AF581750

Advances in Parvovirus Research 2020

Advances in Parvovirus Research 2020

Editor

Giorgio Gallinella

MDPI • Basel • Beijing • Wuhan • Barcelona • Belgrade • Manchester • Tokyo • Cluj • Tianjin

Editor
Giorgio Gallinella
University of Bologna
Italy

Editorial Office
MDPI
St. Alban-Anlage 66
4052 Basel, Switzerland

This is a reprint of articles from the Special Issue published online in the open access journal *Viruses* (ISSN 1999-4915) (available at: https://www.mdpi.com/journal/viruses/special_issues/parvovirus_2020).

For citation purposes, cite each article independently as indicated on the article page online and as indicated below:

LastName, A.A.; LastName, B.B.; LastName, C.C. Article Title. *Journal Name* **Year**, *Volume Number*, Page Range.

ISBN 978-3-0365-2490-0 (Hbk)
ISBN 978-3-0365-2491-7 (PDF)

Contents

About the Editor

Roland Ulber Born on 20.06.1963. Full Professor of Microbiology at the Department of Pharmacy and Biotechnology at the University of Bologna, Italy. He conducts research in the field of virology with a special interest in parvovirus B19.

Preface to "Advances in Parvovirus Research 2020"

The Special Issue of *Viruses* 'Advances in Parvovirus Research 2020' features a series of articles collected in 2020 and 2021. The issue was conceived in continuation of the 2019 Special Issue 'New Insights in Parvovirus Research', which successfully attracted the interest of scientists actively involved in research on the topic of paroviruses. The contributions collectively cover many aspects of basic and translational research on viruses of the family *Parvoviridae*, which are incredibly diverse in their biology and have significant relevance as human and veterinary pathogens, as tools for oncolytic therapy, or as sophisticated gene delivery vectors.

Structural biology is the subject of articles contributed by Mietszch et al. and Chun Yu et al., from the group of late Mavis Agbandje-McKenna, a leading scientist whose achievements will exert a long-lasting influence on her colleagues. Mattola et al. provide a review of current knowledge and advanced experimental techniques for studying the initial phases of parvovirus replicative cycles. Ros et al. summarize known aspects, open issues, and future perspectives related to the minor capsid protein of parvovirus B19. Ferreira et al. report on the uptake mechanism of the oncolytic parvovirus H-1, while Boftsi et al. provide new mechanistic details of post-transcriptional processing in a Minute Virus of Mice virus model. Parvovirus B19 is the subject of contributions from Bua et al. and Ducloux et al., both of which address issues with translational implications for a pathogenic human virus. Other contributions are focused on viruses of veterinary interest, including the review on perspectives related to Aleutian mink disease virus by Markarian et al. and the article on the construction of viral-like particles of tiger feline parvovirus by Jiao et al. Investigations into the molecular phylogenetics of canine parvovirus were the subject of further contributions by Kelman et al., Giraldo-Ramirez et al., and Gainor et al., while Horecka et al. report on the shift in epidemiology connected to the first wave of the COVID-19 pandemic. Finally, Mijanovic et al. present an overview of adeno-associated virus-derived vectors for gene therapy of neurodegenerative diseases, an area of impressive achievements and critical ethical debate.

In the years these papers were submitted, the concomitance of the most critical phases of COVID-19 pandemic hindered normal research activity and diverted relevant energies and resources. Searching for the term 'Parvoviridae' in PubMed reveals the emergence of an inversion in the increasing trend in publications, with an around 30% decrease in published papers on the subject, a fate common to numerous other subject topics in face of the exponential accumulation of works on SARS-CoV2. It is foreseeable and hoped that in the near future, research on parvoviruses will resume its pace and progress toward new knowledge and advanced translational applications.

Giorgio Gallinella
Editor

 viruses

MDPI

Article

Completion of the AAV Structural Atlas: Serotype Capsid Structures Reveals Clade-Specific Features

Mario Mietzsch [1], Ariana Jose [1], Paul Chipman [1], Nilakshee Bhattacharya [2,†], Nadia Daneshparvar [2], Robert McKenna [1] and Mavis Agbandje-McKenna [1,*]

1 Department of Biochemistry and Molecular Biology, Center for Structural Biology, McKnight Brain Institute, College of Medicine, University of Florida, Gainesville, FL 32610, USA; mario.mietzsch@ufl.edu (M.M.); aej51@miami.edu (A.J.); pchipman@ufl.edu (P.C.); rmckenna@ufl.edu (R.M.)
2 Biological Science Imaging Resource, Department of Biological Sciences, Florida State University, Tallahassee, FL 32306, USA; nilakshee.bhattacharya@duke.edu (N.B.); nd15b@my.fsu.edu (N.D.)
* Correspondence: mckenna@ufl.edu
† Current address: Shared Materials Instrumentation Facility, 101 Science Dr, Duke University, Durham, NC 27708, USA.

Abstract: The capsid structures of most Adeno-associated virus (AAV) serotypes, already assigned to an antigenic clade, have been previously determined. This study reports the remaining capsid structures of AAV7, AAV11, AAV12, and AAV13 determined by cryo-electron microscopy and three-dimensional image reconstruction to 2.96, 2.86, 2.54, and 2.76 Å resolution, respectively. These structures complete the structural atlas of the AAV serotype capsids. AAV7 represents the first clade D capsid structure; AAV11 and AAV12 are of a currently unassigned clade that would include AAV4; and AAV13 represents the first AAV2-AAV3 hybrid clade C capsid structure. These newly determined capsid structures all exhibit the AAV capsid features including 5-fold channels, 3-fold protrusions, 2-fold depressions, and a nucleotide binding pocket with an ordered nucleotide in genome-containing capsids. However, these structures have viral proteins that display clade-specific loop conformations. This structural characterization completes our three-dimensional library of the current AAV serotypes to provide an atlas of surface loop configurations compatible with capsid assembly and amenable for future vector engineering efforts. Derived vectors could improve gene delivery success with respect to specific tissue targeting, transduction efficiency, antigenicity or receptor retargeting.

Keywords: AAV; serotype; capsid; cryo-EM; genome packaging; gene delivery

Citation: Mietzsch, M.; Jose, A.; Chipman, P.; Bhattacharya, N.; Daneshparvar, N.; McKenna, R.; Agbandje-McKenna, M. Completion of the AAV Structural Atlas: Serotype Capsid Structures Reveals Clade-Specific Features. *Viruses* **2021**, *13*, 101. https://doi.org/10.3390/v13010101

Academic Editor: Giorgio Gallinella
Received: 22 December 2020
Accepted: 7 January 2021
Published: 13 January 2021

1. Introduction

Adeno-associated viruses (AAV) are single-stranded DNA packaging viruses of the *Parvoviridae* and belong to the genus *Dependoparvovirus* [1]. Vectors based on AAVs are being developed and used as gene delivery biologics to treat a large variety of monogenetic diseases [2]. Thirteen human and primate AAV serotypes, and numerous genomic isolates have been described and have been assigned to six clades A–F or individual clonal isolates [3]. The virions of the AAVs are composed of non-enveloped capsids with T = 1 icosahedral symmetry and diameters of ≈260 Å [4]. They are assembled from 60 viral proteins (VPs): VP1 (≈82 kDa), VP2 (≈73 kDa), and VP3 (≈61 kDa) in an approximate 1:1:10 ratio [5]. The VPs share a common C-terminus that includes the entirety of VP3. Compared to VP3, VP1 and VP2 are extended at their N-termini with a shared ≈65 amino acid (aa) region and additional ≈137 aa N-terminal to VP2 in the case of VP1 (VP1u). The N-terminal regions of VP1 and VP2 contain conserved elements required for AAV infectivity such as a phospholipase A2 (PLA2) domain, a calcium-binding domain, and nuclear localization signals [6,7]. Overall, the VP1 amino acid sequence identity of the AAV serotypes varies between 57 and 99% [8].

The capsid structures of several natural human and primate AAV serotypes, AAV1-AAV6, AAV8, AAV9, AAVhu.37, AAVrh.8, AAVrh.10, and AAVrh.39 have been determined by either X-ray crystallography and/or cryo-electron microscopy (cryo-EM) [9–19]. Regardless of the method of structure determination, only VP3 of the AAVs, except for the first $\approx$15 aa, are structurally ordered. The VP3 structure consists of an anti-parallel, eight-stranded (βB to βI) β-barrel motif, with the BIDG sheet forming the inner surface of the capsid. An additional strand, βA, runs anti-parallel to the βB strand. Furthermore, all AAVs conserve an α-helix (αA) located between βC and βD. Between the individual β-strands, large loops are inserted that are characterized by high sequence and structure variability among the AAVs. These loops form the exterior surface of the capsid and are named after their flanking β-strands. For example, the HI loop is flanked by the βH and βI strands. The sequence variability of different AAVs results in alternative conformations of these loops, which result in AAV serotype-specific capsid surface features. Nine regions of significant diversity at the apex of these loops have been defined as variable regions (VRs) by structural alignments [15]. Despite the structural differences of the VRs, the overall capsid morphology is conserved. These include cylindrical channels at the icosahedral 5-fold symmetry axes, formed by the DE-loops (VR-II), surrounded by a depression largely outlined by the HI-loops. The 5-fold channel is believed to be the route of genomic DNA packaging and VP1u externalization during endo/lysosomal trafficking following cell entry [20,21]. At the 2-fold symmetry axes, depressions are flanked by protrusions surrounding the 3-fold symmetry axes, and raised capsid regions between the 2- and 5-fold axes are termed 2/5-fold walls. The 3-fold region as well as the 2/5-fold wall have been identified as receptor binding sites for many AAV serotypes and serve as determinants of cell and tissue tropism. Among the cellular receptors are sialic acids [22–24], heparan sulfate proteoglycans (HSPG) [25–29], terminal galactose [30,31], sulfated N-acetyl-lactosamine [32], AAVR [33], laminin [34], αvβ1 integrin [35], αvβ5 integrin [36], the hepatocyte growth factor receptor [37], the fibroblast growth factor receptor [38], and platelet-derived growth factor receptor [39]. In addition to receptor binding, the surface of the capsid, including the 5-fold region, displays antigenic sites for antibodies raised by the host immune response [40].

In this study, the structures of the AAV7, AAV11, AAV12, and AAV13 capsids were determined by cryo-EM in an effort to complete the panel of available structures for the defined AAV serotypes. The empty and genome-containing capsid structures of these four AAV serotypes were reconstructed to be between 2.54 to 3.15 Å resolution. All density maps displayed well-defined amino acid side chain densities and showed the characteristic AAV capsid features, including the channels at the 5-fold axes, depressions at the 2-fold and surrounding the 5-fold axes, and protrusions that surround the 3-fold axes. The comparison of the empty (no DNA) and full (genome packaged) capsid structures showed no structural differences of the VP monomer except for an ordered nucleotide at the previously described nucleotide (nt) binding pocket in the case of the full capsids and alternative side chain orientations [17]. Compared to AAV2, significant structural differences were observed primarily at the 3-fold protrusions and the 2/5-fold wall due to aa insertions or deletions as well as sequence differences. This characterization of the structures of AAV7, AAV11 AAV12, and AAV13, completes the library for the defined serotypes. This provides a means to functionally annotate their capsids and a visual platform to aid recombinant DNA vector engineering for improved gene delivery applications.

2. Materials and Methods

2.1. AAV Production and Purification

The AAV7, AAV11, AAV12, and AAV13 producer plasmid *cap* genes were synthesized by GenArt (Thermo Fisher, Waltham, MA, USA) and subcloned into a plasmid with the AAV2 *rep* gene to generate pR2V7, pR2V11, pR2V12, and pR2V13, respectively. Recombinant AAV7, AAV11, AAV12, and AAV13 vectors, with a packaged luciferase gene, were produced by the triple transfection of HEK293 cells, utilizing pTR-UF3-Luciferase, pHelper

(Stratagene, San Diego, CA, USA), and either pR2V7, pR2V11, pR2V12, or pR2V13, and harvested 72 h post transfection as previously described [41]. The cleared lysates containing AAV7, AAV12, and AAV13 capsids were purified by AVB Sepharose and AAV11 by POROS Capture Select AAVX affinity chromatography as previously described [42]. Sample purity and capsid integrity were monitored by SDS-PAGE and negative- stain electron microscopy using a Spirit microscope (FEI, Hillsboro, OR, USA).

2.2. Cryo-Electron Microscopy Data Collection and 3D Image Reconstruction

For each of the purified AAV capsids, 3.5 μL was applied to a glow-discharged Quantifoil copper grid with 2 nm continuous carbon support over holes (Quantifoil R 2/4 400 mesh), blotted, and vitrified using a Vitrobot Mark 4 (FEI, Hillsboro, OR, USA) at 95% humidity and 4 °C. The capsid distribution and ice quality of the grids were screened in-house using an FEI Tecnai G2 F20-TWIN microscope (FEI) operated under low-dose conditions (200 kV, ≈20e$^-$/Å^2). Images were collected on a GatanUltraScan 4000 CCD camera (Gatan, Pleasanton, CA, USA). Grids deemed suitable for high-resolution data collection were used for collecting micrograph movie frames using the Leginon application [43] on a Titan Krios electron microscope. The microscope was operated at 300 kV and data were collected on a Gatan K3 direct electron detector. During data collection, a total dose of ≈60 e$^-$/Å^2 was utilized for 45 to 71 movie frames per micrograph (Table 1). The movie frames were aligned using MotionCor2 with dose weighting [44]. All datasets were collected as part of the NIH "Southeastern Center for Microscopy of MacroMolecular Machines (SECM4)" project. For the three-dimensional image reconstruction, the cisTEM software package was utilized [45] and the data were processed as described previously [46]. The sharpened density maps were inspected using Coot and Chimera [47,48]. The −90 Å^2/0 Å^2 sharpened maps were utilized for assignment of the amino acid main- and side chains. The resolution of the cryo-reconstructed density maps for empty (no DNA) and genome-containing AAV7, AAV11, AAV12, and AAV13 capsids were estimated based on a Fourier Shell Correlation of 0.143 (Table 1).

2.3. Model Building and Structure Refinement

Three-dimensional (3D) homology models of AAV7, AAV11, AAV12, and AAV13 VP3 were generated with the protein structure homology-modeling server Swiss model (https://swissmodel.expasy.org) [49] using their amino acid sequences (NCBI accession numbers YP_077178, AAT46339, ABI16639 and ABZ10812, respectively) and supplying the VP3 structures of AAV8 (PDB accession number: 2QA0) for AAV7, AAV4 (2G8G) for AAV11 and AAV12, and AAV3 (3KIC) for AAV13 as templates [9,14,15]. A T = 1 60-mer capsid coordinate model was generated from the respective VP3 with the VIPERdb2 Oligomer generator subroutine by icosahedral matrix multiplication [50]. The 60-mer capsid models of each AAV were docked into their cryo-reconstructed density maps by rigid body rotations and translations using the 'fit in map' subroutine within UCSF-Chimera [48]. This application uses a correlation coefficient (CC) calculation to assess the quality of the fit between the map generated from the model and the reconstructed map. During the model fitting, the voxel (pixel) size of each reconstructed map was adjusted to optimize the CC between the models and maps. The fitted models were exported relative to the respective map for further use. Each map was resized to the voxel size determined in Chimera using the "e2proc3D.py" subroutine in EMAN2 [51] and then converted to the CCP4 format using the program MAPMAN [52]. A VP monomer was extracted from each 60-mer and the side- and main chains were adjusted into the maps by manual building and the real-space refinement subroutine in Coot [47]. The adjusted capsid model was refined against the map utilizing the rigid body, real space, and B-factor refinement subroutines in Phenix [53]. Capsid model refinement was alternated with visualization and adjustment of VP side- and main chains using Coot while maintaining model geometry as well as rotamer and Ramachandran constraints [47]. The CC and refinement statistics, including root mean square deviations (RMSD) from ideal bond lengths and angles (Table 1), were analyzed using Phenix [53].

Table 1. Summary of data collection, image processing, and refinement statistics.

Cryo-EM Data and Refinement Parameter	AAV7		AAV11		AAV12		AAV13	
	Full	Empty	Full	Empty	Full	Empty	Full	Empty
Total number of micrographs	271		1251		1629		1582	
Defocus range (μm)	0.8–2.0		0.8–3.0		1.0–3.0		0.8–3.0	
Total electron dose ($e^-/Å^2$)	60		60		60		60	
Frames/micrograph	71		45		50		50	
Pixel size (Å/pixel)	1.08		0.85		1.08		1.08	
Capsids used for final map	4695	40,988	10,429	118,351	40,764	220,137	6794	56,962
Resolution of final map (Å)	3.16	2.96	3.15	2.86	2.67	2.54	3.00	2.76
Refinement Statistics								
Map CC	0.871	0.899	0.859	0.863	0.864	0.868	0.856	0.864
RMSD bonds (Å)	0.01	0.01	0.01	0.01	0.01	0.01	0.01	0.01
RMSD angles (°)	0.79	0.83	0.82	0.99	0.91	0.94	0.88	0.95
All-atom clashscore	8.83	7.78	9.29	10.38	8.91	7.96	8.26	8.99
Ramachandran plot (%)								
Outliers	0	0	0	0	0	0	0	0
Allowed	2.3	1.7	2.3	1.7	1.5	1.5	1.9	2.1
Favored	97.7	98.3	97.7	98.3	98.5	98.5	98.1	97.9
Rotamer outliers	0	0	0	0	0	0	0.2	0
C_β deviations	0	0	0	0	0	0	0	0

2.4. AAV Capsid Structure Comparison

The Cα positions of the ordered amino acids within the VP3 atomic coordinates for each of the AAVs were superposed using secondary structure matching (SSM) in Coot [54]. This SSM subroutine also generates a list of the Cα–Cα distances between the aligned structures, which was used to calculate the overall root mean square deviation (RMSD). Deviations between non-overlapping Cα positions, because of residue deletions or insertions, were measured using the distance tool in Coot. Structural identity was determined using PDBeFold (https://www.ebi.ac.uk/msd-srv/ssm/) and calculated as the number of aligned residues (<1.0 Å apart) divided by the total number of residues. Amino acid sequence alignments of the different AAV serotypes were done utilizing the sequence alignment option in VectorNTI (Invitrogen, Carlsbad, CA, USA).

2.5. Structure Accession Numbers

The full and empty AAV7, AAV11, AAV12, and AAV13 cryo-EM reconstructed density maps and models built for their capsids were deposited in the Electron Microscopy Data Bank (EMDB) with accession numbers EMD-23190/PDB ID 7L5U (AAV7 full), EMD-23189/PDB ID 7L5Q (AAV7 empty), EMD-23202/PDB ID 7L6E (AAV11 full), EMD-23203/PDB ID 7L6F (AAV11 empty), EMD-23200/PDB ID 7L6A (AAV12 full), EMD-23201/PDB ID 7L6B (AAV12 empty), EMD-23204/PDB ID 7L6H (AAV13 full), EMD-23205/PDB ID 7L6I (AAV13 empty), respectively.

3. Results and Discussion

3.1. The Structures of AAV7, AAV11, AAV12, and AAV13 Capsids Completes the Serotype List

The capsid structures of AAV serotypes 1–6 and 8–9 have been previously reported [9–15], leaving those of AAV7 and AAV10–13 yet to be determined. AAV10, a member of clade E [3], possesses just a single amino acid (aa) difference (A589T) within VP3 compared to AAVrh.39, for which the capsid structure has been determined [17]. Thus, the AAV10 capsid structure is likely identical to AAVrh.39, especially since the AAVrh.10 capsid, which has several aa differences, is already shown to be structurally identical to AAVrh.39 [17]. In contrast, AAV7 (has an 82 aa difference in VP3 vs. AAV8), AAV11 (109 aa vs. AAV4), AAV12 (109 aa vs. AAV4), and AAV13 (28 aa vs. AAV3) are substantially different to their closest sequence-related AAV serotype, as shown in the parentheses. Thus, to determine their capsid structures recombinant AAV7, AAV11, AAV12, and AAV13 vectors were produced by triple transfection of HEK293 cells followed by purification with AVB affinity chromatography in the case of AAV7, AAV12, and AAV13, and AAVX affinity chromatography in the case of AAV11, as described in the methods. While the affinity purification resulted in highly pure AAV capsid preparations, it did not separate empty (no genome) and genome-containing (full) capsids, and thus, both types of capsids were observed in cryo-EM micrographs (Figure 1A).

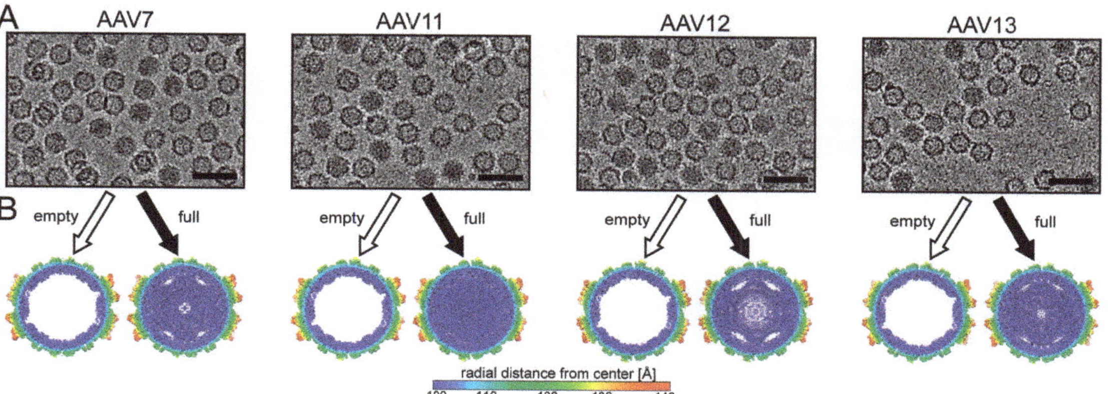

Figure 1. Cryo-electron microscopy (cryo-EM) reconstruction of genome containing (full) and empty AAV7 and AAV11–AAV13 capsids. (**A**) Cryo-electron micrographs showing the presence of full capsids (dark appearance) and empty (light appearance). Scale bar: 50 nm. (**B**) Cross-sectional views of the reconstructed maps determined by cryo-EM reconstruction from full and empty capsids contoured at a sigma (σ) threshold level of 0.9. The reconstructed maps are radially colored (blue to red) according to radial distance to the particle center. This figure was generated using UCSF-Chimera [48].

The distribution of the capsids in the micrographs enabled the independent structural determination of both empty and full capsids using 2D classification, as described previously [46], for each serotype. For AAV7, AAV11, AAV12, and AAV13, the empty/full structures were determined from 40,988/4695, 118,351/10,429, 220,137/40,764, and 56,962/6794 capsids, respectively, to 2.96/3.16, 2.86/3.15, 2.54/2.67, and 2.76/3.00 Å resolution (FSC 0.143), respectively (Table 1). For each of the AAV serotypes, the resolution of the full structures is slightly lower compared to the empty, which is most likely due to fewer capsids used in the reconstructions of the former. Direct comparison of the reconstructed empty and full maps for each AAV serotype in a cross-sectional view clearly showed the electron-dense filled interior of the genome-containing capsids, which is absent from the empty capsids (Figure 1B). Similar to previous observations of full AAV capsid density maps, the majority of the capsid interior is filled except for the region directly underneath the 5-fold channel [17,46]. It has been suggested that the dynamic and flexible VP1/VP2 common region and VP1u could be located in the area under the 5-fold channel in readiness to be externalized through the 5-fold channel, which is a structural rearrangement that is required for its PLA2 enzyme function during the viral life cycle [20].

3.2. The AAV7, AAV11, AAV12, and AAV13 Capsid Structures Conserved the AAV Features

Regardless of whether full or empty maps were analyzed, the different AAV serotypes displayed the characteristic morphological features of other AAVs, e.g., a channel at the 5-fold symmetry axes, trimeric protrusions that surround each 3-fold symmetry axis, and a depression at each 2-fold symmetry axis (Figure 2A). However, the exact morphology of the 3-fold protrusions varies between the different AAV serotypes, with much broader protrusions for AAV11 and AAV12 compared to AAV7 and AAV13. Similarly, the shape and orientation of the 2-fold depression of AAV11 and AAV12 differs from AAV7 and AAV13.

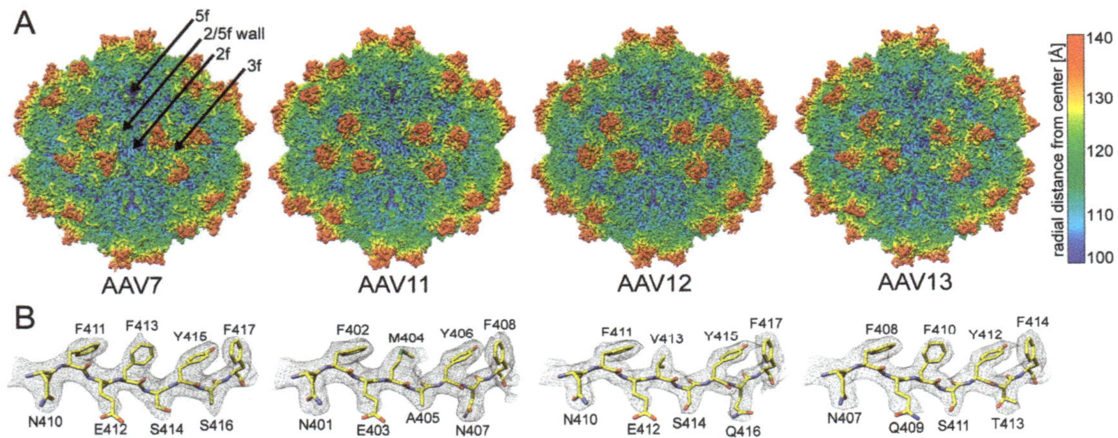

Figure 2. The AAV7 and AAV11–13 capsid structures. (**A**) The capsid surface density maps contoured at a sigma (σ) threshold level of 2.0. The maps are radially colored (blue to red) according to distance to the capsid center, as indicated by the scale bar on the right. The icosahedral 2-, 3-, and 5-fold axes as well as the 2/5-fold wall are indicated on the AAV7 capsid map. (**B**) The AAV7 and AAV11–13 amino acids modeled for the βG strand are shown inside their respective density maps (black mesh). The amino acid residues are as labeled and shown in stick representation and colored according to atom type: C = yellow, O = red, N = blue, S = green. This figure was generated using UCSF-Chimera [48].

The reconstructed maps of the four AAV serotypes, empty and full, showed well-ordered amino acid side-chain densities (Figure 2B) throughout the VP structure starting at aa position 218–220 (AAV7 numbering), which is comparable to the other currently determined AAV serotype capsid structures [9–15]. The only exception was the apex of surface loop VR-IV in AAV7, where aa 455–458 (GGTAG) were disordered, preventing the reliable placement of main- and side-chain residues. A similar disorder was previously observed in AAVrh.10 and AAVrh.39 that share the same or very similar sequence at the apex of the loop, GGTAG and GGTQG, respectively [17]. The glycines on both sides of the apex likely confer the flexibility of this loop and thus the cause of the lack of structural order. AAV11–13 do not possess this accumulation of glycines at this loop; thus, their loops were structurally ordered.

3.3. The Full Capsids of AAV7, AAV11, AAV12, and AAV13 Show Ordered Nucleotides

Similar to previous observations, a structural comparison of empty to the full capsids for the individual AAV serotypes showed them to be largely identical with overall Cα RMSDs ranging from ≈0.2 to 0.3 Å [17,46]. However, a major difference is the observation of weakly ordered density in the interior of the capsid maps in the full structures interpreted as the packaged genome (Figure 1B). This density extends into a pocket underneath the 3-fold symmetry axis and has been interpreted as deoxyadenosine monophosphate (dAMP), which is positioned between conserved prolines 421/632 and histidine 631 (AAV7 numbering) (Figure 3A). We hypothesized that the genome interacts with this 3-fold region of the interior capsid by binding within the pocket to two symmetry-related VP monomers [17]. Due to the imposed icosahedral symmetry during 3D image reconstruction and the fact that the genome cannot follow this symmetry, other nucleotides leading in and out of this pocket are weakly ordered and cannot be reliably modeled. We postulate that as reconstruction methods improve, relaxation of the enforced icosahedral symmetry in future structure determination efforts may allow the observation of a more ordered DNA structure.

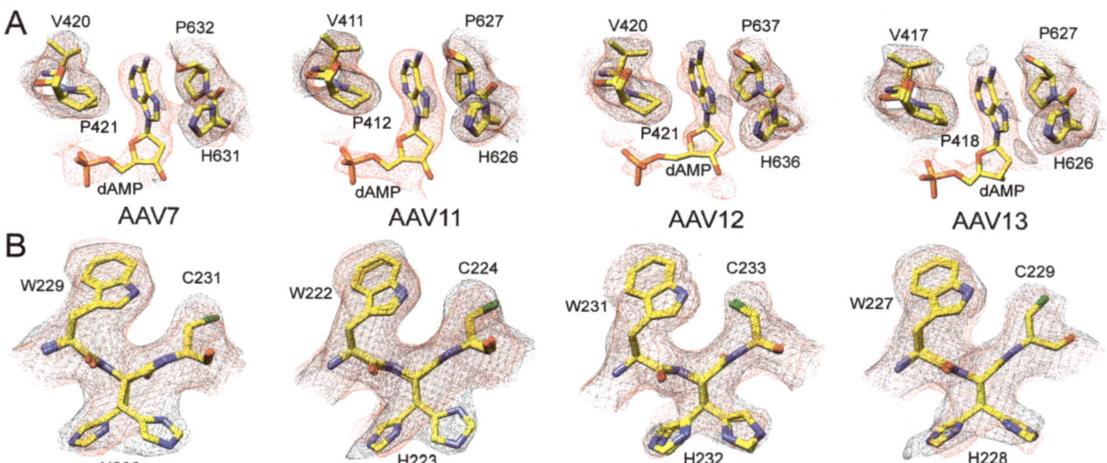

Figure 3. Empty and full Adeno-associated virus (AAV) density map differences. (**A**) The modeled AAV7 and AAV11–13 residues at the nucleotide binding pocket with their respective mesh density maps (black = empty, red = full). The extra density exclusively in the full maps was interpreted as an ordered nucleotide (deoxyadenosine monophosphate, dAMP). (**B**) Dual conformation of histidines (e.g., H230 in AAV7). This histidine adopts alternative side-chain conformations primarily in the absence of packaged DNA with the exception of AAV12. Atom colors: C = yellow, O = red, N = blue, S = green, P = orange. This figure was generated using UCSF-Chimera [48].

While the VP structures of empty and full capsids were largely identical, some alternative side-chain orientations, e.g., histidine 230 (AAV7 numbering), were observed. In the AAV serotype structures determined in this study, the histidine side chain preferred the "left" orientation in full capsids (Figure 3B). However, in AAV12 and AAV13, weak density was also observed toward the "right" orientation. In contrast, both orientations are equally adopted in empty capsids, except for AAV12, where the right orientation appears to be favored (Figure 3B). The dual conformation of H230 was previously observed in empty AAVrh.10 capsids [17]. While the cause of this difference between empty and full capsids is unknown, disordered density at low sigma level in the full maps, likely from the packaged genome, appears to contact the histidine side chain in the "right" orientation and thereby induce this preferred conformation of the side chain. Furthermore, H230 is located near the 5-fold symmetry axis, and the different conformation may be related to the observed differences underneath the 5-fold region in both types of capsids (Figure 1B).

3.4. The AAV7, AAV11, AAV12, and AAV13 Capsid Structures Display Diversity in Surface Loop Conformations

The AAV7, AAV11–13 VP topologies conserve the core eight-stranded anti-parallel β-barrel (βB-βI), with the additional β-strand A and α-helix A (Figure 4A), as described previously for all other AAV structures [9,11–15,17–19,46,55–59]. When superposed, these core regions are homologous for the AAV serotypes (Figure 4A). Between the β-strands, the loops that form the surface of the capsids provide the structural variability among different AAVs. The nine VRs, I-IX previously defined [15], provide serotype-specific functions.

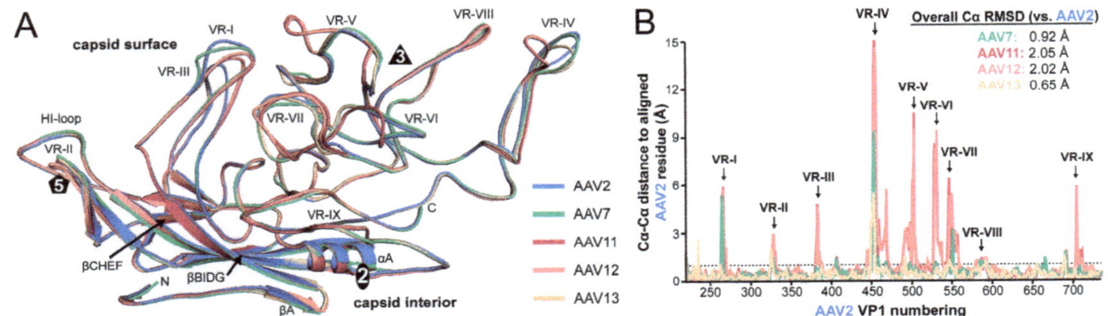

Figure 4. Structural comparison of AAV2, AAV7, and AAV11–13. (**A**) Structural superposition of AAV2 (blue), AAV7 (cyan), AAV11 (burgundy), AAV12 (salmon), and AAV13 (wheat) shown as ribbon diagrams. The positions of β-strands, the N- and C-terminus, the variable regions (VRs), and the icosahedral 2-, 3-, and 5-fold axis are indicated. This figure was generated using PyMol [60]. (**B**) Cα–Cα distance plot (in Å) for the AAV7 and AAV11–13 residues relative to AAV2 of the superposed viral protein (VP) structures. The VRs are indicated and the overall VP Cα-RMSD (root mean square deviation) compared to AAV2 shown. The dashed line marks the Cα–Cα distance variation of 1 Å.

Compared to AAV2, the prototype serotype, some of the AAV7 and AAV11–13 surface loops showed only minor structural differences with Cα distances of ≤1 Å such as VR-VIII and the HI-loop (Figure 4B). AAV7 also shows minor structural differences compared to AAV2 in four additional loops (VR-III, VR-V, VR-VI, and VR-IX) and AAV13 in six additional loops (VR-I, VR-III, VR-V, VR-VI, VR-VII, and VR-IX), respectively. Structural variability (Cα distances of <3 Å) was also observed for the DE-loop/VR-II at the 5-fold symmetry axis for all analyzed AAV serotypes. The absence of major differences in the 5-fold region, which includes the HI-loop, are likely due to the common function these loops have to fulfill such as their role in DNA packaging and VP1u externalization [20,21].

Greater structural variability between AAV7 and AAV2 was seen in VR-I, VR-IV, and VR-VII due to single aa insertions (VR-I and VR-IV) or a single aa deletion (VR-VII) (M). In AAV13, the only significant structural difference is observed in VR-IV, which is slightly shorter due to a single aa deletion compared to AAV2. In contrast, major structural variabilities (vs. AAV2) were seen in VR-I, VR-III, VR-IV, VR-V, VR-VI, VR-VII, and VR-IX for AAV11 and AAV12. Consequently, their overall Cα-RMSD for the entire VP is larger than that of AAV7 and AAV13 (Figure 4B). Most notably is the 5 aa insertion in VR-V for both AAV11 and AAV12, relative to AAV2, but also to AAV7 and AAV13 (Figure 4A). In addition, both AAV serotypes possess a single aa insertion in VR-IV and display a different conformation to the other AAV serotypes with the apex of the loop positioned over part of VR-V. This subloop of VR-V and the alternative conformation of VR-IV are responsible for the broader appearance of the 3-fold protrusions of the AAV11 and AAV12 capsids as described above (Figure 2A). On the side of the 3-fold protrusions VR-VI and VR-VII also showed significant differences with a 1 aa deletion in VR-VI and major structural variabilities in AAV11 and AAV12 (Figure 4A,B). Similarly, structural differences of VR-I, VR-III, and VR-IX lead to morphological differences at the 2/5-fold wall. VR-I takes a different conformation in AAV11 and AAV12 compared to AAV2 due to a 3 aa deletion. This is partially compensated by VR-III with a 2 aa insertion resulting in a broader loop without extending the height of the loop (Figure 4A). Finally, VR-IX of AAV11 and AAV12 displays a differential conformation without amino acid insertions or deletions relative to AAV2. This variation is located near the 2-fold symmetry axis, resulting in a slightly wider depression of the AAV11 and AAV12 capsid (Figure 2A). Their overall RMSD of the Cα coordinates for the entire VP of 2.05 and 2.02 Å (vs. AAV2) is greater than the overall Cα RMSD of AAV2 compared to AAV5, the most divergent AAV serotype, with a Cα RMSD of 1.8 Å [46].

3.5. The AAV7, AAV11, AAV12, and AAV13 Capsid Structures Display Clade-Specific Surface Features

The clades for the AAV serotypes were proposed in 2004 [3] based on more than 100 unique isolates from human and non-human primates that were grouped based on their VP phylogenetical similarity (Figure 5A). AAV serotypes 10–13 were described after this study [28,61,62], and thus, they were originally not grouped into the clades.

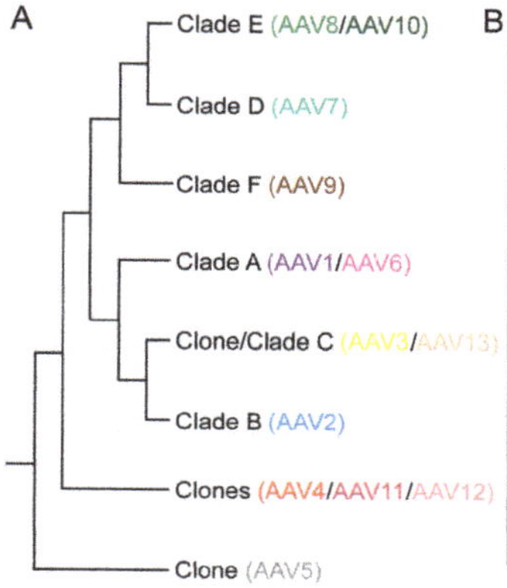

A

Clade E (AAV8/AAV10)
Clade D (AAV7)
Clade F (AAV9)
Clade A (AAV1/AAV6)
Clone/Clade C (AAV3/AAV13)
Clade B (AAV2)
Clones (AAV4/AAV11/AAV12)
Clone (AAV5)

B

	AAV1	AAV2	AAV3	AAV4	AAV5	AAV6	AAV7	AAV8	AAV9	AAV10	AAV11	AAV12	AAV13
AAV1		83/83	87/86	63/58	58/59	99/99	85/82	84/80	82/79	85/82	66/59	60/57	87/86
AAV2	92		88/89	60/56	57/58	83/84	82/82	83/82	82/81	84/84	63/57	60/56	88/90
AAV3	93	91		63/58	58/59	87/86	85/83	86/85	84/80	86/85	65/58	61/56	94/95
AAV4	75	70	68		53/51	63/58	63/58	63/58	62/56	63/58	81/80	78/80	65/58
AAV5	71	69	70	62		58/59	58/58	58/58	57/57	57/57	53/51	52/51	58/58
AAV6	99	91	93	76	72		85/82	84/80	82/79	85/82	66/59	60/57	87/87
AAV7	95	94	89	75	71	94		88/85	82/79	88/86	67/59	62/57	85/83
AAV8	91	94	89	72	71	90	93		85/84	93/93	65/58	62/57	85/84
AAV9	95	94	93	76	70	94	94	93		86/84	64/57	60/56	84/81
AAV10*	89	92	86	70	69	88	92	97	93		66/59	61/57	86/86
AAV11	78	76	76	91	63	78	77	76	79	76		84/87	65/58
AAV12	77	76	74	92	65	75	76	75	79	73	98		60/57
AAV13	95	96	92	75	72	93	96	94	96	92	76	76	

structural identity (% aligned amino acids within 1 Å)

VP1/VP3 amino acid sequence identity (in %)

Figure 5. Relationship of the AAV serotypes. (**A**) Cladogram showing the assignment of the AAV serotypes to their clades as proposed by Gao et al. [3]. (**B**) Amino acid sequence identity of the AAV serotypes given as percentage for VP1 and VP3 or the structural identity as a percentage of aligned amino acids within 1Å when superposed. High values are colored in dark blue or orange and lower values in lighter shades of each color, respectively. * For AAV10, the VP structure of AAVrh.39 was utilized, which varies by a single aa from AAV10.

AAV7 belongs to clade D (Figure 5A) and is closest related to the clade E members AAV8 and AAV10 based on VP1 or VP3 amino acid sequence ranging from 85 to 88% aa sequence identity (Figure 5B). When AAV7 is superposed onto AAV8, the overall Cα RMSD is 0.75 Å with a structural identity of 93% (Figures 5B and 6A), which is slightly lower than the comparison to AAV2 described above at 0.92 Å. However, the 94% structural identity of AAV7 compared with AAV2 is slightly higher than for AAV7 vs. AAV8 (Figure 5B). Compared to AAV8, the AAV7 VP showed different surface loop conformations in VR-I (1 aa deletion), VR-IV (1 aa insertion), and VR-VII (1 aa deletion), respectively. In fact, these AAV7 loops are unique among all the available AAV serotype capsid structures. VR-I of AAV7 is structurally most similar to AAV1 and AAV6, without deletions or insertions but with amino acid variations resulting in Cα distance variation of up to 3 Å. AAV7's VR-VII is the shortest loop among all AAV serotypes with a 1 aa deletion compared to AAV1-AAV4, and AAV6-AAV13 and a 4 aa deletion compared to AAV5. AAV7 vectors were shown to result in high transduction efficiencies of the CNS and spinal cord after delivery into the cerebrospinal fluid or intravenously [63,64]. This indicates that AAV7 might be able to cross the blood–brain barrier (BBB). However, the proposed residues in AAVrh.10 reported to be responsible for this phenotype [17] are only partially conserved in AAV7, e.g., S269, but not N472, where AAV7 has a threonine. More research is needed to determine if AAV7 has the ability to cross the BBB. AAV7 was shown to bind to several AAV8 antibodies, which are termed ADK8, HL2381, and HL2383 [65,66]. These antibodies utilize AAV8's VR-VIII as its epitope [67,68]. The observed cross-reactivities can be explained by the high structural conservation of VR-VIII between AAV7 and AAV8 (Figure 6A).

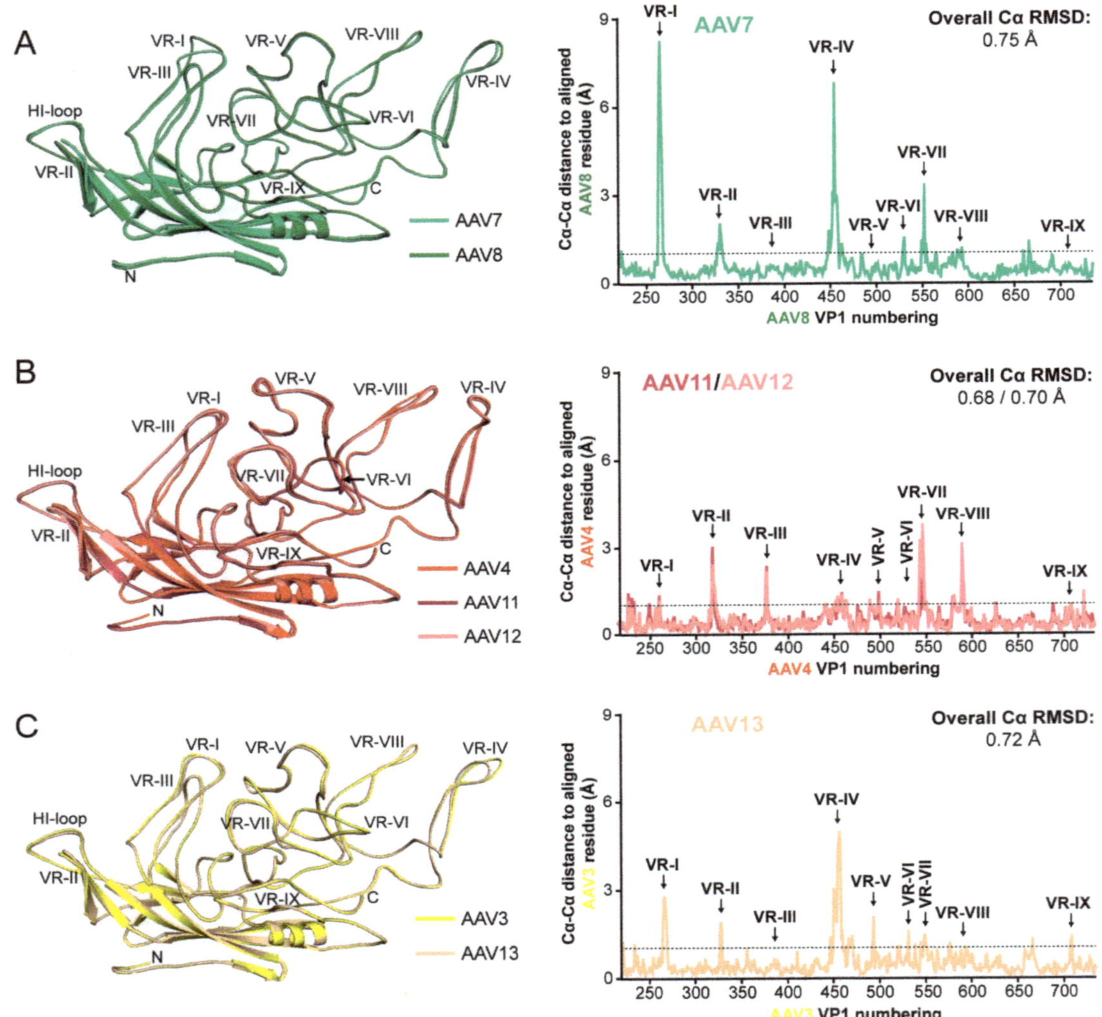

Figure 6. Structural comparison of AAV7 and AAV11–13 to their closest clade member. (**A**) Left—Structural superposition of AAV7 (cyan) and AAV8 (green) shown as ribbon diagrams. The positions of the N- and C-terminus and the variable regions (VRs) are indicated. This figure was generated using PyMol [60]. Right—Cα–Cα distance plot (in Å) for the AAV7 residues relative to AAV8 when the VP structures are superposed. The VRs are indicated, and the overall AAV7 VP Cα–RMSD compared to AAV8 is shown. The dashed line marks a Cα–Cα distance of 1 Å. (**B**) Structural comparison as in (A) for AAV4 (red), AAV11 (burgundy), and AAV12 (salmon). (**C**) Structural comparison as in (A) for AAV3 (yellow) and AAV13 (wheat).

For AAV11 and AAV12, the closest related AAV serotype is AAV4 (Figure 5A) based on VP1 or VP3 amino acid sequence ranging from 78 to 81% aa sequence identity (Figure 5B) However, the sequence identity of AAV11 and AAV12 to each other is slightly higher (84–87%). Surprisingly, the structural identity between AAV11 and AAV12 is 98%, which is only surpassed by AAV1 and AAV6 with 99% sequence and structural identity (Figure 5B) Compared to all other AAV serotypes, the VP3 sequence identity ranges from 51 to 59% (Figure 5B). Consequently, the AAV4 VP is structurally more similar to AAV11 and AAV12 when superposed (Figure 6B) with a structural identity of 91–92% compared to all other

AAV serotypes with structural identities ranging from 63 to 79% (Figure 5B). In particular, AAV4, AAV11, and AAV12 share the insertion in VR-V and the alternative conformation of VR-IV (Figure 6B). The overall Cα RMSD of AAV11 and AAV12 to AAV4 is 0.68 and 0.70 Å with minor loop variations in VR-II, VR-III (1 aa deletion in AAV11 and AAV12), VR-VII, and VR-VIII (Figure 6B). When AAV11 and AAV12 are compared to each other, minor structural differences were observed in VR-II and VR-VII with an overall Cα RMSD of 0.56 Å. We propose that rather than clonal isolates, these three viruses, AAV4, AAV11, and AAV12 (Figure 5A), should be grouped into a new clade G.

While for AAV4, α2–3 linked sialic acid is described as a receptor [22], the receptor for AAV11 and AAV12 is unknown. For AAV12, HSPG and sialic acids were excluded as a receptor [62], and no binding to the available glycans on an array was shown [29]. Amino acids in AAV4 were suggested to be involved in sialic acid binding, which involves residues in VR-V, VR-VI, and VR-VIII [69]. The amino acids in VR-V and VR-VI are conserved structurally and in residue type in AAV11 and AAV12, unlike those in VR-VIII, which may be the reason why AAV11 and AAV12 do not bind sialic acids. Overall, AAV11 and AAV12 vectors have been rarely used for gene delivery purposes; however, AAV11 was described to possess a tropism for the spleen and smooth muscle [61,70], whereas AAV12 was shown to transduce nasal epithelia efficiently [71].

An interesting difference between AAV4, AAV11, and AAV12 for AAV vector production is the requirement for the assembly activating protein (AAP) for capsid assembly [72,73]. While AAV12 is dependent on the presence of AAP, AAV4 and AAV11 are not. An analysis of residues shared between AAV4 and AAV11 but not with AAV12 revealed a total of 23 aa. Previous studies suggested that interior residues are involved in the AAP function [74,75]. Only four of 23 aa differences are located in the interior of the capsid, A301S, A338T, I619V, and R688H (first aa type = AAV4/11, second aa type = AAV12). Of these aa positions, 619 is likely not the determining factor, since AAV11's isoleucine is conserved in most AAV serotypes. A301S and R688H are located near the 2-fold symmetry axis, which is the previously suggested region for AAP binding [74]. More research is needed to confirm the importance of the residues for capsid assembly.

For AAV13, the closest related AAV serotype is AAV3 (Figure 5A) based on VP1 or VP3 amino acid sequence ranging from 94 to 95% aa sequence identity, which is followed by AAV2 with 88–90% (Figure 5B). Nonetheless, when superposed, AAV13 is structurally slightly more similar to AAV2 compared to AAV3 (Cα RMSD: 0.65 Å vs. 0.72 Å and structural identity: 96% vs. 92%) (Figure 4B, Figure 5B, and Figure 6C). In addition to the significant difference in VR-IV caused by 2 aa deletion of AAV13 relative to AAV3, there are also minor variations in VR-I, VR-II, and VR-V (Figure 6C). Common to all of these three AAV serotypes is their ability to bind to HSPG [25,28,29,76] and to the A20 antibody [77]. For AAV13, a critical aa for this binding is K528, which is not present in AAV2 or AAV3 [28]. This residue is located on the side of the 3-fold protrusion within VR-VI. The mutation of K528 to glutamic acid results in the inability to bind HSPG [28,29]. Interestingly, this residue is in a structural equivalent position to AAV6-K531 reported to be important for HSPG binding of AAV6 [78]. AAV13 vectors have currently not been used for gene delivery purposes. Thus, more research is needed to determine its tropism and transduction efficiency.

4. Conclusions

This study determined the capsid structures of AAV7, AAV11, AAV12, and AAV13, thereby completing the panel of available structures for all currently defined AAV serotypes. While these capsids conserve the AAV capsid features such as the 5-fold channels, protrusions around the 3-fold symmetry axes, depressions at the 2-fold axes, as well as nucleotide binding pocket, they also display surface loops that are not found in any other AAV serotype structure. These separate AAV7 from its closest related AAV members contained within clade E (such as AAV8, AAV10, or AAVrh.10).

AAV11 and AAV12 share structural similarity to AAV4 with loop conformations that are also not found in other AAV serotypes. Thus, while not defined as such, AAV4, AAV11, and AAV12 might form a separate clade. Lastly, AAV13 with AAV3 as its closest related AAV serotype shares structural similarity to both AAV2 and AAV3. It likely belongs to clade C, which was previously described to contain AAV2–AAV3 hybrid members.

The definition of clades suggests antigenic specificity, with members being cross-reactive [3]. However, recent data shows that members of different clades cross-react, and thus the clade definition requires revisiting [66]. The completion of the AAV serotype structural atlas, providing visualization of the conserved and variable regions, shows that the serotypes can also be grouped based on structural morphology. These structures provide a template for engineering the AAV capsids for targeted tissue tropism and the escape of recognition by host antibodies toward improved vector efficacy.

Author Contributions: M.M. was responsible for sample production and purification, cryo-reconstruction, structure refinement and analysis, model building and refinement, and manuscript preparation. A.J. was responsible for sample production and purification. P.C. vitrified sample and screened cryo-EM grids. N.B. and N.D. collected cryo-EM data. R.M. supervised the project and contributed to manuscript preparation. M.A.-M. conceived and supervised the project, analyzed all results, and contributed to manuscript preparation. All authors have read and agreed to the published version of the manuscript.

Funding: The TF20 cryo-electron microscopes was provided by the UF College of Medicine (COM) and Division of Sponsored Programs (DSP). Data collection at Florida State University was made possible by NIH grants S10 OD018142-01 Purchase of a direct electron camera for the Titan-Krios at FSU (PI Taylor), S10 RR025080-01 Purchase of a FEI Titan Krios for 3-D EM (PI Taylor), and U24 GM116788 The Southeastern Consortium for Microscopy of MacroMolecular Machines (PI Taylor). The University of Florida COM and NIH GM082946 provided funds for the research efforts at the University of Florida.

Acknowledgments: The authors thank the UF-ICBR Electron microscopy core for access to electron microscopes utilized for cryo-electron micrograph screening.

Conflicts of Interest: MAM is a SAB member for Voyager Therapeutics, Inc., and AGTC, has a sponsored research agreement with Voyager Therapeutics and Intima Biosciences, Inc. and is a consultant for Intima Biosciences, Inc. MAM is a co-founder of StrideBio, Inc. This is a biopharmaceutical company with interest in developing AAV vectors for gene delivery application.

References

1. Cotmore, S.F.; Agbandje-McKenna, M.; Canuti, M.; Chiorini, J.A.; Eis-Hubinger, A.-M.; Hughes, J.; Mietzsch, M.; Modha, S.; Ogliastro, M.; Pénzes, J.J.; et al. ICTV virus taxonomy profile: Parvoviridae. *J. Gen. Virol.* **2019**, *100*, 367–368. [CrossRef] [PubMed]
2. Wang, D.; Tai, P.W.; Gao, G. Adeno-associated virus vector as a platform for gene therapy delivery. *Nat. Rev. Drug Discov.* **2019**, *18*, 358–378. [CrossRef] [PubMed]
3. Gao, G.; Vandenberghe, L.H.; Alvira, M.R.; Lu, Y.; Calcedo, R.; Zhou, X.; Wilson, J.M. Clades of adeno-associated viruses are widely disseminated in human tissues. *J. Virol.* **2004**, *78*, 6381–6388. [CrossRef]
4. Mietzsch, M.; Pénzes, J.J.; Agbandje-McKenna, M. Twenty-five years of structural parvovirology. *Viruses* **2019**, *11*, 362. [CrossRef] [PubMed]
5. Snijder, J.; Van De Waterbeemd, M.; Damoc, E.; Denisov, E.; Grinfeld, D.; Bennett, A.; Agbandje-McKenna, M.; Makarov, A.; Heck, A.J.R. Defining the stoichiometry and cargo load of viral and bacterial nanoparticles by orbitrap mass spectrometry. *J. Am. Chem. Soc.* **2014**, *136*, 7295–7299. [CrossRef] [PubMed]
6. Girod, A.; Wobus, C.E.; Zádori, Z.; Ried, M.; Leike, K.; Tijssen, P.; Kleinschmidt, J.A.; Hallek, M. The VP1 capsid protein of adeno-associated virus type 2 is carrying a phospholipase A2 domain required for virus infectivity. *J. Gen. Virol.* **2002**, *83 Pt 5*, 973–978. [CrossRef]
7. Popa-Wagner, R.; Porwal, M.; Kann, M.; Reuss, M.; Weimer, M.; Florin, L.; Kleinschmidt, J.A. Impact of VP1-specific protein sequence motifs on adeno-associated virus type 2 intracellular trafficking and nuclear entry. *J. Virol.* **2012**, *86*, 9163–9174. [CrossRef] [PubMed]
8. Daya, S.; Berns, K.I. Gene therapy using adeno-associated virus vectors. *Clin. Microbiol. Rev.* **2008**, *21*, 583–593. [CrossRef]
9. Nam, H.-J.; Lane, M.D.; Padron, E.; Gurda, B.L.; McKenna, R.; Kohlbrenner, E.; Aslanidi, G.; Byrne, B.; Muzyczka, N.; Zolotukhin, S.; et al. Structure of adeno-associated virus serotype 8, a gene therapy vector. *J. Virol.* **2007**, *81*, 12260–12271. [CrossRef]

10. Govindasamy, L.; DiMattia, M.A.; Gurda, B.L.; Halder, S.; McKenna, R.; Chiorini, J.A.; Muzyczka, N.; Zolotukhin, S.; Agbandje-McKenna, M. Structural insights into adeno-associated virus serotype 5. *J. Virol.* **2013**, *87*, 11187–11199. [CrossRef]
11. Xie, Q.; Bu, W.; Bhatia, S.; Hare, J.; Somasundaram, T.; Azzi, A.; Chapman, M.S. The atomic structure of adeno-associated virus (AAV-2), a vector for human gene therapy. *Proc. Natl. Acad. Sci. USA* **2002**, *99*, 10405–10410. [CrossRef] [PubMed]
12. DiMattia, M.A.; Nam, H.-J.; Van Vliet, K.; Mitchell, M.; Bennett, A.; Gurda, B.L.; McKenna, R.; Olson, N.H.; Sinkovits, R.S.; Potter, M.; et al. Structural insight into the unique properties of adeno-associated virus serotype 9. *J. Virol.* **2012**, *86*, 6947–6958. [CrossRef] [PubMed]
13. Ng, R.; Govindasamy, L.; Gurda, B.L.; McKenna, R.; Kozyreva, O.G.; Samulski, R.J.; Parent, K.N.; Baker, T.S.; Agbandje-McKenna, M. Structural characterization of the dual glycan binding adeno-associated virus serotype 6. *J. Virol.* **2010**, *84*, 12945–12957. [CrossRef] [PubMed]
14. Lerch, T.F.; Xie, Q.; Chapman, M.S. The structure of adeno-associated virus serotype 3B (AAV-3B): Insights into receptor binding and immune evasion. *Virology* **2010**, *403*, 26–36. [CrossRef] [PubMed]
15. Govindasamy, L.; Padron, E.; McKenna, R.; Muzyczka, N.; Kaludov, N.; Chiorini, J.A.; Agbandje-McKenna, M. Structurally mapping the diverse phenotype of adeno-associated virus serotype 4. *J. Virol.* **2006**, *80*, 11556–11570. [CrossRef]
16. Lerch, T.F.; O'Donnell, J.K.; Meyer, N.L.; Xie, Q.; Taylor, K.A.; Stagg, S.M.; Chapman, M.S. Structure of AAV-DJ, a retargeted gene therapy vector: Cryo-electron microscopy at 4.5 Å resolution. *Structure* **2012**, *20*, 1310–1320. [CrossRef]
17. Mietzsch, M.; Barnes, C.; Hull, J.A.; Chipman, P.; Xie, J.; Bhattacharya, N.; Sousa, D.; McKenna, R.; Gao, G.; Agbandje-McKenna, M. Comparative analysis of the capsid structures of AAVrh.10, AAVrh.39, and AAV8. *J. Virol.* **2020**, *94*, 6. [CrossRef]
18. Halder, S.; Van Vliet, K.; Smith, J.K.; Duong, T.T.P.; McKenna, R.; Wilson, J.M.; Agbandje-McKenna, M. Structure of neurotropic adeno-associated virus AAVrh.8. *J. Struct. Biol.* **2015**, *192*, 21–36. [CrossRef]
19. Kaelber, J.T.; Yost, S.A.; Webber, K.A.; Firlar, E.; Liu, Y.; Danos, O.; Mercer, A.C. Structure of the AAVhu.37 capsid by cryoelectron microscopy. *Acta Crystallogr. Sect. F Struct. Biol. Commun.* **2020**, *76 Pt 2*, 58–64. [CrossRef]
20. Venkatakrishnan, B.; Yarbrough, J.; Domsic, J.; Bennett, A.; Bothner, B.; Kozyreva, O.G.; Samulski, R.J.; Muzyczka, N.; McKenna, R.; Agbandje-McKenna, M. Structure and dynamics of adeno-associated virus serotype 1 VP1-unique N-terminal domain and its role in capsid trafficking. *J. Virol.* **2013**, *87*, 4974–4984. [CrossRef]
21. Bleker, S.; Sonntag, F.; Kleinschmidt, J.A. Mutational analysis of narrow pores at the fivefold symmetry axes of adeno-associated virus type 2 capsids reveals a dual role in genome packaging and activation of phospholipase A2 activity. *J. Virol.* **2005**, *79*, 2528–2540. [CrossRef] [PubMed]
22. Kaludov, N.; Brown, K.E.; Walters, R.W.; Zabner, J.; Chiorini, J.A. Adeno-associated virus serotype 4 (AAV4) and AAV5 both require sialic acid binding for hemagglutination and efficient transduction but differ in sialic acid linkage specificity. *J. Virol.* **2001**, *75*, 6884–6893. [CrossRef] [PubMed]
23. Wu, Z.; Miller, E.; Agbandje-McKenna, M.; Samulski, R.J. Alpha2,3 and alpha2,6 N-linked sialic acids facilitate efficient binding and transduction by adeno-associated virus types 1 and 6. *J. Virol.* **2006**, *80*, 9093–9103. [CrossRef] [PubMed]
24. Walters, R.W.; Yi, S.M.P.; Keshavjee, S.; Brown, K.E.; Welsh, M.J.; Chiorini, J.A.; Zabner, J. Binding of adeno-associated virus type 5 to 2,3-linked sialic acid is required for gene transfer. *J. Biol. Chem.* **2001**, *276*, 20610–20616. [CrossRef]
25. Summerford, C.; Samulski, R.J. Membrane-associated heparan sulfate proteoglycan is a receptor for adeno-associated virus type 2 virions. *J. Virol.* **1998**, *72*, 1438–1445. [CrossRef]
26. Handa, A.; Muramatsu, S.-I.; Qiu, J.; Mizukami, H.; Brown, K.E. Adeno-associated virus (AAV)-3-based vectors transduce haematopoietic cells not susceptible to transduction with AAV-2-based vectors. *J. Gen. Virol.* **2000**, *81 Pt 8*, 2077–2084. [CrossRef]
27. Halbert, C.L.; Allen, J.M.; Miller, A.D. Adeno-associated virus type 6 (AAV6) vectors mediate efficient transduction of airway epithelial cells in mouse lungs compared to that of AAV2 vectors. *J. Virol.* **2001**, *75*, 6615–6624. [CrossRef] [PubMed]
28. Schmidt, M.; Govindasamy, L.; Afione, S.; Kaludov, N.; Agbandje-McKenna, M.; Chiorini, J.A. Molecular characterization of the heparin-dependent transduction domain on the capsid of a novel adeno-associated virus isolate, AAV(VR-942). *J. Virol.* **2008**, *82*, 8911–8916. [CrossRef] [PubMed]
29. Mietzsch, M.; Broecker, F.; Reinhardt, A.; Seeberger, P.H.; Heilbronn, R.; Imperiale, M.J. Differential adeno-associated virus serotype-specific interaction patterns with synthetic heparins and other glycans. *J. Virol.* **2014**, *88*, 2991–3003. [CrossRef]
30. Bell, C.L.; Vandenberghe, L.H.; Bell, P.; Limberis, M.P.; Gao, G.-P.; Van Vliet, K.; Agbandje-McKenna, M.; Wilson, J.M. The AAV9 receptor and its modification to improve in vivo lung gene transfer in mice. *J. Clin. Investig.* **2011**, *121*, 2427–2435. [CrossRef]
31. Shen, S.; Bryant, K.D.; Brown, S.M.; Randell, S.H.; Asokan, A. Terminal N-linked galactose is the primary receptor for adeno-associated virus 9. *J. Biol. Chem.* **2011**, *286*, 13532–13540. [CrossRef] [PubMed]
32. Hahm, H.S.; Broecker, F.; Kawasaki, F.; Mietzsch, M.; Heilbronn, R.; Fukuda, M.; Seeberger, P.H. Automated glycan assembly of Oligo-N-acetyllactosamine and keratan sulfate probes to study virus-glycan interactions. *Chem* **2017**, *2*, 114–124. [CrossRef]
33. Pillay, S.; Meyer, N.L.; Puschnik, A.S.; Davulcu, O.; Diep, J.; Ishikawa, Y.; Jae, L.T.; Wosen, J.E.; Nagamine, C.M.; Chapman, M.S.; et al. An essential receptor for adeno-associated virus infection. *Nature* **2016**, *530*, 108–112. [CrossRef]
34. Akache, B.; Grimm, D.; Pandey, K.; Yant, S.R.; Xu, H.; Kay, M.A. The 37/67-kilodalton laminin receptor is a receptor for adeno-associated virus serotypes 8, 2, 3, and 9. *J. Virol.* **2006**, *80*, 9831–9836. [CrossRef]
35. Asokan, A.; Hamra, J.B.; Govindasamy, L.; Agbandje-McKenna, M.; Samulski, R.J. Adeno-associated virus type 2 contains an integrin alpha5beta1 binding domain essential for viral cell entry. *J. Virol.* **2006**, *80*, 8961–8969. [CrossRef] [PubMed]

36. Summerford, C.; Bartlett, J.S.; Samulski, R.J. AlphaVbeta5 integrin: A co-receptor for adeno-associated virus type 2 infection. *Nat. Med.* **1999**, *5*, 78–82. [CrossRef]
37. Kashiwakura, Y.; Tamayose, K.; Iwabuchi, K.; Hirai, Y.; Shimada, T.; Matsumoto, K.; Nakamura, T.; Watanabe, M.; Oshimi, K.; Daida, H. Hepatocyte growth factor receptor is a coreceptor for adeno-associated virus type 2 infection. *J. Virol.* **2005**, *79*, 609–614. [CrossRef]
38. Blackburn, S.D.; Steadman, R.A.; Johnson, F.B. Attachment of adeno-associated virus type 3H to fibroblast growth factor receptor 1. *Arch. Virol.* **2005**, *151*, 617–623. [CrossRef]
39. Di Pasquale, G.; Davidson, B.L.; Stein, C.S.; Martins, I.; Scudiero, D.; Monks, A.; Chiorini, J.A. Identification of PDGFR as a receptor for AAV-5 transduction. *Nat. Med.* **2003**, *9*, 1306–1312. [CrossRef]
40. Emmanuel, S.N.; Mietzsch, M.; Tseng, Y.S.; Smith, J.K.; Agbandje-McKenna, M. Parvovirus capsid-antibody complex structures reveal conservation of antigenic epitopes across the family. *Viral Immunol.* **2020**. [CrossRef]
41. Jose, A.; Mietzsch, M.; Smith, J.K.; Kurian, J.; Chipman, P.; McKenna, R.; Chiorini, J.; Agbandje-McKenna, M. High-resolution structural characterization of a new adeno-associated virus serotype 5 antibody epitope toward engineering antibody-resistant recombinant gene delivery vectors. *J. Virol.* **2018**, *93*, e01394-18. [CrossRef] [PubMed]
42. Mietzsch, M.; Smith, J.K.; Yu, J.C.; Banala, V.; Emmanuel, S.N.; Jose, A.; Chipman, P.; Bhattacharya, N.; McKenna, R.; Agbandje-McKenna, M. Characterization of AAV-specific affinity ligands: Consequences for vector purification and development strategies. *Mol. Ther. Methods Clin. Dev.* **2020**, *19*, 362–373. [CrossRef] [PubMed]
43. Suloway, C.; Pulokas, J.; Fellmann, D.; Cheng, A.; Guerra, F.; Quispe, J.; Stagg, S.; Potter, C.S.; Carragher, B. Automated molecular microscopy: The new Leginon system. *J. Struct. Biol.* **2005**, *151*, 41–60. [CrossRef] [PubMed]
44. Zheng, S.Q.; Palovcak, E.; Armache, J.-P.; Verba, K.A.; Cheng, Y.; Agard, D. MotionCor2: Anisotropic correction of beam-induced motion for improved cryo-electron microscopy. *Nat. Methods* **2017**, *14*, 331–332. [CrossRef]
45. Grant, T.; Rohou, A.; Grigorieff, N. cisTEM, user-friendly software for single-particle image processing. *eLife* **2018**, *7*, e35383. [CrossRef]
46. Mietzsch, M.; Li, Y.; Kurian, J.; Smith, J.K.; Chipman, P.; McKenna, R.; Yang, L.; Agbandje-McKenna, M. Structural characterization of a bat Adeno-associated virus capsid. *J. Struct. Biol.* **2020**, *211*, 107547. [CrossRef]
47. Emsley, P.; Cowtan, K. Coot: Model-building tools for molecular graphics. *Acta Crystallogr. Sect. D Biol. Crystallogr.* **2004**, *60 Pt 12*, 2126–2132. [CrossRef]
48. Pettersen, E.F.; Goddard, T.D.; Huang, C.C.; Couch, G.S.; Greenblatt, D.M.; Meng, E.C.; Ferrin, T.E. UCSF Chimera—A visualization system for exploratory research and analysis. *J. Comput. Chem.* **2004**, *25*, 1605–1612. [CrossRef]
49. Biasini, M.; Bienert, S.; Waterhouse, A.; Arnold, K.; Studer, G.; Schmidt, T.; Kiefer, F.; Cassarino, T.G.; Bertoni, M.; Bordoli, L.; et al. SWISS-MODEL: Modelling protein tertiary and quaternary structure using evolutionary information. *Nucleic Acids Res.* **2014**, *42*, W252–W258. [CrossRef]
50. Carrillo-Tripp, M.; Shepherd, C.M.; Borelli, I.A.; Venkataraman, S.; Lander, G.; Natarajan, P.; Johnson, J.E.; Brooks, C.L., III; Reddy, V.S. VIPERdb2: An enhanced and web API enabled relational database for structural virology. *Nucleic Acids Res.* **2009**, *37*, D436–D442. [CrossRef]
51. Tang, G.; Peng, L.; Baldwin, P.R.; Mann, D.S.; Jiang, W.; Rees, I.; Ludtke, S.J. EMAN2: An extensible image processing suite for electron microscopy. *J. Struct. Biol.* **2007**, *157*, 38–46. [CrossRef] [PubMed]
52. Kleywegt, G.J.; Jones, T.A. xdlMAPMAN and xdlDATAMAN—Programs for reformatting, analysis and manipulation of biomacromolecular electron-density maps and reflection data sets. *Acta Crystallogr. Sect. D Biol. Crystallogr.* **1996**, *52 Pt 4*, 826–828. [CrossRef] [PubMed]
53. Adams, P.D.; Afonine, P.V.; Bunkóczi, G.; Chen, V.B.; Davis, I.W.; Echols, N.; Headd, J.J.; Hung, L.-W.; Kapral, G.J.; Grosse-Kunstleve, R.W.; et al. PHENIX: A comprehensive Python-based system for macromolecular structure solution. *Int. Tables Crystallogr.* **2010**, *66 Pt 2*, 539–547. [CrossRef]
54. Krissinel, E.; Henrick, K. Secondary-structure matching (SSM), a new tool for fast protein structure alignment in three dimensions. *Acta Crystallogr. Sect. D Biol. Crystallogr.* **2004**, *60 Pt 12*, 2256–2268. [CrossRef]
55. Mikals, K.; Nam, H.-J.; Van Vliet, K.; Vandenberghe, L.H.; Mays, L.E.; McKenna, R.; Wilson, J.M.; Agbandje-McKenna, M. The structure of AAVrh32.33, a novel gene delivery vector. *J. Struct. Biol.* **2014**, *186*, 308–317. [CrossRef] [PubMed]
56. Burg, M.; Rosebrough, C.; Drouin, L.M.; Bennett, A.; Mietzsch, M.; Chipman, P.; McKenna, R.; Sousa, D.; Potter, M.; Byrne, B.; et al. Atomic structure of a rationally engineered gene delivery vector, AAV2.5. *J. Struct. Biol.* **2018**, *203*, 236–241. [CrossRef] [PubMed]
57. Bennett, A.; Keravala, A.; Makal, V.; Kurian, J.; Belbellaa, B.; Aeran, R.; Tseng, Y.-S.; Sousa, D.; Spear, J.; Gasmi, M.; et al. Structure comparison of the chimeric AAV2.7m8 vector with parental AAV2. *J. Struct. Biol.* **2019**, *209*, 107433. [CrossRef] [PubMed]
58. Tan, Y.Z.; Aiyer, S.; Mietzsch, M.; Hull, J.A.; McKenna, R.; Grieger, J.; Samulski, R.J.; Baker, T.S.; Agbandje-McKenna, M.; Lyumkis, D. Sub-2 Å Ewald curvature corrected structure of an AAV2 capsid variant. *Nat. Commun.* **2018**, *9*, 3628. [CrossRef]
59. Guenther, C.M.; Brun, M.J.; Bennett, A.D.; Ho, M.L.; Chen, W.; Zhu, B.; Lam, M.; Yamagami, M.; Kwon, S.; Bhattacharya, N.; et al. Protease-activatable adeno-associated virus vector for gene delivery to damaged heart tissue. *Mol. Ther.* **2019**, *27*, 611–622. [CrossRef]
60. DeLano, W.L. *The PyMOL Molecular Graphics System*; DeLano Scientific: San Carlos, CA, USA, 2002.
61. Mori, S.; Takeuchi, T.; Enomoto, Y.; Kondo, K.; Sato, K.; Ono, F.; Sata, T.; Kanda, T. Tissue distribution of cynomolgus adeno-associated viruses AAV10, AAV11, and AAVcy.7 in naturally infected monkeys. *Arch. Virol.* **2008**, *153*, 375–380. [CrossRef]

62. Schmidt, M.; Voutetakis, A.; Afione, S.; Zheng, C.; Mandikian, D.; Chiorini, J.A. Adeno-associated virus type 12 (AAV12): A novel AAV serotype with sialic acid- and heparan sulfate proteoglycan-independent transduction activity. *J. Virol.* **2008**, *82*, 1399–1406. [CrossRef]
63. Samaranch, L.; Salegio, E.A.; San Sebastian, W.; Kells, A.P.; Bringas, J.R.; Forsayeth, J.; Bankiewicz, K.S. Strong cortical and spinal cord transduction after AAV7 and AAV9 delivery into the cerebrospinal fluid of nonhuman primates. *Hum. Gene Ther.* **2013**, *24*, 526–532. [CrossRef] [PubMed]
64. Zhang, H.; Yang, B.; Mu, X.; Ahmed, S.S.; Su, Q.; He, R.; Wang, H.; Mueller, C.; Sena-Esteves, M.; Brown, R.; et al. Several rAAV vectors efficiently cross the blood–brain barrier and transduce neurons and astrocytes in the neonatal mouse central nervous system. *Mol. Ther.* **2011**, *19*, 1440–1448. [CrossRef] [PubMed]
65. Mietzsch, M.; Grasse, S.; Zurawski, C.; Weger, S.; Bennett, A.; Agbandje-McKenna, M.; Muzyczka, N.; Zolotukhin, S.; Heilbronn, R. OneBac: Platform for scalable and high-titer production of adeno-associated virus serotype 1–12 vectors for gene therapy. *Hum. Gene Ther.* **2014**, *25*, 212–222. [CrossRef] [PubMed]
66. Tseng, Y.-S.; Van Vliet, K.; Rao, L.; McKenna, R.; Byrne, B.J.; Asokan, A.; Agbandje-McKenna, M. Generation and characterization of anti-adeno-associated virus serotype 8 (AAV8) and anti-AAV9 monoclonal antibodies. *J. Virol. Methods* **2016**, *236*, 105–110. [CrossRef] [PubMed]
67. Gurda, B.L.; Raupp, C.; Popa-Wagner, R.; Naumer, M.; Olson, N.H.; Ng, R.; McKenna, R.; Baker, T.S.; Kleinschmidt, J.A.; Agbandje-McKenna, M. Mapping a neutralizing epitope onto the capsid of adeno-associated virus serotype 8. *J. Virol.* **2012**, *86*, 7739–7751. [CrossRef]
68. Havlik, L.P.; Simon, K.E.; Smith, J.K.; Klinc, K.A.; Tse, L.V.; Oh, D.K.; Fanous, M.M.; Meganck, R.M.; Mietzsch, M.; Agbandje-McKenna, M.; et al. Co-evolution of AAV capsid antigenicity and tropism through a structure-guided approach. *J. Virol.* **2020**, *94*, e00976-20. [CrossRef]
69. Shen, S.; Troupes, A.N.; Pulicherla, N.; Asokan, A. Multiple roles for sialylated glycans in determining the cardiopulmonary tropism of adeno-associated virus 4. *J. Virol.* **2013**, *87*, 13206–13213. [CrossRef]
70. Westhaus, A.; Cabanes-Creus, M.; Rybicki, A.; Baltazar, G.; Navarro, R.G.; Zhu, E.; Drouyer, M.; Knight, M.; Albu, R.F.; Ng, B.H.; et al. High-throughput in vitro, ex vivo, and in vivo screen of adeno-associated virus vectors based on physical and functional transduction. *Hum. Gene Ther.* **2020**, *31*, 575–589. [CrossRef]
71. Quinn, K.; Quirion, M.R.; Lo, C.-Y.; Misplon, J.A.; Epstein, S.L.; Chiorini, J.A. Intranasal administration of adeno-associated virus type 12 (AAV12) leads to transduction of the nasal epithelia and can initiate transgene-specific immune response. *Mol. Ther.* **2011**, *19*, 1990–1998. [CrossRef]
72. Sonntag, F.; Schmidt, K.; Kleinschmidt, J.A. A viral assembly factor promotes AAV2 capsid formation in the nucleolus. *Proc. Natl. Acad. Sci. USA* **2010**, *107*, 10220–10225. [CrossRef] [PubMed]
73. Earley, L.F.; Powers, J.M.; Adachi, K.; Baumgart, J.T.; Meyer, N.L.; Xie, Q.; Chapman, M.S.; Nakai, H. Adeno-associated virus (AAV) assembly-activating protein is not an essential requirement for capsid assembly of AAV serotypes 4, 5, and 11. *J. Virol.* **2017**, *91*, e01980-16. [CrossRef] [PubMed]
74. Naumer, M.; Sonntag, F.; Schmidt, K.; Nieto, K.; Panke, C.; Davey, N.E.; Popa-Wagner, R.; Kleinschmidt, J.A. Properties of the adeno-associated virus assembly-activating protein. *J. Virol.* **2012**, *86*, 13038–13048. [CrossRef] [PubMed]
75. Maurer, A.C.; Cepeda Diaz, A.K.; Vandenberghe, L.H. Residues on AAV capsid lumen dictate interactions and compatibility with the assembly-activating protein. *J. Virol.* **2019**, *93*, e02013-18. [CrossRef]
76. Lerch, T.F.; Chapman, M.S. Identification of the heparin binding site on adeno-associated virus serotype 3B (AAV-3B). *Virology* **2012**, *423*, 6–13. [CrossRef]
77. Wobus, C.E.; Hügle-Dörr, B.; Girod, A.; Petersen, G.; Hallek, M.; Kleinschmidt, J.A. Monoclonal antibodies against the adeno-associated virus type 2 (AAV-2) capsid: Epitope mapping and identification of capsid domains involved in AAV-2–cell interaction and neutralization of AAV-2 infection. *J. Virol.* **2000**, *74*, 9281–9293. [CrossRef]
78. Bennett, A.D.; Wong, K.; Lewis, J.; Tseng, Y.-S.; Smith, J.K.; Chipman, P.; McKenna, R.; Samulski, R.J.; Kleinschmidt, J.; Agbandje-McKenna, M. AAV6 K531 serves a dual function in selective receptor and antibody ADK6 recognition. *Virology* **2018**, *518*, 369–376. [CrossRef]

Article

Characterization of the GBoV1 Capsid and Its Antibody Interactions

Jennifer Chun Yu [1], Mario Mietzsch [1], Amriti Singh [1,†], Alberto Jimenez Ybargollin [1], Shweta Kailasan [1,‡], Paul Chipman [1], Nilakshee Bhattacharya [2,§], Julia Fakhiri [3,‖], Dirk Grimm [3], Amit Kapoor [4], Indrė Kučinskaitė-Kodzė [5], Aurelija Žvirblienė [5], Maria Söderlund-Venermo [6], Robert McKenna [1] and Mavis Agbandje-McKenna [1,*]

1 Department of Biochemistry and Molecular Biology, Center for Structural Biology, College of Medicine, University of Florida, Gainesville, FL 32610, USA; jennifer.yu@ufl.edu (J.C.Y.); mario.mietzsch@ufl.edu (M.M.); amritisingh@ufl.edu (A.S.); albertojimenezy@cop.ufl.edu (A.J.Y.); shwetakailasan@gmail.com (S.K.); pchipman@ufl.edu (P.C.); rmckenna@ufl.edu (R.M.)
2 Biological Science Imaging Resource, Department of Biological Sciences, Florida State University, Tallahassee, FL 32306, USA; nilakshee.bhattacharya@duke.edu
3 Department of Infectious Diseases/Virology, Medical Faculty, BioQuant, University of Heidelberg, 69120 Heidelberg, Germany; julia.fakhiri@hotmail.com (J.F.); dirk.grimm@bioquant.uni-heidelberg.de (D.G.)
4 Center for Vaccines and Immunity, The Research Institute at Nationwide Children's Hospital, Columbus, OH 43220, USA; Amit.Kapoor@Nationwidechildrens.org
5 Department of Immunology and Cell Biology of the Institute of Biotechnology of Vilnius University, 10257 Vilnius, Lithuania; indkuc@ibt.lt (I.K.-K.); azvirb@ibt.lt (A.Ž.)
6 Department of Virology, University of Helsinki, 00014 Helsinki, Finland; maria.soderlund-venermo@helsinki.fi
* Correspondence: mckenna@ufl.edu
† Current address: 410 Salt Meadow Circle, Apt 306, Bradenton, FL 34208, USA.
‡ Current address: Integrated Biotherapeutics, Inc., 4 Research Ct, Suite 300, Rockville, MD 20850, USA.
§ Current address: Shared Materials Instrumentation Facility, 101 Science Dr, Duke University, Durham, NC 27708, USA.
‖ Current address: Roche Diagnostics, Pharmaceutical Research and Early Development, Department for Cell Technologies, Nonnenwald 2, 82377 Penzberg, Germany.

Citation: Yu, J.C.; Mietzsch, M.; Singh, A.; Jimenez Ybargollin, A.; Kailasan, S.; Chipman, P.; Bhattacharya, N.; Fakhiri, J.; Grimm, D.; Kapoor, A.; et al. Characterization of the GBoV1 Capsid and Its Antibody Interactions. *Viruses* **2021**, *13*, 330. https://doi.org/10.3390/v13020330

Academic Editor: Giorgio Gallinella

Received: 15 January 2021
Accepted: 15 February 2021
Published: 20 February 2021

Abstract: Human bocavirus 1 (HBoV1) has gained attention as a gene delivery vector with its ability to infect polarized human airway epithelia and 5.5 kb genome packaging capacity. Gorilla bocavirus 1 (GBoV1) VP3 shares 86% amino acid sequence identity with HBoV1 but has better transduction efficiency in several human cell types. Here, we report the capsid structure of GBoV1 determined to 2.76 Å resolution using cryo-electron microscopy (cryo-EM) and its interaction with mouse monoclonal antibodies (mAbs) and human sera. GBoV1 shares capsid surface morphologies with other parvoviruses, with a channel at the 5-fold symmetry axis, protrusions surrounding the 3-fold axis and a depression at the 2-fold axis. A 2/5-fold wall separates the 2-fold and 5-fold axes. Compared to HBoV1, differences are localized to the 3-fold protrusions. Consistently, native dot immunoblots and cryo-EM showed cross-reactivity and binding, respectively, by a 5-fold targeted HBoV1 mAb, 15C6. Surprisingly, recognition was observed for one out of three 3-fold targeted mAbs, 12C1, indicating some structural similarity at this region. In addition, GBoV1, tested against 40 human sera, showed the similar rates of seropositivity as HBoV1. Immunogenic reactivity against parvoviral vectors is a significant barrier to efficient gene delivery. This study is a step towards optimizing bocaparvovirus vectors with antibody escape properties.

Keywords: bocavirus; capsid; parvovirus; cryo-EM; gene therapy; antigenicity

1. Introduction

Gorilla bocavirus 1 (GBoV1) is a member of the genus *Bocaparvovirus* in the *Parvoviridae* that contain single-stranded DNA (ssDNA) packaging viruses [1]. The family is divided into three subfamilies, including the subfamily *Parvovirinae* whose members infect vertebrate hosts [1]. Bocaparvoviruses represents the largest genus in this subfamily, with 21 classified species that infect a variety of hosts, including cows, rabbits, rodents, humans and non-human primates [2–10]. Bovine parvovirus (BPV) is the first discovered member of the genus, isolated from cattle in 1959 [3]. The first discovered member infecting humans is human bocavirus 1 (HBoV1), isolated in 2005 from nasopharyngeal aspirates of children under 2 years of age with acute respiratory infections [7]. Enteric strains, HBoV2-4, were then described from children with acute gastroenteritis [8,9]. GBoV1 was isolated from gorillas with enteritis and is the first identified non-human primate bocavirus [10].

The bocaviruses have a ~5.5 kb genome that consist of three open reading frames (ORFs), the *ns* gene on the left end, the *cap* gene on the right end and the *np* gene between *ns* and *cap* [7]. The three ORFs are flanked by two non-identical hairpin structures and are transcribed from a single promoter (p5) to generate a single pre-mRNA that is alternatively spliced [11,12]. The *ns* gene encodes non-structural proteins that are essential for viral DNA replication [11]. The *np* gene encodes the NP1 protein, which was shown to be play a role in pre-mRNA processing and subsequent capsid protein expression [11,13]. The structural proteins that form the viral capsid, VP1, VP2 and VP3, are encoded by the *cap* gene [14]. Sixty copies of VP1, VP2 and VP3 assemble one T = 1 icosahedral capsid in an approximate 1:1:10 ratio in 2-, 3-, 5-related symmetries [14,15]. The three VPs share the same C-terminus. VP3 is the major capsid protein and is the smallest of the three VPs. VP2 shares a common region with VP1 at the N-terminus, called the VP1/2 common region [16]. The unique region of the minor VP1 protein N-terminus (VP1u) contains a phospholipase activity (PLA2) shown to be responsible for endosomal escape during trafficking to the nucleus and is absolutely required for infectivity [17–19]. The VP1u is hypothesized to externalize through a channel located at the 5-fold axis of the capsid to activate its PLA2 activity [20,21]. While VP1 incorporation is essential for viral infectivity, the VP3 protein alone was shown to form intact capsids, termed VP3-only capsids, with similar antigenicity to wild-type capsids [22,23]. It is the capsid that interacts with the host environment and is a determinant of host and cell recognition, host immune response and cell entry [24].

Previously, the capsid structures of HBoV1-4 have been reported that showed conserved features across the *Parvoviridae* family, as well as features unique to the genus [23,25]. The bocavirus VP monomer has a conserved eight-stranded β-barrel motif (βB to βI), forming the interior of the capsid, with a βA strand that runs antiparallel to βB and an α-helix (αA) located between strands βC and βD of the β-barrel [16]. The loops between the β strands form the surface of the capsid and these surface loops are labeled based on the flanking β strands, for example, the loop between βD and βE is the DE loop. Within the loops are defined variable regions (VRs), ranging from VR-I to VR-IX. These variable regions (VRs) were previously defined for the bocaviruses based on comparisons to the structure of bovine parvovirus (BPV), the first capsid structure solved from the *Bocaparvoviridae* genus [26]. In addition to the conserved β-barrel motif, βA and αA, the bocaviruses were reported to possess a unique α-helix (αB) located near VR-III as well as a basket-like structure beneath the 5-fold channel [23].

Recently, HBoV1 and GBoV1, along with enteric human strains HBoV2-4, were proposed as gene therapy delivery vectors. The interest in HBoV1 stems from its specific tropism for the apical side of polarized human airway epithelia (pHAE), which is optimal for the treatment of cystic fibrosis [27,28]. Gene therapy has been the "gold standard" for the treatment of monogenetic diseases such as cystic fibrosis but Adeno-associated virus serotype 2 (AAV2), a vector used for treatment was shown to be inefficient at delivery to the lung and has tropism for the basolateral side of pHAE [29–31]. In addition to the favorable tropism of HBoV1, the expanded genome capacity of bocaviruses (5.5 kb), compared to AAV, allows the packaging of the full-length 4.7 kb *CFTR* gene [7,28]. Due to these advantages, various studies have aimed to optimize a recombinant (r)AAV2/HBoV1 pseudotyped vector, a HBoV1 capsid-based vector that packages a transgene with rAAV2 inverted terminal repeats (ITRs), for delivery of the *CFTR* gene. This vector was shown to have tropisms similar to the HBoV1 wild-type virus [28,32,33].

Despite the advantages of HBoV1 as a gene therapy vector, the capsid has high seroprevalence in the human population, a hurdle to therapeutic gene delivery. GBoV1 is an alternative to HBoV1, as it is capable of also infecting the apical side of pHAE efficiently, package 5.5 kb and less susceptible to neutralization by pooled intravenous immunoglobulin (IVIgs) [33]. The goal of this study was to characterize the GBoV1 capsid and to better understand the functional regions including antigenicity of the capsid. We report the high-resolution structure of the GBoV1 capsid, determined by cryo-electron microscopy (cryo-EM) and 3D single-particle reconstruction. The GBoV1 VP3 monomer conserves features common to parvoviruses and contains features unique to bocaviruses, for example, α-helix αB. Within the capsid interior is a 5-fold basket-like density, which appears smaller when compared to HBoV1-4. The main differences between the HBoV1 and GBoV1 monomer are localized to VR-I, VR-III and VR-V. Low-resolution structures of the GBoV1 capsid complexed with mouse monoclonal antibodies (mAbs) 15C6 and 12C1, originally generated against HBoV1, show epitopes localized to the 5-fold and 3-fold axes, respectively, in agreement with reports for HBoV1 [22]. However, two other mAbs, targeted at the HBoV1 3-fold axis, did not recognize GBoV1 indicating structural variation at this capsid region between the two viruses. Both viruses showed comparable high seroprevalence against human sera suggesting a high degree of cross-reactivity for the sera tested and a conserved capsid region, such as the 5-fold region, as forming the epitopes. These observations begin to unravel the antigenic properties of GBoV1 and provide information that could aid engineering vectors with reduced antigenic reactivity and, thus, therapeutic efficacy.

2. Methods

2.1. Virus-Like Particle Production and Purification

The Bac-to-Bac baculovirus system was used for the expression of HBoV1, HBoV4, GBoV1, AAV2 and AAV5 virus-like particles (VLPs) as described previously for HBoV1 (VP3 only), AAV2 and AAV5 [22,34]. For the generation of GBoV1 expressing VP1, VP2 and VP3 (termed wild-type), the entire *cap* gene (NCBI accession no. NC_014358.1) was cloned and inserted into the pFastBac plasmid with an ACG start codon for VP1. For the generation of GBoV1 VP3-only VLPs, the VP3-encoding sequence (without the VP1 unique region and the VP1/2 common region) was directly cloned and inserted into the pFastbac plasmid for transposition into the baculovirus expression vector. Based on the standard manufacturer's protocol, the baculovirus expression vectors were then used to generate recombinant baculovirus stocks expressing GBoV1 wild-type and VP3-only VLPs [35]. Briefly, *Sf*9 insect cells, maintained in SFM Sf9-900 medium (Thermo Fisher, Waltham, MA, USA) with 10% fetal bovine serum (FBS) and antibiotics, were infected at a multiplicity of infection (MOI) of 5 and harvested 72 h post-infection. Sucrose cushion and sucrose density gradients were performed for the purification of VLPs after three rounds of freeze-thaw cycles and benzonase (Millipore) treatment of the cell pellets, as described previously [15]. Purified samples were dialyzed into 1× phosphate-buffered saline (PBS) (2.8 mM KCl, 137 mM NaCl, 10 mM Na_2HPO_4, 1.8 mM KH_2PO_4) and concentrated to 0.5–2 mg/mL using Apollo concentrators (Orbital Biosciences, Topsfield, MA, USA). Purity and capsid integrity of the VLPs was confirmed by sodium dodecyl sulfate polyacrylamide gel electrophoresis (SDS-PAGE) and negative stain electron microscopy analysis using a Tecnai G2 Spirit electron microscope (FEI, Hillsboro, OR, USA) at 120 kV, respectively, as described previously [22].

2.2. Dot Immunoblot Analysis

In dot immunoblots HBoV1 VP3-specific mAbs were probed against GBoV1 (VP3 only and WT), HBoV1 (VP3 only), HBoV4 (VP3 only), AAV2 and AAV5 native capsids. To determine the seroprevalence, human sera from healthy donors (Valley Biomedical, Winchester, VA, USA) were utilized in dot immunoblots against GBoV1, HBoV1 (VP3 only), AAV2 and AAV5. H1-H1 polyclonal antibody (rabbit double-immunized with HBoV1 VP3 VLPs) was used against denatured capsids as a positive BoV control [36]. Dot blots were performed on nitrocellulose membrane dipped in 1× PBS. Denatured (incubating capsids at 100 °C for 5 min) and non-denatured VLPs were applied directly to the membrane at approximately 10 ng^{-1} µg using a vacuum manifold, letting the sample incubate for 10 min. The membranes were blocked in 6% milk in 1× PBS overnight at 4 °C. For the primary antibody, mAb 15C6 was added at a 1:2000 dilution; mAbs 12C1, 4C2 and 9G12 [22] were added at a 1:1000 dilution; H1-H1 was added at a 1:1000 dilution; human sera samples were added at a 1:500 dilution in 6% milk in PBS-T (PBS with 0.1% Tween) and incubated on the membrane for 1.5 h at room temperature (RT). Secondary anti-mouse and anti-rabbit antibodies were applied to the membrane in a 1:5000 dilution and anti-human IgG was applied in a 1:50,000 dilution in 6% milk in PBS-T for 1 h at RT. Five min washes were performed three times before and after incubation with secondary antibody. Blocking, primary and secondary antibody steps were all performed on a shaker. Finally, luminol substrate was applied to the membrane and incubated in the dark for 1 min, before the membrane was exposed to X-ray radiography film and developed.

2.3. Generation of Fab Fragments

Production of the 15C6, 12C1, 4C2 and 9G12 mAbs from BALB/c mice injected with HBoV1 VLPs and subsequent purification of IgGs were previously described [22]. IgG from 15C6 and 12C1 at a concentration of approximately 1–2 mg/mL was buffer-exchanged into 20 mM sodium phosphate, pH 7.0, 10 mM EDTA for papain cleavage. Papain was added to the IgG samples and incubated for 16–20 h at 37 °C with rotation. The papain-IgG mixture was centrifuged at a low speed (1000× *g*) and the supernatant containing cleaved Fab fragments was loaded onto a protein A column. The Fc portion of the cleaved IgG was captured in the protein A column and flowthrough containing the desired Fab fragments was collected. This flowthrough was concentrated to ~0.5 mg/mL in an Apollo concentrator with a 9 kDa molecular mass cutoff. sodium dodecyl sulfate-polyacrylamide gel electrophoresis (SDS-PAGE) was performed to confirm the purity of the Fab fragment sample.

2.4. Preparation of GBoV1-Fab Complexes and GBoV1 VLPs for Cryo-EM Data Collection

GBoV1 VLPs were mixed with the 15C6 and 12C1 Fab fragments at a 1:120 to 1:180 (VLP:Fab) ratio to ensure binding-site saturation. The VLP: antibody complexes were incubated on ice for 30 min to 1 h and 3 µL of sample was vitrified onto C-flat holey carbon grids (Protochips, Inc.) using the Vitrobot Mark IV (FEI Co.). For the GBoV1 VLPs (wild-type and VP3-only), 3 µL of sample was prepared as described for the complexes.

2.5. Cryo-EM Data Collection

For the GBoV1-Fab complexes, micrographs were collected from frozen grids using a Tecnai G2 F20-TWIN transmission electron microscope (FEI) with a 200 kV voltage under low-dose conditions (20 $e^-/Å^2$) at a magnification of 82,500× on a 16-megapixel charge coupled device (CCD) camera with pixel size 15 µm, resulting in micrographs with a pixel size of 1.82 Å. This microscope and camera were also used for screening grids for ice quality and particle distribution of GBoV1 prior to high-resolution data collection and also for collecting a low-resolution data set for comparison to the complex structures. For the GBoV1 VLP-alone high-resolution studies, holey carbon grids with vitrified VLPs were used to collect micrograph movie frames on a Titan Krios electron microscope (FEI Co.) operated at 300 kV with a K3 DED using the Leginon application. High-resolution data collection was performed with a total dose of 60 to 67 $e^-/Å^2$ for up to 50 movie frames per micrograph. The movie frames collected on the K3 detector were aligned using MotionCor2 with dose weighting as previously described [37]. Data sets were collected as part of the NIH project "Southeastern Center for Microscopy of MacroMolecular Machines" (SECM4).

2.6. 3D Particle Reconstruction

The cisTEM software package was used for three-dimensional (3D) image reconstruction of both, wild-type and antibody-complex structures [38]. The aligned micrographs were first imported and their microscope-based contrast transfer function (CTF) estimated. Suboptimal-quality micrographs were eliminated. Capsids on the remaining micrographs were automatically selected using a particle radius of 125 Å. The selected capsids were subjected to 2D classification and undesirable classes, such as ice or impurities, were removed from the dataset. Both, *ab-initio* 3D reconstruction and automatic refinement was performed under default settings. *Ab-initio* 3D reconstruction generated an initial low-resolution model with 10% of the total boxed particles with imposed icosahedral symmetry and automatically refined with the entire dataset. Map sharpening of the high-resolution structure used a pre-cut off B-factor value of 90 $Å^2$ and variable post-cut off B-factor values such as 0, 20 and 40 $Å^2$. Using the UCSF-Chimera software, the sharpened density maps were analyzed and the -90 $Å^2/0$ $Å^2$ map was used for further model building and structure refinement. The final resolution of the structures was estimated based on a Fourier shell correlation (FSC) threshold criterion of 0.143 (Table 1).

Table 1. Summary of data collection, processing and refinement statistics.

Parameter	GBoV1
Total no. of micrographs	1411
Defocus range (μm)	1.08–3.19
Electron dose ($e^-/Å^2$)	60
No. of frames/micrograph	50
Pixel size (Å/pixel)	1.08
No. of capsids used for final map	168,565
Resolution of final map (Å)	2.76
PHENIX model refinement statistics	
Residue range	33–542
Map CC	0.877
RMSD (Å)	
Bonds	0.01
Angles	0.89
All-atom clash score	10.61
Ramachandran plot (%)	
Favored	98.4
Allowed	1.6
Outliers	0.0
Rotamer outliers	0.0
No. of Cβ deviations	0

2.7. Model Building and Structure Refinement

The 3D model for GBoV1 wild-type VP3 monomer was generated from the protein sequence (NCBI accession ADK34012.1) in the online program SWISS-MODEL using the structure of HBoV1 (Research Collaboratory for Structural Bioinformatics [RCSB] PDB code 5URF) as a template [39]. This reference monomer model was used to generate a 60mer (based on 60 copies of the VP3 protein) with the VIPERdb2 oligomer generator [40] and docked as rigid bodies into the GBoV1 density map using the "fit in map" subroutine in UCSF-Chimera [41]. The docked VP monomer model was adjusted to better fit the GBoV1 wild-type cryo-reconstructed density map with manual model-building tools and real-space-refine options in Coot [42]. Further refinement was performed on the model with PHENIX, using the real-space-refinement subroutine under default settings for five macrocycles [43]. This refined model was inspected in Coot and amino acid side chains are adjusted, if needed, for favorable statistics. After another round of refinement in PHENIX, an icosahedral model was generated from 60 copies of the refined VP3 monomer with the

VIPERdb2 oligomer generator. The 60-mer VP3 monomer was further refined in PHENIX using B-factor refinement options.

2.8. Antibody Epitope Mapping

The high-resolution 60-mer GBoV1 structure was rigid body-docked into the antibody-complex density maps, using the "fit in map" subroutine in UCSF-Chimera. A generic Fab (PDB ID: 2FBJ) was fitted into the density of the Fab using the same subroutine in UCSF-Chimera [41]. The resulting pseudo-atomic model was used to generate a roadmap using RIVEM [44]. The contact residues, residues in the interface between the capsid and antibody structure, were identified via manual inspection in the program Coot [42]. Occluded residues were also identified manually by generating a roadmap in RIVEM using the GBoV1 structure and generic Fab.

2.9. Sequence and Structural Comparison

The VP3 models of HBoV1-4 and GBoV1 were analyzed in Coot and using the superposition tool. Overall paired root mean squared deviations (RMSD) were calculated between Cα positions. The distances between Cα positions of regions with insertions or deletions were manually measured in Coot with the distance tool. Regions with two or more adjacent amino acids and a greater than 2 Å difference determined by Coot are considered to be structurally diverse and are assigned to previously described VRs [42].

2.10. Structure Accession Numbers

The GBoV1 WT cryo-EM reconstructed density map and model built for the capsid were deposited in the Electron Microscopy Data Bank (EMDB) with accession numbers EMD-23460 and PDB ID 7LNK, respectively.

3. Results & Discussion

3.1. GBoV1 Shares Conserved Capsid Features with the HBoVs

The GBoV1 VLPs were produced using recombinant baculovirus expressing either GBoV1 VP1, VP2 and VP3 (GBoV1 WT) or only VP3 (GBoV1 VP3-only) in *Sf*9 insect cells. The purified GBoV1 WT sample was analyzed on SDS-PAGE to confirm the presence and purity of VP1, VP2 and VP3 with a corresponding molecular weight of approximately 60, 65 and 80 kDa. For GBoV1 VP3-only, only one band is present at 60 kDa (Figure 1a). Cryo-EM micrographs confirmed the presence of intact capsids with a diameter of approximately 250 Å without the presence of contaminants (Figure 1a). Thus, the samples were deemed suitable for data collection for high-resolution structure determination and movie frame micrographs were collected. 3D image reconstruction of 168,565 GBoV1 WT and 218,746 GBoV1 VP3-only capsids resulted in structures with an estimated resolution of 2.76 Å for both types of capsids based on an FSC threshold of 0.143 (Figure 1b–d, Table 1).

The GBoV1 capsid structures were identical, despite being assembled from different VP compositions (Figure 1a). They share the conserved features of the *Parvoviridae* subfamily, with a channel at the 5-fold symmetry axis, protrusions around the 3-fold axis, the 2/5-fold wall, located between the depressions at the 2- and 5-fold axes and depressions at the 2-fold as well as around the 5-fold axis (Figure 1b,c) [16]. The basket-like structure beneath the 5-fold channel previously reported within the capsids of HBoV1-HBoV4 and BPV [15,23,25,26], was less pronounced in GBoV1 suggesting less order at the N-terminus of the VP (Figure 1c). This basket contains residues located at the N-terminus of VP3 and is part of a glycine-rich region hypothesized, for parvoviruses, to act as a hinge for the externalization of the VP1u to utilize its PLA2 activity. The location of this density below the 5-fold axes is consistent with the suggested use of this channel for the VP1u externalization [19,20]. Interestingly, this region of GBoV1, residues 1–32 (VP3 numbering), is similar to the analogous regions of HBoV2-4 (aa1-32) with a sequence identity ranging from 69–79% but shares only a 50% sequence identity with HBoV1 despite the higher sequence identity across the entire VP3 for all five viruses (Table 2).

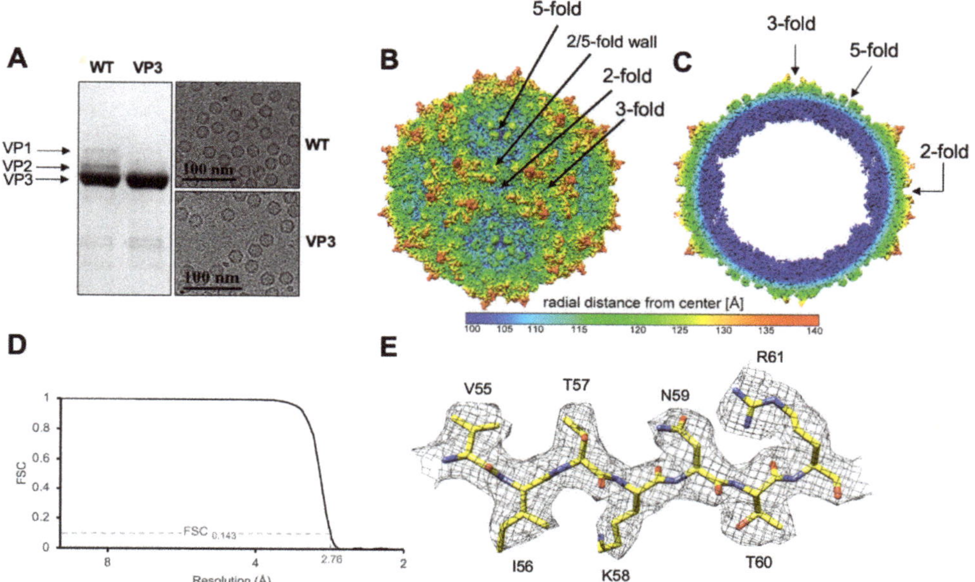

Figure 1. The capsid structure of Gorilla bocavirus 1 (GBoV1). (**A**) sodium dodecyl sulfate-polyacrylamide gel electrophoresis (SDS-PAGE) of GBoV1 WT and VP3 only samples confirming the presence of VP1, VP2 and VP3 (~80, 65, 60 kDa) and cryo-electron micrograph showing intact viral particles. (**B**) Capsid density map of GBoV1 WT contoured at sigma (σ) threshold of 1.0. The radial distance from the center measured in Å is colored as shown. Arrows point to the 5-fold, 3-fold or 2-fold symmetry axis and the 2/5-fold wall. (**C**) Cross-sectional view of GBoV1 WT density map. (**D**) Fourier shell correlation (FSC) plot for the cryo-reconstruction with an estimated resolution of 2.76 Å at an FSC threshold of 0.143. Resolution (Å) is presented using a log scale. (**E**) Atomic model of amino acids 55–61 (βB) represented within their density map contoured at a σ threshold level of 1. C = yellow, O = red and N = blue. Panels B, C and E were made using UCSF-Chimera [41].

Table 2. HBoV1-4 and GBoV1 primary VP3 sequence identity (bottom left) and structural identity (top right).

	HBoV1	HBoV2	HBoV3	HBoV4	GBoV1
HBoV1		94.7	93.9	93.8	94.1
HBoV2	77.5		98.8	97.4	98.2
HBoV3	77.2	89.2		97.3	96.5
HBoV4	77.2	88.7	90.2		95.3
GBoV1	86.3	79.7	79.9	79.4	

In both the GBoV1 structures, excluding residues 1–32, residue 33 to the last C-terminal residue, aa542 (VP3 numbering), were structurally ordered and models could be built into the cryo-reconstructed density maps (e.g., in Figure 1e). The densities of the amino acid side chains were well-defined for the β-strands and most of the surface loops. Some acidic residue side chain densities were less defined. This observation is caused by a high sensitivity of these residue types to radiation damage as has been reported in other high-resolution cryo-constructed maps [45]. The GBoV1 VP3 structure conserved the parvovirus features, including the eight-stranded β-barrel motif, α-helix A and β-strand A (Figure 2a). The structure also featured α-helix B, a region unique to the bocaparvoviruses and the ten defined VRs, VR-I to VRVIIIB and VR-IX (Figure 2a). The GBoV1 WT model refinement statistics (Table 1) are consistent or better than for structures reported at this resolution by cryo-reconstruction for other bocaparvoviruses as well as other parvoviruses [16,23]. The root mean squared deviation (RMSD) between the VP models built into the GBoV1 WT and VP3 reconstructed density maps is 0.32 Å. Due to this high similarity, only the GBoV1 WT model will be used for further analysis.

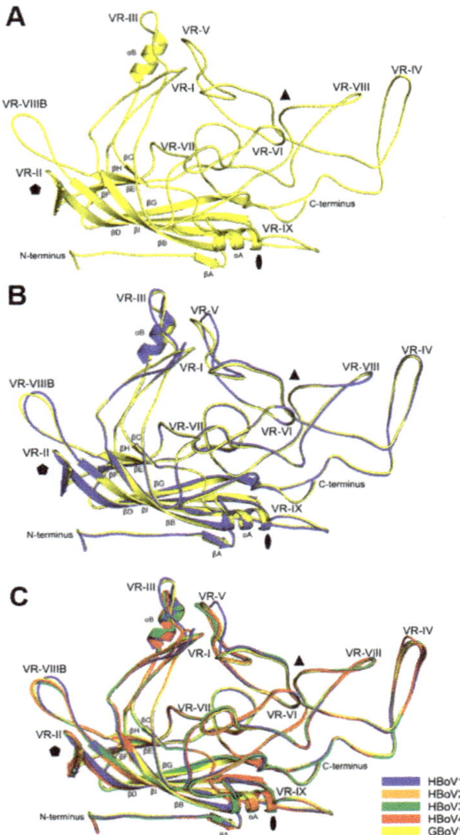

Figure 2. Structural comparison of GBoV1 to the HBoVs. (**A**) The VP3 monomer structure of GBoV1 shown as a ribbon diagram, with the secondary structure elements, N- and C-terminus and VRs labeled. The approximate positions of the icosahedral 2-, 3- and 5-fold axes are indicated as filled oval, triangle and pentagon, respectively. (**B**) VP3 monomer structures of HBoV1 (blue) and GBoV1 (yellow) superposed. The labels are as in panel (**A**). (**C**) VP3 monomer structures of HBoV1 (blue), HBoV2 (orange), HBoV3 (green), HBoV4 (red) and GBoV1 (yellow) superposed. The labels are as in panel (**A**). The color for each model is as given beside panel (**C**). Images were superposed in the Coot program [42] and visualized in the PyMol program [46].

3.2. Structural Differences between GBoV1 and the HBoVs Are Localized to the Variable Regions

The GBoV1 VP3 monomer has high primary sequence identity to the HBoVs, with 86.3% for HBoV1 and ~79% for HBoV2-4 (Table 2). The structural identity was determined by superposing the model of GBoV1 onto the previously published models of HBoV1-4 in Coot (Figure 2a–c) [23,42]. A structure-based sequence alignment was generated using these measured Cα distances, revealing that secondary structures (βI-G, βC-F core, αA and αB) are conserved and the surface loops between these regions are characterized by amino acid substitutions, insertions and deletions that result in structural differences (Figures 2 and 3).

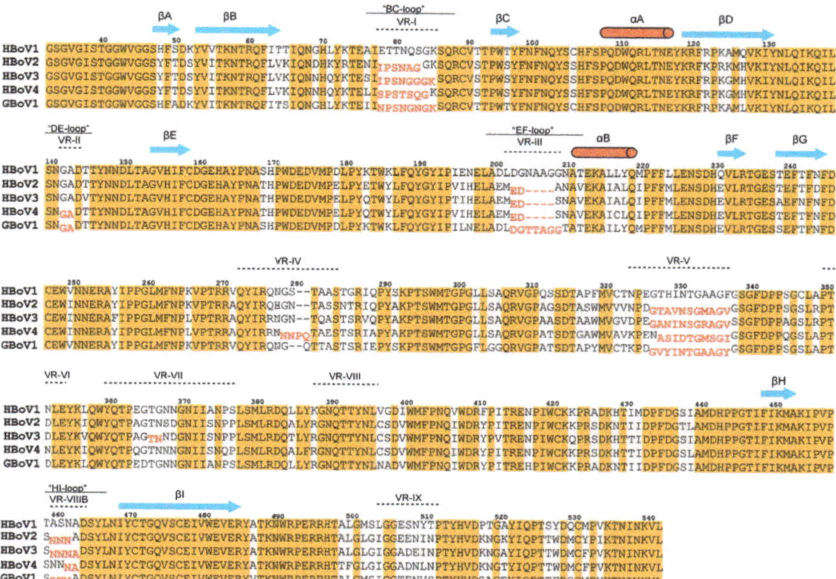

Figure 3. Structure-based sequence alignment of Human bocavirus 1 (HBoV1)-4 and GBoV1. The structure-based sequence alignment, starting from the first ordered residue (aa33), was generated using distance values from the Coot [42] superpose tool. Secondary structural elements, β-strands and α-helices, are indicated by blue arrows and red cylinders, respectively. Regions highlighted with orange indicate sequence identity between HBoV1-4 and GBoV1. The locations of the VRs are also indicated based on the previously defined VRs [23]. Amino acid number, based on HBoV1, is shown above the sequences. Structural variability, defined by amino acids whose Cα atoms are >2 Å apart, are offset low and highlighted in red.

The highest structural variabilities between the VP3s of the five viruses compared are localized to VR-I, VR-III and VR-V, with an RMSD of up to 3.3 Å, 4.5 Å and 3 Å, respectively (Figure 3, Table 3). VR-I and VR-III, along with VR-VII and VR-IX, form the 2/5-fold wall, a region of the capsid reported to be important for antigenic reactivity and receptor binding in parvoviruses [16]. VR-I has high structural variability as a result of the primary sequence differences at residues 78–85 (Figure 2b,c and Figure 3). For VR-III.

GBoV1 and HBoV1 are both structurally divergent to HBoV2-4 with a four amino acid insertion at the apex of the EF loop (Figure 3). Due to the four amino acid insertion in HBoV1 (a respiratory virus) that is not present in HBoV2-4 (gastroenteric virus) (Figure 3), VR-III was previously suggested as a region that determines tissue tropism [23]. The GBoV1 VR-III contains two amino acid substitutions relative to HBoV1 at residues 205–206 (VR-III), where NA is switched to TT (Figures 2c and 3). With these substitutions, the RMSD at VR-III for GBoV1 to HBoV1 (4.5 Å) is higher than that of HBoV2-4 (3.1–3.3 Å) (Table 2). The high structural variability between GBoV1 and HBoV1-4 suggests VR-III may also play a role in host tropism. In other parvoviruses, the region analogous to VR-III serves as a determinant for tissue tropism, pathogenicity, transduction efficiency and antigenicity [24]. VR-V along with VR-IV and VR-VIII form the protrusions around the 3-fold axis. The 3-fold protrusions have been shown to be part of an antigenic footprint for HBoV1, as well as to be important for both, antigenicity and transduction efficiency in parvoviruses [16]. Interestingly, the GBoV1 VR-V is more structurally similar to the VR-V of HBoV2-4, compared to HBoV1 (Table 3).

Table 3. Local root mean squared deviations (RMSDs) in angstroms (Å) for aligned HBoV1-4 and GBoV1 VRs. Higher values are shaded darker.

	VR-I	VR-III	VR-V	VR-II	VR-VIIIB	VR-IV	VR-VI	VR-VII	VR-VIII	VR-IX
HBoV1 vs. HBoV2	3.1	3.1	2.9	1.9	1.8	1.1	0.7	1.1	0.7	1.0
HBoV1 vs. HBoV3	2.9	3.2	3.6	1.3	2.9	0.9	0.4	1.3	0.6	1.3
HBoV1 vs. HBoV4	2.8	3.3	3.0	2.6	2.5	2.1	1.1	1.2	1.2	1.1
HBoV1 vs. GBoV1	3.3	4.5	3.0	2.1	2.5	0.9	0.4	1.1	0.7	1.1
HBoV2 vs. HBoV3	1.8	1.0	1.5	1.4	1.6	0.8	0.8	0.8	0.9	0.6
HBoV2 vs. HBoV4	2.9	1.2	1.5	1.2	1.6	1.6	0.5	0.8	1.1	0.7
HBoV2 vs. GBoV1	0.8	3.4	1.2	0.6	0.6	0.9	0.6	0.7	0.7	0.2
HBoV3 vs. HBoV4	2.9	0.9	1.5	1.4	1.3	2.0	0.7	0.9	1.3	0.6
HBoV3 vs. GBoV1	2.0	3.3	1.8	1.8	1.7	0.9	0.5	0.9	0.9	0.8
HBoV4 vs. GBoV1	3.2	3.8	1.2	0.9	1.5	1.8	0.6	0.8	0.9	0.3

When comparing the GBoV1 and HBoV1 structures, RMSDs of 2.1 Å and 2.5 Å are observed for VR-II and VR-VIIIB (Figure 2b, Table 3). Differences between GBoV1 and the other viruses range from 0.6 to 2.9 Å (Table 3). VR-II is located at the apex of the DE loop, five of which form the 5-fold channel. The 5-fold channel has been proposed to be important for genome packaging and VP1u externalization [24]. VR-II is highly conserved and is part of a cross-reactive epitope between HBoV1, HBoV3 and HBoV4 [22]. The slight structural difference within the VR-II is attributed to the need for this region to be flexible to allow the proposed externalization through this 5-fold channel. VR-VIIIB or the HI loop, is located on the depressions around the 5-fold channel. While HBoV1 structurally is the most divergent compared to HBoV2-4 and GBoV1, this loop is part of a cross-reactive epitope including HBoV1, HBoV2 and HBoV4 [22]. This suggests that only a few residues within the two loops are important for the antibody recognition of the cross-reactive antibody. This region is also reported as being important for genome packaging and capsid assembly of other parvoviruses [47–49].

For VR-IV, VR-VI, VR-VII, VR-VIII and VR-IX, there is the least (RMSD of 1.1–0.4 Å) structural divergence of GBoV1 compared to HBoV1 (Table 3, Figure 2b). VR-IV forms the protrusions around the icosahedral 3-fold axis [23]. HBoV4 differs from the other four viruses in that it has a two amino acid insertion within this loop, conferring a different conformation to the protrusions around the 3-fold compared to the other bocaparvoviruses (Figures 2c and 3). VR-VI, VR-VII and VR-VIII, all located on the side of the 3-fold protrusions, have minor to no amino acid sequence differences (Figure 3). These VRs have been shown to play a role in antigenicity and also parvovirus infectivity [16,24]. Lastly, VR-IX is located at the 2/5-fold wall. This region is structurally identical for HBoV1-4 and GBoV1, while having amino acid sequence variations and has been implicated as host tropism determinant for bocaparvoviruses [23]. As an example, BPV has a 7 amino acid deletion in VR-IX compared to the HBoVs [26]. GBoV1's VR-IX is structurally identical to HBoV1-4 and it was shown to be capable of infecting human cell lines [33], suggesting that this region may govern primate and human cell tropisms.

3.3. The GBoV1 Capsid Differs Antigenically to the HBoV1 Capsid

Antibodies 15C6, 12C1, 4G2 and 9G12 were generated, in mice, against HBoV1 capsids, using the hybridoma technology, in a previous study [22]. These antibodies were tested for reactivity against the GBoV1 capsid using native dot immunoblots (Figure 4). 15C6 and 12C1 were cross-reactive between HBoV1 and GBoV1, whereas 4C2 and 9G12 were specific for HBoV1. Previously, the 15C6 binding footprint was mapped to capsid surface features surrounding the icosahedral 5-fold axis, whereas 12C1, 4C2 and 9G12 were shown to recognize the protrusions surrounding the 3-fold axis [22]. As expected, the reactivity of these HBoV1-specific mAbs were the same for GBoV1 WT and VP3-only capsids (Figure 4). This is consistent with the observation that the epitopes of the two cross-reacting antibodies are located on the capsid surface which is formed by the VP3 common region. The antibody H1-H1, a polyclonal rabbit antibody generated against HBoV1 VP3 VLPs, served as a positive control, detecting denatured VLPs [36].

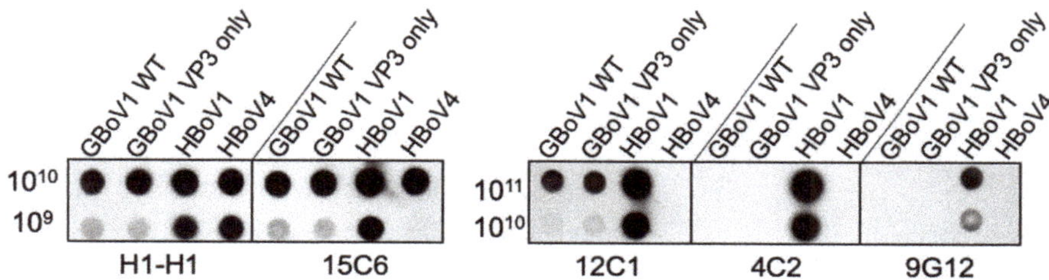

Figure 4. Cross-reactivity of GBoV1 capsids with HBoV1 antibodies via native dot blot. 10^{11}, 10^{10} or 10^{9} viral capsids were loaded onto a nitrocellulose membrane and tested against H1-H1 (positive control for denatured virus-like particles (VLPs)) and HBoV1 antibodies 15C6, 12C1, 4C2 and 9G12 (detecting conformational epitopes). 10^{11} not shown for 15C6 and H1-H1 due to overexposure.

Cryo-EM and 3D image reconstruction were used to determine the structures of the 15C6 and 12C1 Fabs complexed with the GBoV1 WT capsid (Figure 5). A total of 1895 individual capsid complexes were used for the reconstruction of the GBoV1:15C6 and 4108 of the GBoV1:12C1 complexes, with estimated resolutions of 6.4 Å and 6.2 Å, respectively, based on a FSC threshold level of 0.143 (Figure 5a). For direct comparison, a low resolution GBoV1 WT-capsid structure was determined to 5.3 Å from 17,284 capsids (Figure 5a,b,e).

The density maps of the GBoV1:15C6 complex showed density corresponding to the bound 15C6 Fabs surrounding the 5-fold channel (Figure 5c). For the GBoV1:12C1 complex, the 12C1 Fab was bound on the protrusions surround the 3-fold axis (Figure 5d). A 0.5 σ threshold density map was used to visualize both, the complimentary-determining regions (CDR) and the constant regions of the Fab (Figure 5d). At 1σ threshold, five copies of the Fab are visible of the GBoV1:15C6 complex but only the CDR for the GBoV1:12C1 complex (not shown). The visible surface features, 3-fold protrusions and 2/5-fold wall, for the GBoV1:15C6 complex is consistent with the low-resolution GBoV1 WT structure. The 5-fold channel and depressions surrounding the channel is also consistent in the GBoV1:12C1 complex with the low-resolution GBoV1 WT structure (Figure 5b). Underneath the 5-fold channel, a basket-like density can be seen in all three structures (Figure 5e–g), consistent with improved ordering at lower resolution as observed in the low- and high-resolution structures of BPV and HBoV1-4 [15,23].

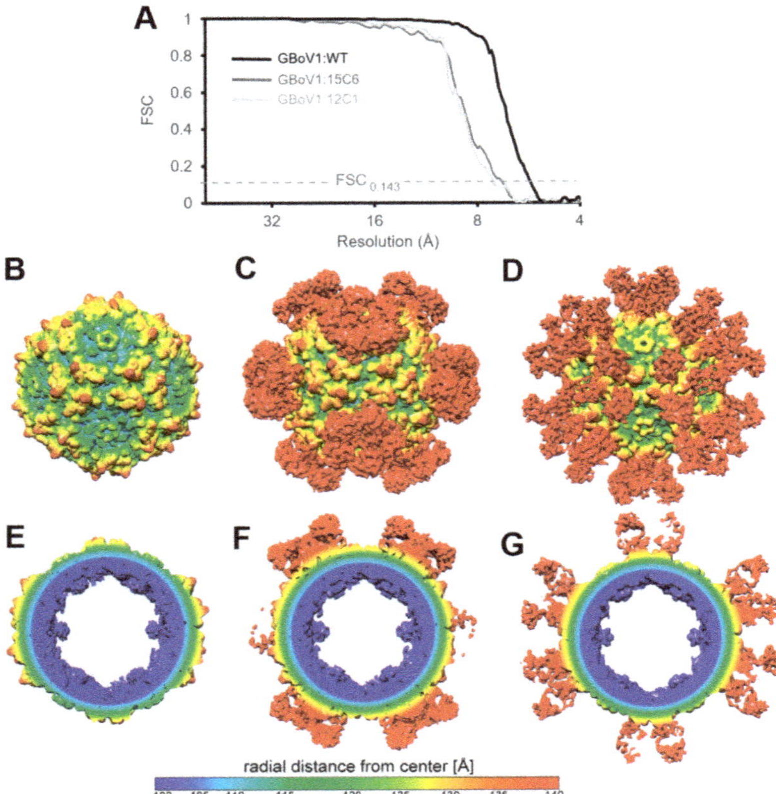

Figure 5. Antibody epitopes on GBoV1 capsid localized to 5- and 3-fold axes for 15C6 and 12C1. (**A**) FSC plots for the cryo-reconstruction with an estimated resolution value of 5.3Å, 6.4 Å and 6.2 Å, at an FSC threshold value of 0.143 for GBoV1, GBoV1:15C6 and GBoV1:12C1, respectively. Resolution (Å) is presented using a $\log_2$ scale. (**B**) Capsid density map of GBoV1 contoured at σ threshold of 1.0. (**C**) Capsid density map of GBoV1 complexed with 15C6 (GBoV1:15C6) contoured at σ threshold of 1.0. (**D**) Capsid density map of GBoV1 complexed with 15C6 contoured at σ threshold of 0.5. (**E**) Cross-sectional view of the GBoV1 complex density map. (**F**) Cross-sectional view of the GBoV1:15C6 complex density map. (**G**) Cross-sectional view of the GBoV1:12C1 capsid density map.

Rigid-body docking of the refined 60-mer model of GBoV1 WT was performed for the GBoV1:15C6 and GBoV1:12C1 density maps, along with a generic Fab (PDB ID: 2FBJ), with a CC of 0.93 and 0.90, respectively (Figure 6). Steric clashes at the 5-fold axis for the individual 15C6 Fabs likely resulted in local disorder for the Fab (Figure 6). For 12C1 it appears that enough space is available at the 3-fold protrusions (Figure 5) but disorder also is observed. A 2D stereographic projection (roadmap) representation of the complex, generated based on the fitted models, identified the contact and occluded (within Fab footprint but not contacting capsid) residues (Figure 7) [44]. The 15C6 epitope lines the 5-fold channel, encompassing the residues that form VR-II (residues 142-GAD-144) and VR-VIIIB (residues 460-STNA-463), which are the DE and HI loops, respectively (Figure 7a). These residues are conserved between GBoV1 and HBoV1, with the exception of residue S460, which is A460 in HBoV1. The high conservation within this sequence region as well as high structural similarity explains the cross-reactivity between the GBoV1 and HBoV1 capsid for the 15C6 antibody. This antibody also cross-reacts with HBoV2 and HBoV4 [22]. The 12C1 epitope sits on the protrusions around the 3-fold axis and contains contact residues from VR-I (80-SNGN-83), VR-IV (276-IRQNGQTTA-284) and VR-VIII (390-NQTT-393) (Figure 7b). Compared to HBoV1, the GBoV1 VR-IV and VR-VIII are structurally identical despite having amino acid differences (Figure 3). This structural identity likely dictates the cross-reactivity of 12C1 for the GBoV1 capsid. Interestingly, antibodies 4G2 and 9G12, also recognizing the 3-fold protrusions, were not cross-reactive despite the 94.3% structural identity that is shared between the GBoV1 and HBoV1 VP3 monomers [22]. VR-I, VR-III and VR-V are the most structurally divergent loops between HBoV1 and GBoV1 and contain residues outside the 15C6 and 12C1 epitopes (Figure 7c). These residues outside these epitopes, particularly at the 3-fold protrusion, are potentially responsible for this difference in antigenic reactivity with respect to 4G2 and 9G12.

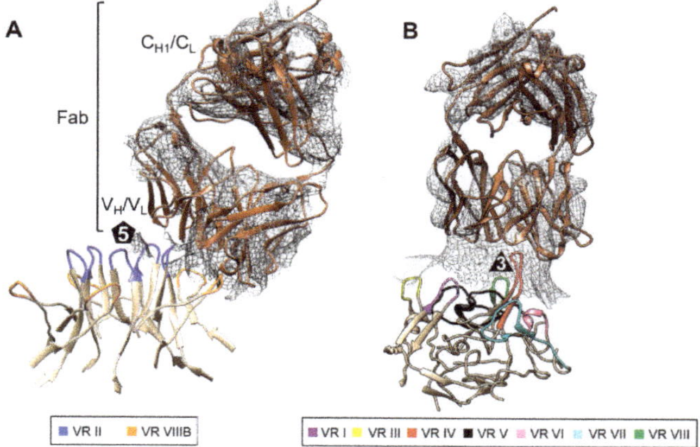

Figure 6. GBoV1-Fab binding interfaces. (**A**) Close-up view of the GBoV1 WT structure docked to a generic Fab (PDB ID 2FBJ) within the cryo-reconstructed density of GBoV1-15C6 (represented as a gray mesh, contoured at 0.5σ) and (**B**) GBoV1-12C1. Highlighted VRs are colored as shown in key. Generic Fab (dark brown) consists of a heavy and light chain, each with constant and variable regions. The Fab variable region interacts with the surface of the capsid. The GBoV1 capsid is also colored in tan.

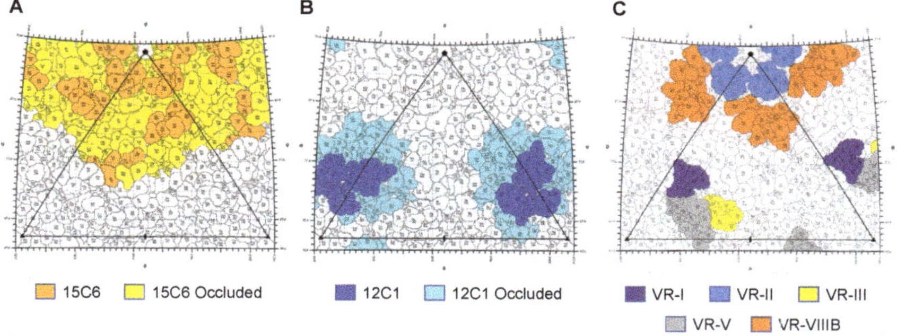

Figure 7. The GBoV1 15C6 and 12C1 epitopes. (**A**) Roadmap surface representation of the GBoV1 15C6 epitope. Colored in orange are the modeled contact residues between the GBoV1 capsid and the 15C6 Fab model. Colored in yellow are the residues occluded by the bound 15C6 Fab. (**B**) Roadmap surface representation of the GBoV1 12C1 epitope. Colored in blue are the modeled contact residues between the GBoV1 capsid and the 12C1 Fab model. Colored in cyan are the residues occluded by the 12C1 Fab. (**C**) Position of VR-I, VR-II, VR-III and VR-V, VR-VIIIB on the GBoV1 capsid. Amino acid residues that are exposed on the capsid surface are labeled with their 3-letter code and residue number. The 5-fold, 3-fold and 2-fold axis are indicated by a filled pentagon, triangle and ellipse, respectively. The roadmaps were generated with the RIVEM program [44].

3.4. HBoV1 and GBoV1 Share Similar Rates of Seropositivity

Forty human serum samples from adult donors were screened by native dot immunoblot against HBoV1, GBoV1, AAV2 and AAV5 capsids. Approximately 98.3% of the samples reacted to HBoV1 capsids, 88.3% against GBoV1, 25% against AAV2 and 14.2% against AAV5, respectively (Figure 8). This suggests that HBoV infections are prevalent in North America, the source of the analyzed human sera. This seroprevalence is comparable to results from a previous study analyzing sera from adults in Finland (95%) and Pakistan (99%) [50]. Nevertheless, seroprevalences are obtained without consideration of HBoV2-4 cross-reacting IgGs, so the true rates of HBoV1-specific and GBoV1-specific seropositivity will be skewed by potential HBoV2-4 IgGs. The high GBoV1 response is thus likely caused by anti-HBoV antibodies. In addition, generally weaker signal intensities were observed for GBoV1 compared to HBoV1 across the forty samples (Figure 8). The minor difference of 10% in seropositivity indicates variation in antigenic reactivity. Interestingly, AAV2 has been reported to be have a 72% seroprevalence (59% neutralizing) in French adults, a significant difference compared to the data here, suggesting location may be a large variable in these epidemiological studies [51]. As an example, another study reported 25–30% neutralizing antibodies against AAV2 when 100 human serum samples from North American adults were tested [52]. It is important to note that native dot immunoblots do not report the neutralization potential of the antibodies from the forty samples but rather the antibodies capable of recognizing the capsid surface. Further study of neutralizing factors and cross-reactivity within the human sera will be needed to determine the effect of such samples on the transduction efficiency in human cells and tissue.

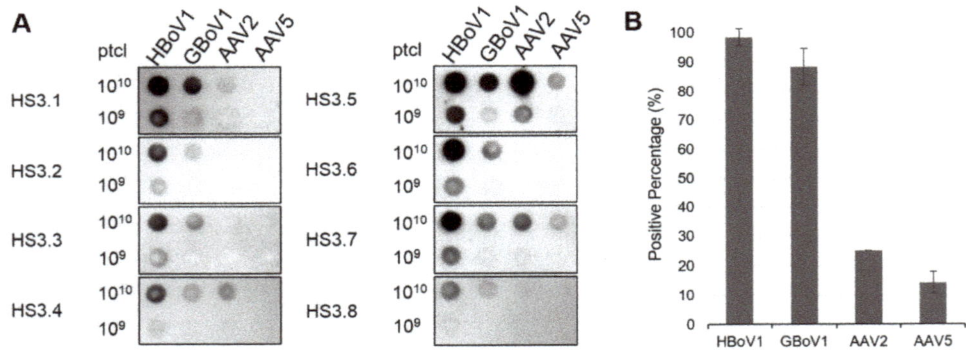

Figure 8. Dot immunoblot analysis of HBoV1 and GBoV1 against human sera. (**A**) Representative native dot immunoblots of HBoV1 and GBoV1 against human sera with 10^{10} or 10^9 loaded capsid particles. AAV2 and AAV5 are used as controls. Samples tested are as labeled. (**B**) Bar graph representation of the percentage of positive signal based visual inspection of the 40 dot immunoblots reactivities. $n = 3$.

4. Conclusions

This study reports the first high-resolution structure of the GBoV1 WT capsid, resolved to 2.76 Å resolution. Compared to other members of the genus, the GBoV1 capsid shares similar surface features, such as the channel at the 5-fold symmetry axis, protrusions around the 3-fold axis, the 2/5-fold wall, located between the depressions at the 2- and 5-fold and depressions at the 2-fold as well as around the 5-fold axes. In addition to the high-resolution WT structure, the structures of two capsid-antibody complex structures are reported, highlighting antigenic epitopes on the GBoV1 capsid surface. Both, GBoV1 and HBoV1 share a high sequence and structural identity, with major structural differences localized to VR-I, VR-III and VR-V. VR-I and VR-III are both part of the 2/5-fold wall of the capsid and VR-V is located on the protrusions around the icosahedral 3-fold. These VRs contain residues that are within the epitopes of HBoV1 cross-reactive monoclonal antibodies

12C1, 4C2 and 9G12. All three antibodies share the same antibody epitope, as previously reported [27]. Interestingly, native dot immunoblots show that the GBoV1 capsid is capable of escaping 4C2 and 9G12 but not 12C1, suggesting that minor structural differences at these VRs are responsible for the GBoV1 capsid's ability to escape antibodies 4C2 and 9G12. Overall, the GBoV1 and HBoV1 capsid are antigenically similar at the icosahedral 5-fold axis, a region that is most conserved amongst parvoviruses, yet differ at the 3-fold axis. The reported capsid structures and epitopes can guide strategies for vector engineering and aid to develop the GBoV1 capsid structure as a viral vector. In addition, the HBoV1 and GBoV1 capsid seropositivity rates against human sera points to high cross-reactivity between the two viruses. Potential binding sites are likely the 5-fold region that is highly conserved. This, however, remains to be determined. Interestingly, the HBoV1 and GBoV1 seropositivity rates with human sera were significantly higher than those for AAV2 and AAV5 in the North American adult samples tested. This observation further emphasizes the need to understand the antigenic reactivity of the bocaparvoviruses if they are to be developed as vectors for clinical gene delivery.

Author Contributions: J.C.Y. was responsible for virus production and purification, native dot immunoblot analysis, cryo-reconstruction, structure refinement and analysis, model building and refinement and manuscript preparation. M.M. contributed to cryo-reconstruction, structure building, model building and refinement analysis and manuscript preparation. A.S. was responsible for aiding in virus production and purification. A.J.Y. contributed to the dot immunoblot analysis for the human sera. S.K. produced and purified HBoV4. J.F., D.G. and A.K. were responsible for plasmid design. I.K.-K. and A.Ž. developed the HBoV mAbs. M.S.-V. provided the H1-H1 antibody and contributed to interpretation of the results and manuscript preparation. P.C. vitrified sample and screened cryo-EM grids. N.B. collected cryo-EM data. R.M. and M.A.-M. conceived and designed this project, analyzed all results and contributed to manuscript preparation. All authors have read and agreed to the published version of this manuscript.

Funding: This project was possible due to the National Health Institute grant NIH R01 GM02946, the Sigrid Jusélius Foundation and the Life and Health Medical Foundation and the Cystic Fibrosis Foundation (CFF) grant GRIMM15XX0.

Acknowledgments: The authors thank the University of Florida (UF) Interdisciplinary Center for Biotechnology Research (ICBR) EM lab for providing negative-stain EM and cryo-EM services. We also thank Florida State University for providing cryo-EM data collection services. This research was made possible by the NIH grant R01 GM02946.

Conflicts of Interest: The authors declare no conflict of interest.

References

1. Cotmore, S.F.; Agbandje-McKenna, M.; Canuti, M.; Chiorini, J.A.; Eis-Hubinger, A.M.; Hughes, J.; Mietzsch, M.; Modha, S.; Ogliastro, M.; Pénzes, J.J.; et al. ICTV virus taxonomy profile: Parvoviridae. *J. Gen. Virol.* **2019**, *100*, 367–368. [CrossRef]
2. Bates, R.C.; Storz, J.; Reed, D.E. Isolation and comparison of bovine parvoviruses. *J. Infect. Dis.* **1972**, *126*, 531–536. [CrossRef] [PubMed]
3. Abixanti, F.R.; Warfield, M.S. Recovery of a hemadsorbing virus (HADEN) from the gastrointestinal tract of calves. *Virology* **1961**, *14*, 288–289. [CrossRef]
4. Lanave, G.; Martella, V.; Farkas, S.L.; Marton, S.; Fehér, E.; Bodnar, L.; Lavazza, A.; Decaro, N.; Buonavoglia, C.; Bányai, K. Novel bocaparvoviruses in rabbits. *Vet. J.* **2015**, *206*, 131–135. [CrossRef]
5. Lau, S.K.P.; Yeung, H.C.; Li, K.S.M.; Lam, C.S.F.; Cai, J.P.; Yuen, M.C.; Wang, M.; Zheng, B.J.; Woo, P.C.Y.; Yuen, K.Y. Identification and genomic characterization of a novel rat bocavirus from brown rats in China. *Infect. Genet. Evol.* **2017**, *47*, 68–76. [CrossRef] [PubMed]
6. Zhang, C.; Song, F.; Xiu, L.; Liu, Y.; Yang, J.; Yao, L.; Peng, J. Identification and characterization of a novel rodent bocavirus from different rodent species in China. *Emerg. Microbes Infect.* **2018**, *7*, 48. [CrossRef]
7. Allander, T.; Tammi, M.T.; Eriksson, M.; Bjerkner, A.; Tiveljung-Lindell, A.; Andersson, B. Cloning of a human parvovirus by molecular screening of respiratory tract samples. *Proc. Natl. Acad. Sci. USA* **2005**, *102*, 12891–12896. [CrossRef]
8. Kapoor, A.; Simmonds, P.; Slikas, E.; Li, L.; Bodhidatta, L.; Sethabutr, O.; Triki, H.; Bahri, O.; Oderinde, B.S.; Baba, M.M.; et al. Human Bocaviruses Are Highly Diverse, Dispersed, Recombination Prone and Prevalent in Enteric Infections. *J. Infect. Dis.* **2010**, *201*, 1633–1643. [CrossRef]

9. Arthur, J.L.; Higgins, G.D.; Davidson, G.P.; Givney, R.C.; Ratcliff, R.M. A Novel Bocavirus Associated with Acute Gastroenteritis in Australian Children. *PLoS Pathog.* **2009**, *5*, e1000391. [CrossRef]
10. Kapoor, A.; Mehta, N.; Esper, F.; Poljsak-Prijatelj, M.; Quan, P.L.; Qaisar, N.; Delwart, E.; Lipkin, W.I. Identification and characterization of a new bocavirus species in gorillas. *PLoS ONE* **2010**, *5*, 11948. [CrossRef]
11. Huang, Q.; Deng, X.; Yan, Z.; Cheng, F.; Luo, Y.; Shen, W.; Lei-Butters, D.C.M.; Chen, A.Y.; Li, Y.; Tang, L.; et al. Establishment of a Reverse Genetics System for Studying Human Bocavirus in Human Airway Epithelia. *PLoS Pathog.* **2012**, *8*, 1002899. [CrossRef]
12. Dijkman, R.; Koekkoek, S.M.; Molenkamp, R.; Schildgen, O.; van der Hoek, L. Human Bocavirus Can Be Cultured in Differentiated Human Airway Epithelial Cells. *J. Virol.* **2009**, *83*, 7739–7748. [CrossRef]
13. Zou, W.; Cheng, F.; Shen, W.; Engelhardt, J.F.; Yan, Z.; Qiu, J. Nonstructural Protein NP1 of Human Bocavirus 1 Plays a Critical Role in the Expression of Viral Capsid Proteins. *J. Virol.* **2016**, *90*, 4658–4669. [CrossRef]
14. Qiu, J.; Söderlund-Venermo, M.; Young, N.S. Human parvoviruses. *Clin. Microbiol. Rev.* **2017**, *30*, 43–113. [CrossRef] [PubMed]
15. Gurda, B.L.; Parent, K.N.; Bladek, H.; Sinkovits, R.S.; DiMattia, M.A.; Rence, C.; Castro, A.; McKenna, R.; Olson, N.; Brown, K.; et al. Human Bocavirus Capsid Structure: Insights into the Structural Repertoire of the Parvoviridae. *J. Virol.* **2010**, *84*, 5880–5889. [CrossRef] [PubMed]
16. Mietzsch, M.; Pénzes, J.J.; Agbandje-Mckenna, M. Twenty-Five Years of Structural Parvovirology. *Viruses* **2019**, *11*, 362. [CrossRef]
17. Zádori, Z.; Szelei, J.; Lacoste, M.C.; Li, Y.; Gariépy, S.; Raymond, P.; Allaire, M.; Nabi, I.R.; Tijssen, P. A Viral Phospholipase A2 Is Required for Parvovirus Infectivity. *Dev. Cell* **2001**, *1*, 291–302. [CrossRef]
18. Qu, X.W.; Liu, W.P.; Qi, Z.Y.; Duan, Z.J.; Zheng, L.S.; Kuang, Z.Z.; Zhang, W.J.; Hou, Y. De Phospholipase A2-like activity of human bocavirus VP1 unique region. *Biochem. Biophys. Res. Commun.* **2008**, *365*, 158–163. [CrossRef] [PubMed]
19. Stahnke, S.; Lux, K.; Uhrig, S.; Kreppel, F.; Hösel, M.; Coutelle, O.; Ogris, M.; Hallek, M.; Büning, H. Intrinsic phospholipase A2 activity of adeno-associated virus is involved in endosomal escape of incoming particles. *Virology* **2011**, *409*, 77–83. [CrossRef]
20. Bleker, S.; Sonntag, F.; Kleinschmidt, J.A. Mutational Analysis of Narrow Pores at the Fivefold Symmetry Axes of Adeno-Associated Virus Type 2 Capsids Reveals a Dual Role in Genome Packaging and Activation of Phospholipase A2 Activity. *J. Virol.* **2005**, *79*, 2528–2540. [CrossRef]
21. Mani, B.; Baltzer, C.; Valle, N.; Almendral, J.M.; Kempf, C.; Ros, C. Low pH-Dependent Endosomal Processing of the Incoming Parvovirus Minute Virus of Mice Virion Leads to Externalization of the VP1 N-Terminal Sequence (N-VP1), N-VP2 Cleavage and Uncoating of the Full-Length Genome. *J. Virol.* **2006**, *80*, 1015–1024. [CrossRef] [PubMed]
22. Kailasan, S.; Garrison, J.; Ilyas, M.; Chipman, P.; McKenna, R.; Kantola, K.; Söderlund-Venermo, M.; Kučinskaitė-Kodzė, I.; Žvirblienė, A.; Agbandje-McKenna, M. Mapping Antigenic Epitopes on the Human Bocavirus Capsid. *J. Virol.* **2016**, *90*, 4670–4680. [CrossRef] [PubMed]
23. Mietzsch, M.; Kailasan, S.; Garrison, J.; Ilyas, M.; Chipman, P.; Kantola, K.; Janssen, M.E.; Spear, J.; Sousa, D.; McKenna, R.; et al. Structural Insights into Human Bocaparvoviruses. *J. Virol.* **2017**, *91*. [CrossRef] [PubMed]
24. Halder, S.; Ng, R.; Agbandje-Mckenna, M. Parvoviruses: Structure and infection. *Future Virol.* **2012**, *7*, 253–278. [CrossRef]
25. Luo, M.; Mietzsch, M.; Chipman, P.; Song, K.; Xu, C.; Spear, J.; Sousa, D.; McKenna, R.; Soderlund-Venermo, M.; Agbandje-Mckenna, M. pH-Induced Conformational Changes of Human Bocavirus Capsids. *J. Virol.* **2021**. [CrossRef] [PubMed]
26. Kailasan, S.; Halder, S.; Gurda, B.; Bladek, H.; Chipman, P.R.; McKenna, R.; Brown, K.; Agbandje-McKenna, M. Structure of an Enteric Pathogen, Bovine Parvovirus. *J. Virol.* **2015**, *89*, 2603–2614. [CrossRef]
27. Deng, X.; Yan, Z.; Luo, Y.; Xu, J.; Cheng, F.; Li, Y.; Engelhardt, J.F.; Qiu, J. In Vitro Modeling of Human Bocavirus 1 Infection of Polarized Primary Human Airway Epithelia. *J. Virol.* **2013**, *87*, 4097–4102. [CrossRef] [PubMed]
28. Yan, Z.; Keiser, N.W.; Song, Y.; Deng, X.; Cheng, F.; Qiu, J.; Engelhardt, J.F. A novel chimeric adenoassociated virus 2/human bocavirus 1 parvovirus vector efficiently transduces human airway epithelia. *Mol. Ther.* **2013**, *21*, 2181–2194. [CrossRef]
29. Aitken, M.L.; Moss, R.B.; Waltz, D.A.; Dovey, M.E.; Tonelli, M.R.; McNamara, S.C.; Gibson, R.L.; Ramsey, B.W.; Carter, B.J.; Reynolds, T.C. A phase I study of aerosolized administration of tgAAVCF to cystic fibrosis subjects with mild lung disease. *Hum. Gene Ther.* **2001**, *12*, 1907–1916. [CrossRef]
30. Moss, R.B.; Milla, C.; Colombo, J.; Accurso, F.; Zeitlin, P.L.; Clancy, J.P.; Spencer, L.T.; Pilewski, J.; Waltz, D.A.; Dorkin, H.L.; et al. Repeated aerosolized AAV-CFTR for treatment of cystic fibrosis: A randomized placebo-controlled phase 2B trial. *Hum. Gene Ther.* **2007**, *18*, 726–732. [CrossRef]
31. Mueller, C.; Flotte, T.R. Gene therapy for cystic fibrosis. *Clin. Rev. Allergy Immunol.* **2008**, *35*, 164–178. [CrossRef]
32. Yan, Z.; Zou, W.; Feng, Z.; Shen, W.; Park, S.Y.; Deng, X.; Qiu, J.; Engelhardt, J.F. Establishment of a High-Yield Recombinant Adeno-Associated Virus/Human Bocavirus Vector Production System Independent of Bocavirus Nonstructural Proteins. *Hum. Gene Ther.* **2019**, *30*, 556–570. [CrossRef] [PubMed]
33. Fakhiri, J.; Schneider, M.A.; Puschhof, J.; Stanifer, M.; Schildgen, V.; Holderbach, S.; Voss, Y.; El Andari, J.; Schildgen, O.; Boulant, S.; et al. Novel Chimeric Gene Therapy Vectors Based on Adeno-Associated Virus and Four Different Mammalian Bocaviruses. *Mol. Ther. Methods Clin. Dev.* **2019**, *12*, 202–222. [CrossRef]
34. Rayaprolu, V.; Kruse, S.; Kant, R.; Venkatakrishnan, B.; Movahed, N.; Brooke, D.; Lins, B.; Bennett, A.; Potter, T.; McKenna, R.; et al. Comparative Analysis of Adeno-Associated Virus Capsid Stability and Dynamics. *J. Virol.* **2013**, *87*, 13150–13160. [CrossRef]
35. Invitrogen Bac-to-Bac® Baculovirus Expression System. *Manual* **2015**, 1–78. [CrossRef]
36. Li, X.; Kantola, K.; Hedman, L.; Arku, B.; Hedman, K.; Söderlund-Venermo, M. Original antigenic sin with human bocaviruses 1–4. *J. Gen. Virol.* **2015**, *96*, 3099–3108. [CrossRef]

37. Zheng, S.Q.; Palovcak, E.; Armache, J.P.; Verba, K.A.; Cheng, Y.; Agard, D.A. MotionCor2: Anisotropic correction of beam-induced motion for improved cryo-electron microscopy. *Nat. Methods* **2017**, *14*, 331–332. [CrossRef] [PubMed]
38. Grant, T.; Rohou, A.; Grigorieff, N. CisTEM, user-friendly software for single-particle image processing. *Elife* **2018**, *7*. [CrossRef]
39. Schwede, T.; Kopp, J.; Guex, N.; Peitsch, M.C. SWISS-MODEL: An automated protein homology-modeling server. *Nucleic Acids Res.* **2003**, *31*, 3381–3385. [CrossRef]
40. Ho, P.T.; Montiel-Garcia, D.J.; Wong, J.J.; Carrillo-Tripp, M.; Brooks, C.L.; Johnson, J.E.; Reddy, V.S. VIPERdb: A Tool for Virus Research. *Annu. Rev. Virol.* **2018**, 477–488. [CrossRef]
41. Pettersen, E.F.; Goddard, T.D.; Huang, C.C.; Couch, G.S.; Greenblatt, D.M.; Meng, E.C.; Ferrin, T.E. UCSF Chimera - A visualization system for exploratory research and analysis. *J. Comput. Chem.* **2004**, *25*, 1605–1612. [CrossRef]
42. Emsley, P.; Cowtan, K. Coot: Model-building tools for molecular graphics. *Acta Crystallogr. Sect. D Biol. Crystallogr.* **2004**, *60*, 2126–2132. [CrossRef]
43. Adams, P.D.; Afonine, P.V.; Bunkóczi, G.; Chen, V.B.; Davis, I.W.; Echols, N.; Headd, J.J.; Hung, L.W.; Kapral, G.J.; Grosse-Kunstleve, R.W.; et al. PHENIX: A comprehensive Python-based system for macromolecular structure solution. *Acta Crystallogr. Sect. D Biol. Crystallogr.* **2010**, *66*, 213–221. [CrossRef] [PubMed]
44. Xiao, C.; Rossmann, M.G. Interpretation of electron density with stereographic roadmap projections. *J. Struct. Biol.* **2007**, *158*, 182–187. [CrossRef] [PubMed]
45. Bartesaghi, A.; Matthies, D.; Banerjee, S.; Merk, A.; Subramaniam, S. Structure of β-galactosidase at 3.2-Å resolution obtained by cryo-electron microscopy. *Proc. Natl. Acad. Sci. USA* **2014**, *111*, 11709–11714. [CrossRef] [PubMed]
46. The PyMOL Molecular Graphics System 2020. Available online: https://pymol.org/2/ (accessed on 25 January 2021).
47. Gurda, B.L.; DiMattia, M.A.; Miller, E.B.; Bennett, A.; McKenna, R.; Weichert, W.S.; Nelson, C.D.; Chen, W.-j.; Muzyczka, N.; Olson, N.H.; et al. Capsid Antibodies to Different Adeno-Associated Virus Serotypes Bind Common Regions. *J. Virol.* **2013**, *87*, 9111–9124. [CrossRef]
48. DiPrimio, N.; Asokan, A.; Govindasamy, L.; Agbandje-McKenna, M.; Samulski, R.J. Surface Loop Dynamics in Adeno-Associated Virus Capsid Assembly. *J. Virol.* **2008**, *82*, 5178–5189. [CrossRef] [PubMed]
49. Fakhiri, J.; Linse, K.-P.; Mietzsch, M.; Xu, M.; Schneider, M.A.; Meister, M.; Schildgen, O.; Schnitzler, P.; Soderlund-Venermo, M.; Agbandje-McKenna, M.; et al. Impact of Natural or Synthetic Singletons in the Capsid of Human Bocavirus 1 on Particle Infectivity and Immunoreactivity. *J. Virol.* **2020**, *94*. [CrossRef]
50. Kantola, K.; Hedman, L.; Arthur, J.; Alibeto, A.; Delwart, E.; Jartti, T.; Ruuskanen, O.; Hedman, K.; Söderlund-Venermo, M. Seroepidemiology of human bocaviruses 1-4. *J. Infect. Dis.* **2011**, *204*, 1403–1412. [CrossRef]
51. Benveniste, O.; Boutin, S.; Monteilhet, V.; Veron, P.; Leborgne, C.; Montus, M.F.; Masurier, C. Prevalence of Serum IgG and Neutralizing Factors Against Adeno-Associated Virus (AAV) Types 1,2,5,6,8 and 9 in the Healthy Population: Implications for Gene Therapy Using AAV Vectors. *Hum. Gene Ther.* **2010**, *21*, 704–712. [CrossRef]
52. Calcedo, R.; Vandenberghe, L.H.; Gao, G.; Lin, J.; Wilson, J.M. Worldwide epidemiology of neutralizing antibodies to adeno-associated viruses. *J. Infect. Dis.* **2009**, *199*, 381–390. [CrossRef] [PubMed]

Review

Concepts to Reveal Parvovirus–Nucleus Interactions

Salla Mattola [1], Satu Hakanen [1], Sami Salminen [1], Vesa Aho [1], Elina Mäntylä [2], Teemu O. Ihalainen [2], Michael Kann [3,4] and Maija Vihinen-Ranta [1,*]

[1] Department of Biological and Environmental Science, University of Jyvaskyla, 40500 Jyvaskyla, Finland; salla.m.mattola@jyu.fi (S.M.); satu.a.hakanen@jyu.fi (S.H.); Sami.j.salminen@jyu.fi (S.S.); Vesa.p.aho@jyu.fi (V.A.)
[2] BioMediTech, Faculty of Medicine and Health Technology, Tampere University, 33520 Tampere, Finland; elina.mantyla@tuni.fi (E.M.); teemu.ihalainen@tuni.fi (T.O.I.)
[3] Department of Infectious Diseases, Institute of Biomedicine, Sahlgrenska Academy, University of Gothenburg, 41390 Gothenburg, Sweden; michael.kann@gu.se
[4] Department of Clinical Microbiology, Sahlgrenska University Hospital, 41345 Gothenburg, Sweden
* Correspondence: maija.vihinen-ranta@jyu.fi

Abstract: Parvoviruses are small single-stranded (ss) DNA viruses, which replicate in the nucleoplasm and affect both the structure and function of the nucleus. The nuclear stage of the parvovirus life cycle starts at the nuclear entry of incoming capsids and culminates in the successful passage of progeny capsids out of the nucleus. In this review, we will present past, current, and future microscopy and biochemical techniques and demonstrate their potential in revealing the dynamics and molecular interactions in the intranuclear processes of parvovirus infection. In particular, a number of advanced techniques will be presented for the detection of infection-induced changes, such as DNA modification and damage, as well as protein–chromatin interactions.

Keywords: parvoviruses; nucleus; imaging of viral interactions and dynamics; analysis of protein–protein interactions; analysis of virus–chromatin interactions

Citation: Mattola, S.; Hakanen, S.; Salminen, S.; Aho, V.; Mäntylä, E.; Ihalainen, T.O.; Kann, M.; Vihinen-Ranta, M. Concepts to Reveal Parvovirus–Nucleus Interactions. *Viruses* **2021**, *13*, 1306. https://doi.org/10.3390/v13071306

Academic Editor: Giorgio Gallinella

Received: 1 June 2021
Accepted: 2 July 2021
Published: 5 July 2021

1. Introduction

Parvoviruses are not only significant pathogens causing diseases in humans and animals but also promising candidates in gene therapy, in oncolytic therapy, in vaccine development, and as passive immunization vectors [1–7]. Compared to some other viruses that only need a few viral particles for infection, parvoviruses are extremely inefficient. In infection and disease development, this incapability is compensated by high replication. Finding new ways to treat parvoviral diseases and to facilitate the development of parvovirus-based therapies requires deepening the understanding of infection and propagation in their host cells.

Although parvoviruses and their infection have been extensively studied throughout the past decades, there is still a lack of molecular level understanding of the virus–host cell interactions. Due to their low particles to infectious unit ratio, the identification and tracking of virus-induced events, which contribute to viral propagation, is a key challenge. Furthermore, the small size of parvovirus (~20 nm in diameter) hinders the attachment of fluorescent probes, which limits capsid detection by single-virus tracking.

Parvoviruses are divided into two classes: autonomous parvoviruses, such as canine parvovirus (CPV), minute virus of mice (MVM), and rat parvovirus (H-1PV), and dependoparvoviruses, such as adeno-associated viruses (AAV), which require coinfection with either adenoviruses or herpes simplex virus in their late stages of infection [8]. Parvoviruses are composed of two to three capsid proteins (viral proteins, VPs; VP1, 2, and 3). They enclose a c. 5 kb-long ssDNA genome, which consists of two overlapping open reading frames. The expression is controlled by two promoters, the early P4 and late P38. The former guides the expression of viral nonstructural proteins 1 and 2 (NS1 and

NS2), while the latter controls the expression of capsid proteins [9–11]. In the infectious virion, which has a diameter of 18–26 nm, the genome is covalently bound to the NS1 (Rep78 in AAV) protein [12–15]. This protein is cytotoxic and has central roles in viral replication attributed to its helicase, endonuclease, ATPase, and site-specific DNA-binding activities [16,17]. NS2 plays a role in viral replication [12,18], development of viral replication centres [19], viral mRNA translation [20], and the assembly [21] and nuclear egress of capsids [22–26]. In gene therapy, which is mostly based on AAV, the single-stranded genome is replaced by a double-stranded self-complementary genome, which does not allow replication [15].

After the cellular entry and cytoplasmic release, parvoviral capsids enter the nucleus through the nuclear pore complexes (NPCs) and/or via disruption of the nuclear envelope (NE) [27–34]. The VP1 capsid protein bears nuclear localization signals (NLSs) within its VP1-unique region in the N-terminal domain [35–41], which are thought to allow nuclear import by interaction with nuclear transport factors of the importin family [30,42,43]. In assembled capsids, this domain is hidden.

Once arriving in the nucleus, the genome replicates via a rolling circle mechanism, during which the genome concatemer is cleaved to monomers by NS1 [44]. The gene expression of parvoviruses is coupled to the S-phase of the cell cycle, and it leads to the formation of distinct replication centre foci where viral gene transcription and productive replication occur [19,45,46]. As the infection proceeds, the replication centres expand [27,28,47], which is accompanied by changes in the cellular chromatin structure and chromatin marginalization to the nuclear periphery at later stages of infection [45,47]. Besides the dramatic morphological changes, parvovirus infections are known to induce substantial damage to the host DNA [48–50], and MVM replication centres have been shown to associate with the sites of cellular DNA damage [51,52]. This allows the virus to recruit cellular DNA replication and DNA damage response proteins, which promote viral replication and gene expression [45,49,53]. NS1 of MVM is responsible for nicking the host DNA, which subsequently results in S phase cell cycle arrest [54]. However, during human parvovirus B19 (B19V) infection, a G2/M arrest is induced by the NS1 protein through a p53-independent pathway, which does not depend on the DNA damage response [50]. In addition to evoking disturbances in the cell cycle, parvoviruses are known to cause apoptosis of the infected cells, another hallmark of DNA damage [55,56].

These nuclear changes are followed by progeny capsid assembly in the nucleus, which is combined with the encapsidation of viral genomes covalently bound to NS1. The progeny virions leave the cell by lysis, probably after export from the nucleus [57–60]. This lytic viral release, in conjunction with the S-phase-dependent replication, enables the use of autonomous parvoviruses in oncotherapy for the destruction of rapidly dividing cancer cells [61].

2. Imaging of Viral Interactions and Dynamics in the Cytoplasm and Nucleus

To date, a broad variety of microscopy-based imaging and spectroscopy applications have enlightened the steps in the early infection of several parvoviruses (Figure 1). Upon nuclear import, CPV can pass the NE [27,28,62,63], which was confirmed by single-particle tracking analyses of fluorophore-labelled AAV capsids (Figure 1, boxes 1 and 2) [64]. Similar analyses have also been used to study the receptor binding of canine parvovirus [65,66] as well as the cytoplasmic trafficking [67] and nuclear import of AAV [27,28,64,68].

The schematic represents the fluorescent microscopy methodology for the imaging of the parvoviral life cycle in the nuclear region. (1) Analysis of fluorescent virus particle dynamics by single-particle tracking and high-speed super-resolution microscopy verified the import of viral capsids through the nuclear pore complex. Image correlation analysis using the pair correlation function (pCF) revealed the importin β-mediated nuclear transport of capsids. Confocal microscopy combined with EM characterized an alternative nuclear entry pathway for parvoviruses through virus-induced nuclear envelope ruptures. (2) Tracking of fluorescent capsids after their nuclear entry demonstrated that they moved by diffusion in the nucleoplasm. Furthermore, image correlation using the autocorrelation function (ACF) indicated that the capsids were disintegrated after their nuclear import. (3) Super-resolution microscopy analysis indicated that viral replication centres were located close to sites of cellular DNA damage. Fluorescence recovery after photobleaching (FRAP) studies showed that infection affected the diffusion of nuclear proteins, such as transcription-associated proteins. (4) Fluorescent tagging of progeny capsids (green) has allowed for analyses of capsid dynamics in living cells. Images were created with BioRender.com.

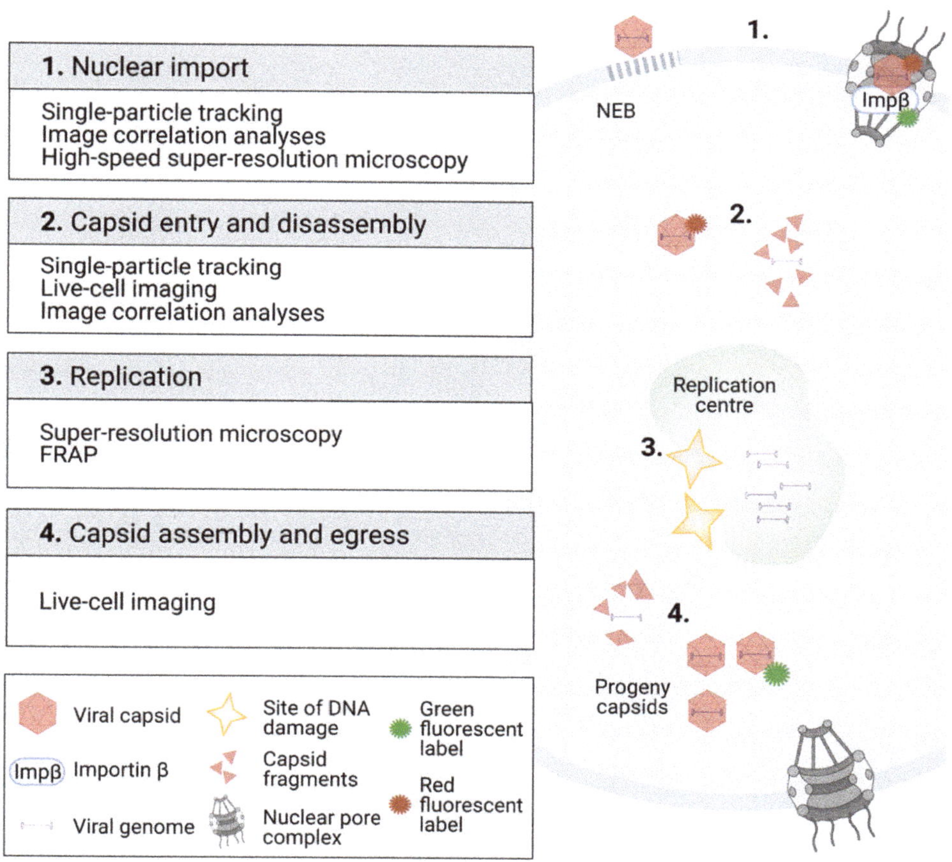

Figure 1. Imaging of viruses in the nucleus of infected cells.

Imaging of autonomous parvovirus capsids has partially been hampered by the limited possibilities to express recombinant viruses that contain fluorescent proteins, as the enlarged genome size leads to poor viral genome packaging. Therefore, little is known about virus–nucleus interactions following the assembly of viral capsid. However, AAV-2 studies have shown that large peptides can be inserted into the VP2 protein with a minimal effect on viral assembly or infectivity [69]. This has allowed the creation of fluorescent protein-tagged AAV particles for live cell analysis of intranuclear dynamics [70]. The loop regions of AAV capsid proteins exposed to the capsid surface have been used for the insertion of shorter peptides, which enables the labelling of viral particles with a fluorescent dye [71,72].

Tracking of individual viruses is a powerful tool to examine the mechanisms of their intracellular transport, and it is straightforward, for example, to conclude whether the motion is directed or random diffusion. For active processes, such as transport along microtubules, the dynamics can be deduced from a low number of particles. However, insight into the parvoviral life cycle has revealed the diffusive dynamics of events. For example, following of the trajectories of Cy5-labelled AAV capsids in the cytoplasm and nucleus showed that the majority of capsids move by regular diffusion, but a smaller fraction of the capsids exhibits anomalous subdiffusion [64]. The analysis of a small number of randomly moving diffusing particles is challenging, but when the motions of typically hundreds or thousands of particles are averaged, their movement can be characterized. The mean squared displacement (MSD) of the particles follows the law $MSD = 2dDt$, where D is the diffusion coefficient of the particle, d is the dimensionality of the motion, and t is the time. Measuring the MSD allows for the determination of the particle diffusion coefficient, which can then be further connected to the particle radius r, temperature T, and viscosity η of the medium by the Stokes–Einstein equation:

$$D = \frac{k_B T}{6\pi\eta r}.$$

Recently, image correlation spectroscopy has been used to verify the nuclear capsid import and intranuclear disassembly of capsids in living cells (Figure 1, boxes 1 and 2) [30]. Image correlation methods are based on the principles of fluorescence correlation spectroscopy (FCS), which measures fluctuations of fluorescence intensity in a small volume by using the focused excitation laser beam. The recorded fluctuations in photon counts, collected as a time series, are used to calculate the time autocorrelation function (ACF) to resolve the dynamics of fluorescently tagged proteins. The ACF represents the correlation of the fluorescent signal between the starting time point ($t = t_0$) and following time points ($t = t_0 + \Delta t$) of the experiment, thus yielding information on fluorescent molecule diffusion time in the focal spot. In parvovirus studies, the ACF calculated for a time series of laser scanning microscopy images containing temporal information of the intensity fluctuations and spatial distribution maps of the fluorescent viral particles has enabled the analysis of fast and slow diffusion, or even immobile viral particles [30].

To obtain more information about the possible directed movement of fluorescent particles, pair correlation function (pCF) analysis can also be used. The pCF measures the correlation over time and space and thus can distinguish directed movement or obstacles to diffusion. In parvovirus studies, pCF revealed a positive correlation between pixels across the NE within an image series, thereby demonstrating the nuclear import of capsid through the NE [30,73–75]. In addition, pCF analysis detected a spatiotemporal correlation between the fluorescent viral capsid and importin β, suggesting that importin β mediates capsid translocation through the nuclear pore complex [30]. An alternative or parallel existing nuclear entry pathway has been derived from studies using fluorescence and electron microscopy. The experiments have demonstrated that the NE undergoes substantial damage at early times during parvovirus H1, CPV, and AAV2 infection, indicating an NPC-independent nuclear entry of capsids [31,33].

The theoretical nuclear diffusion coefficient of capsids obtained from the Stokes-Einstein law, assuming that the viscosity of the nucleoplasm is approximately four times higher than in water [76,77], is in the order of 10 $\mu m^2/s$. This is in accordance with the experimental finding of 5 $\mu m^2/s$ obtained for the mobile population of virus-like particles of parvovirus [30,47]. In the cellular scale, this is a relatively fast diffusion rate, and it means that on average, the virus particles are able to diffuse a 10 μm distance in a time scale of a few seconds, when not restricted by physical barriers or by interactions.

Studies of nucleoplasmic capsid diffusion coefficients by ACF, which improved temporal resolution from the millisecond to microsecond scale, have revealed distinct diffusion dynamics for intact capsids and potential capsid fragments, suggesting that capsids are disintegrated in the nucleoplasm after their import [30]. The detailed mechanisms by which the viral genome is released into the nucleoplasm remain to be determined. However, fluorescence microscopy analyses have shown that capsids are already modified prior to nuclear import and nuclear disassembly when VP1 N-terminus is exposed during the endocytic entry [41,78–80]. According to immunoprecipitation analyses, B19V capsid uncoating is enhanced by cytoplasmic divalent cations [81]. Previously published studies have demonstrated that at least for MVM, the nuclear release of DNA occurs without a complete disassembly of the capsids [78,82–85]. In summary, it can be concluded that parvoviral capsids enter the nucleus either via NPC or by passing through transient holes in the NE, which allow the entry of intact capsids. Intact capsids entering the nucleus may undergo structural change which leads to viral genome release at some distance from the NE [30,86].

As outlined before, progressing parvovirus infection leads to the development of viral replication centres [46,87] and relocation of host chromatin to the nuclear periphery [45,47–49,88]. Recently, super-resolution microscopy has demonstrated that viral replication centres originate close to DNA damage sites (Figure 1, box 3) [52]. The introduction of photobleaching experiments in the analyses of intranuclear mobility and kinetics of viral and cellular proteins has allowed a better monitoring of nuclear changes upon parvoviral infection (Figure 1, box 3). In these studies, a high-intensity laser is used to photobleach the fluorescence of a fluorescent molecule, typically a fluorescent fusion protein, from a defined area of the cell. In fluorescence recovery, after photobleaching (FRAP), a region of interest is bleached, and the recovery of fluorescence in the bleached region is measured. The rate of fluorescence recovery is determined by the exchange of fluorescent molecules between the bleached region and the surrounding unbleached area, thereby allowing the analysis of protein dynamics and interactions. In fluorescence loss in photobleaching (FLIP), an area of the cell is continuously photobleached with laser pulses, and images taken between the pulses measure the response in the entire pool of fluorescent molecules. Similar to FRAP, the rate of fluorescence loss is related to the mobility of the fluorescent molecules.

In CPV infection, FRAP experiments (Figure 1, box 3) have revealed that the dynamics of transcription-associated protein change during infection [89] and further demonstrated that infection leads to an increased protein mobility in the nucleoplasm, which potentially alters protein–protein and protein–DNA binding reactions during viral replication [47]. Additionally, FRAP has been used to study the kinetics of NS1-EYFP in noninfected cell nuclei. The results have shown that NS1-EYFP mobility is not consistent with free diffusion and suggested transient binding to nuclear components [90]. Shown by FLIP, the nucleocytoplasmic shuttling of NS1-EYFP has been discovered [90].

Further central questions in the late stages of the nuclear life cycle of parvoviruses, such as capsid assembly and nuclear egress, have been addressed using fluorescent microscopy of immunostained cells. These studies, in combination with biochemical characterizations, showed that MVM capsids assemble in the nucleus from VP1/VP2 trimers [60,91], and these trimers expose a structured nuclear localization motif [58]. For AAV-2, the subcellular localization of capsid assembly to nucleoli was identified with immunofluorescence and in situ hybridization microscopy techniques. Viral genome sequence analysis and mutational studies revealed that the capsid assembly is mediated by the viral assembly associated protein (AAP) [92,93]. Moreover, X-ray crystallography and cryo-EM analyses of MVM capsids demonstrated that viral DNA is packed through a fivefold packaging channel [94,95]. Studies have also revealed that MVM capsids leave the nucleus prior to cell lysis and NE breakdown [96], suggesting that capsids have to exit the nucleus through the NPCs [22,23]. A similar combination of techniques was used to show that MVM capsids egress the nucleus dependent upon chromosomal region maintenance 1 (CRM1, also known as exportin 1) protein [96], which is a nuclear export factor for various proteins and different cellular RNAs (snRNA, rRNA, some mRNAs) [97]. Notably, the nuclear exit was limited to genome-containing capsids phosphorylated in the unordered domain of VP2, while empty capsids exhibited nuclear accumulation [96]. By combining classical immunofluorescence microscopy with surface plasmon resonance spectroscopy, it has been shown that the CRM1-dependent nuclear export of MVM capsids is mediated by the supraphysiological NES in NS2 [22].

3. Screening and Validation of Protein–Protein Interactions

The nuclear import of intact parvovirus capsids is not limited by the NPC diameter, which is able to transport particles with a diameter of ~39 nm [98]. There is accumulating evidence that the nuclear entry of the parvovirus capsid depends on the host machinery for nuclear import, requiring coordinated interaction with different host proteins. Earlier studies have shown that the capsid proteins of MVM and CPV, in addition to AAV capsids, have basic regions containing NLSs or a structured nuclear localization motif in their capsid proteins. [35–41,60,79] During endocytic entry, the acidification of capsid leads to NLS exposure, and after reaching the cytoplasm, this would thus allow the attachment of nuclear import factors. Studies including coimmunoprecipitation assays (Co-IP) have verified that CPV and AAV2 capsids interact with Imp β [42,99]. However, these assays elucidate neither the localization of the interaction in the cell environment nor the phase of the infection. The proximity ligation assay (PLA) has allowed comprehensive imaging and quantitation of interactions within the host cell. This antibody-based technique enables the detection of two proteins that are in close proximity to each other (~40 nm) [100]. Therefore, PLA is capable of visualizing protein–protein interactions beyond the diffraction limit (Figure 2A). For CPV, in situ proximity ligation analysis, combined with confocal microscopy and image analysis, has demonstrated that capsids are able to recruit cytoplasmic Imp β for nuclear transport [42]. Coimmunoprecipitation analyses have indicated that entering H-1PV and AAV2 capsids interact with nucleoporins, which are proteins of the NPC [31].

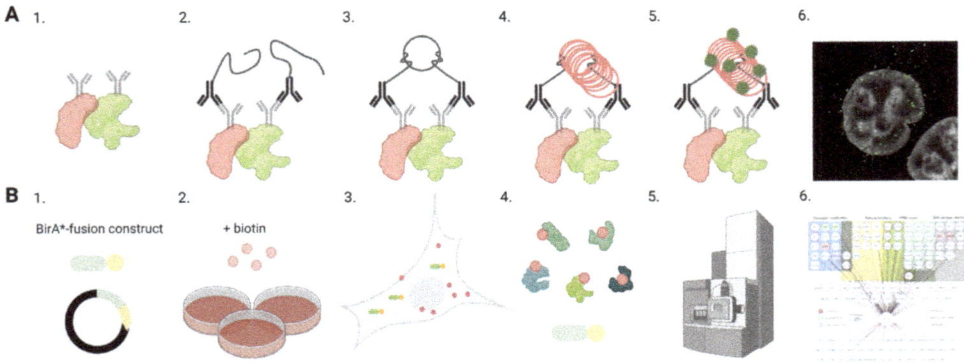

Figure 2. Analyses of protein–protein interactions in infection. Schematic overviews of proximity ligation assay (PLA) and proximity-dependent biotin identification (BioID) methods to identify and localize interactions between viral and host proteins. (**A**) The schematic representation of PLA assay. (1) Primary antibodies are used to target proteins of interest shown in red and green. (2) Secondary antibodies with PLA oligonucleotide probes bind to the primary antibodies. (3) Closely located PLA probes are ligated together, and (4) the formed circular DNA is amplified. (5) The amplified DNA (red) is labelled by fluorescent probes (green). (6) Confocal microscopy image shows the intracellular distribution of the PLA signals (green). Nuclei were stained with DAPI (grey). (**B**) Outlines of the BioID workflow. (1) Transfection of cells with BirA*-viral protein-fusion constructs and the generation of a stable inducible cell line. (2) Addition of biotin to the culture media and viruses if infection is required. (3) Cell culture period during which biotin ligase activity of BirA* fusion protein induces proximity-dependent biotinylation of neighbouring endogenous and viral proteins. (4) Cell lysis and the streptavidin-affinity purification of biotinylated proteins from cell lysates. (5) Mass spectroscopy and analyses of protein associations. (6) Interaction network indicating interaction partners of viral protein and biological processes involved. Images were created with BioRender.com.

Knowledge of viral protein interactions with cellular proteins is essential for understanding the intranuclear processes such as viral replication, capsid assembly, and nuclear egress. Affinity purification-mass spectrometry proteomics approaches have been traditionally used to analyse protein–protein interactions in infection [101–103]. Recently, many new screening methods have been generated to recognize protein–protein associations [104–106]. One of the methods is the proximity-dependent biotin identification (BioID) assay combined with mass spectrometry [107–109] (Figure 2B). BioID is a proximity-tagging method that utilizes a fusion of promiscuous biotin ligase, BirA, to a protein of interest to identify protein–protein associations and proximate proteins. The working radius for biotinylation via BirA is 10–40 nm, depending on the used application. Mass spectrometry-based proteomics applications such as BioID are able to recognize highly transient protein–protein interactions during the viral lifecycle. BioID studies of parvovirus human bocavirus 1 (HBoV1) have revealed interaction between viral nuclear protein 1 (NP1) and factors mediating nuclear import and mRNA processing [110]. A BioID analysis of AAV2 Rep proteins has revealed their association with cellular proteins, such as the transcriptional corepressor KAP1, which assist the viral genome in resisting epigenetic silencing, thereby allowing the lytic replication of AAV [111]. BioID has also been used to recognize interactions between viral proteins and DNA damage-related proteins. BioID has revealed an AAV Rep protein interaction with the Mre11 part of the MRN complex, an important initiator of the AMT response [111]. Overall, BioID has allowed for identifying associations of the viral protein of interest in a wide variety of nuclear processes, which, for CPV NS2, include DNA damage response and chromatin modification [112].

4. Detection of DNA Damage, DNA Repair, and Virus–DNA Interactions

Progression of parvovirus infection depends upon the induction of a cell cycle arrest and cell lysis. It leads to the activation of DNA damage response (DDR) [19,45], which promotes the infection and viral reproduction [113,114]. Ataxia telangiectasia and Rad3(ATR)-mediated DDR activation is linked to replication fork stalling, whereas the activation of the Ataxia-telangiectasia mutated (ATM)-mediated route is the initial response to a double-stranded DNA break (DSB) [115,116]. The activation of the ATR route has been observed for MVM, B19, and HBoV1 [51,111,112], and the ATM route for MVM, HBoV1, and AAV [45,117–119] (Figure 3A). Recognition of DNA damage induces the recruitment of proteins responsible for DNA damage repair to the site of the damage. During parvovirus infection, the emergence of DNA damage can be observed either indirectly by the accumulation of DDR proteins to the damage site or by observing the formation of actual DNA breakages. MVM infection has been shown to cause accumulation of proteins of the ATM signalling route (e.g., phosphorylated H2AX (γ-H2AX), Nbs1, RPA32, Chk2, p53, MDC1, MRN) to the replication start sites together with the viral replication protein NS1 [19,45]. During viral replication, at least newly synthesized viral DNA is bound to RPA, a known activator of ATR [120]. However, in MVM infection, this does not lead to the full activation of the ATR response since checkpoint kinase1 (Chk1) is not activated [49,51] (Figure 3A).

Recently, a high-throughput viral chromosome conformation capture sequencing assay (V3C-seq) has been applied to study the association of MVM viral genomes with host chromatin [121] (Figure 3C). V3C-seq is based on the chromosome conformation capture sequencing technology (3C-seq) [122] used to study chromosome arrangement in the nucleus by crosslinking the sites of genomic associations and identifying these regions with sequencing. 3C-seq studies have revealed that MVM genomes become associated with DNA damage sites during early stages of infection [121]. These sites of DNA damage with associated viral genomes increase as the infection proceeds. Nuclear localization of this association was further verified with fluorescent in situ hybridization (FISH) and super-resolution stochastic optical reconstruction microscopy (STORM). The introduction of externally induced DNA damage sites with laser irradiation or with CRISPR-Cas9 to a specific genomic locus resulted in parvoviral genome association with these regions. V3C-seq analyses have also revealed that the viral genome association sites and DNA damage sites overlap with self-interacting genetic regions, also known as topologically associating domains (TADs) [52]. Recently, it has been shown that the localization of viral genomes to the DNA damage sites is mediated by viral NS1 [121].

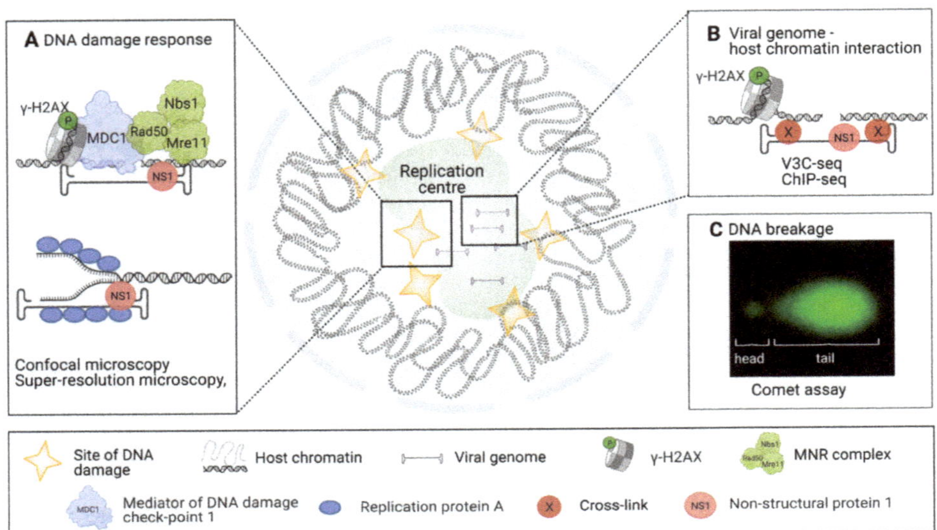

Figure 3. Approaches revealing virus-induced DNA damage. The schematic diagram of diverse methods for the analyses of DNA damage response (DDR), viral and host DNA interactions, and DNA damage in infection. (**A**) Analyses of ATM and ATR-mediated DNA damage signalling pathways by confocal and super-resolution microscopy, ATM-mediated cellular response to DNA damage functions through phosphorylation of proteins related to DNA damage and DNA damage repair such as γ-H2AX, MDC1, Rad50, Nbs1, and Mre11. In MVM infection these proteins are found in replication start sites together with viral NS1. In parvovirus-infected cells, the ATR-mediated response depends on RPA and viral NS1 interaction. (**B**) Elucidation of interactions between viral genome and host cell chromatin by using high-throughput viral chromosome conformation capture sequencing assay (V3C-seq). Moreover, association of DNA damage site MVM genomes has been shown by ChIP-seq. This analysis has been used to verify the association between NS1-mediated viral genome replication and DDR. (**C**) Studies of host cell chromatin disintegration by comet assay. Images were created with BioRender.com.

Classical DNA damage analyses in viral infection are qPCR or agarose gel electrophoresis, which do not allow investigations on the single-cell level. This obstacle was solved by comet assay—also known as single-cell gel electrophoresis—which is a sensitive, quantitative, and relatively simple imaging-based method to observe DNA breakages (Figure 3C) [123–125]. Scraped or trypsinized cells are cast into low-density agarose gel and lysed, after which the remaining nucleoids are placed in an electric field and stained. DNA lesions, both single and double stranded, result in a relaxation of DNA supercoiling. The relaxed DNA loops migrate towards the positively charged pole during electrophoresis, forming the characteristic comet tail pattern. The relative DNA content in the comet tail versus the head thus reflects the number of DNA lesions. Unlike the various DDR pathway markers, which might be activated in response to viral genomes or proteins [126], this method relies on the physical properties of damaged host DNA. Comet assay studies and ChIP-seq analysis have demonstrated that MVM infection causes host DNA damage, which increases as the infection proceeds [52]. In contrast, the comet assay has revealed no significant DNA damage in cells infected by the bocavirus minute virus of canine [127], nor in cells infected by human B19V [127]. The potential nucleolytic activity of parvoviral NS1 protein against host DNA has been investigated in expression studies for HBoV1 [117] and human B19V [127], but these studies did not find significant host DNA damage in NS1-expressing cells.

To benefit from host cell responses such as the DDR, viral proteins or viral genomes are required to interact directly with DNA or DNA-modifying proteins. The interactions of cellular DNA-binding proteins and viral proteins with host chromatin and viral genomes in MVM and CPV infections have been studied by ChIP-seq methods [52,88,121]. These studies have shown the acetylation of histones bound to CPV genome and MVM genome association with cellular γ-H2AX sites and the viral NS1 protein [52,88,121] (Figure 3B). Furthermore, the studies of the genomic reactivation of latent AAV genome by ChIP and ChIP coupled to qPCR have revealed the mechanism by which cellular proteins induce viral genome repression [111].

5. Recent Methods for Future Studies of Parvovirus–Nucleus Interactions

Despite of decades of research, many detailed mechanisms of virus–host interactions are not well understood, and many new observations raise further questions, requiring the use of newly developed techniques. Next-generation sequencing (NGS) and fluorescence imaging technologies are currently advancing rapidly [128,129], offering excellent opportunities for detailed analysis of infection-induced changes in the host chromatin organization and high-resolution imaging of parvovirus infection. For example, these methods combined with spatial transcriptomics allow analyses of the spatial heterogeneity of the gene expression within the sample [130–133].

NGS is a modern sequencing methodology where massive parallel sequencing is used to map the sequences of millions of small DNA fragments. Bioinformatics is then used to combine the acquired sequencing data, which can be then compared to reference genome(s). Various approaches allow for obtaining information about expressed genes [134], genome accessibility [135], binding regions of different DNA interacting proteins [136–138], or chromatin–chromatin interactions and organization [139]. As an example, the assay for transposase-accessible chromatin with sequencing (ATAC-seq) is based on hyperactive Tn5 transposase mutants [135]. In this assay, the hyperactive Tn5 is used to tagment the accessible chromatin by conjugating short and specific DNA oligomers into the accessible regions. These regions of the genome are then isolated and sequenced, yielding a high-resolution map of the accessible regions of the genome. Thus, ATAC-seq has great potential in studies on how parvoviral infection changes the host cell chromatin organization or in studies of viral genome packaging or release. This is exemplified by recent results showing that baculovirus infection induces significant changes in the organization of host genome, such as an increase in chromatin accessibility, relocation close to the NE, and nucleosome disassembly [140]. Moreover, ATAC-seq analysis of Epstein–Barr virus (EBV), a member of the herpesvirus family, has demonstrated that B cell chromatin undergoes significant remodelling during infection, which leads to the regulation of cell cycle, apoptosis pathways, and interferon regulatory factors [141]. Another example of a similar DNA-tagging method is DNA adenine methyltransferase identification (DamID)-sequencing [142]. Here, DNA adenine methyltransferase (Dam) is fused to a protein of interest, and this fusion protein is expressed in cells. The Dam enzyme recognizes DNA sequence GATC and methylates the adenine in the close vicinity of the fusion protein. These methylated regions of chromatin can then be sequenced and mapped. Thus, these sequences correspond to the chromatin that has been in close vicinity to the expressed fusion protein. This DamID-seq has been used to map the chromatin interacting with the nuclear lamina and lamina-associated domains [143]. In addition to sequencing, both ATAC-seq and DamID-seq can be combined with high-resolution fluorescence imaging. In the case of ATAC-seq, fluorescent oligomers are used together with hyperactive Tn5, and therefore, the tagmented and accessible chromatin can be visualized by fluorescence microscopy. This ATAC-see method [144] allows imaging the accessible chromatin regions and would be directly applicable to parvoviral studies regarding host cell chromatin or viral genome organization. DamID can be used together with methylated DNA-recognizing fluorescent ^{m6}A-tracer fusion protein.

^{m6}A-tracer binds to the GATC sequence when adenine is methylated by Dam methylase. By fusing ^{m6}A-tracer to a fluorescent protein, the fluorescent signal localizes to the methylated DNA [145]. The great advantage of the DamID ^{m6}A-tracer system is the possibility to use it in living cells. Thus, one can follow the chromatin dynamics by live cell microscopy. We envision that the system could be used to follow parvovirus infection-induced dynamic reorganization of the host genome.

Imaging and sequencing approaches are directly combined in spatial transcriptomics, where transcriptomes are resolved by high resolution microscopy or by capturing, so that spatial information about the location is also recorded. In microscopy-based spatially resolved transcriptomics or genomics, the different RNA and DNA species are labelled via sequential fluorescence in situ *hybridization* and barcoding. This approach offers the highest resolution, and recently, the imaging of 3660 chromosomal loci together with 17 chromatin marks in single cells has been reported [146].

6. Concluding Remarks

Conventional confocal microscopy approaches, including the imaging of fluorescent viral capsids and proteins and their interplay with cellular components within the host cell, have been successfully used in parvovirus studies. The development of live cell imaging and super-resolution microscopy, combined with image data analysis, together with the development of new screening tools for analyses of protein–protein and DNA–protein interactions, has further enhanced our understanding of virus–nucleus interactions and the nuclear dynamics of infection. In the near future, combining fluorescence data and ultrastructural information from electron micrographs will allow answering detailed questions regarding the mechanisms of intranuclear events in viral infection. Moreover, the advances in super-resolution microscopy applications will enable us to probe cell–virus interactions and dynamics in previously unattainable detail.

Author Contributions: S.M., S.H., S.S., V.A., E.M., T.O.I., M.K. and M.V.-R. wrote the paper. All authors have read and agreed to the published version of the manuscript.

Funding: This research was funded by the Jane and Aatos Erkko Foundation (M.V.-R.), Academy of Finland, grant numbers n330896 (M.V.-R.), 308315 (T.O.I.), 314106 (T.O.I.), and 332615 (E.M.); the Biocenter Finland, viral gene transfer (M.V.-R.); the Graduate School of the University of Jyvaskyla (S.M.); and a starting grant of the University of Gothenburg (M.K.).

Institutional Review Board Statement: Not applicable.

Informed Consent Statement: Not applicable.

Data Availability Statement: Data available in a publicly accessible repository.

Conflicts of Interest: The authors declare no conflict of interest.

References

1. Heegaard, E.D.; Brown, K.E. Human Parvovirus B19. *Clin. Microbiol. Rev.* **2002**, *15*, 485–505. [CrossRef] [PubMed]
2. Nandi, S.; Kumar, M. Canine Parvovirus: Current Perspective. *Indian J. Virol.* **2010**, *21*, 31–44. [CrossRef] [PubMed]
3. Kotterman, M.A.; Schaffer, D.V. Engineering Adeno-Associated Viruses for Clinical Gene Therapy. *Nat. Rev. Genet.* **2014**, *15*, 445–451. [CrossRef] [PubMed]
4. Li, C.; Samulski, R.J. Engineering Adeno-Associated Virus Vectors for Gene Therapy. *Nat. Rev. Genet.* **2020**, *21*, 255–272. [CrossRef] [PubMed]
5. Marchini, A.; Bonifati, S.; Scott, E.M.; Angelova, A.L.; Rommelaere, J. Oncolytic Parvoviruses: From Basic Virology to Clinical Applications. *Virol. J.* **2015**, *12*, 1–16. [CrossRef]
6. Zabaleta, N.; Dai, W.; Bhatt, U.; Chichester, J.A.; Estelien, R.; Sanmiguel, J.; Michalson, K.T.; Diop, C.; Maciorowski, D.; Qi, W.; et al. Immunogenicity of an AAV-Based, Room-Temperature Stable, Single Dose COVID-19 Vaccine in Mice and Non-Human Primates. *bioRxiv* **2021**. [CrossRef]
7. Tse, L.V.; Meganck, R.M.; Graham, R.L.; Baric, R.S. The Current and Future State of Vaccines, Antivirals and Gene Therapies Against Emerging Coronaviruses. *Front. Microbiol.* **2020**, *11*, 658. [CrossRef]

8. Pénzes, J.J.; Söderlund-Venermo, M.; Canuti, M.; Eis-Hübinger, A.M.; Hughes, J.; Cotmore, S.F.; Harrach, B. Reorganizing the Family Parvoviridae: A Revised Taxonomy Independent of the Canonical Approach Based on Host Association. *Arch. Virol.* **2020**, *165*, 2133–2146. [CrossRef]
9. Cotmore, S.F.; Tattersall, P. Dna Replication in the Autonomous Parvoviruses. *Semin. Virol.* **1995**, *6*, 271–281. [CrossRef]
10. Li, X.; Rhode, S.L. Mutation of Lysine 405 to Serine in the Parvovirus H-1 NS1 Abolishes Its Functions for Viral DNA Replication, Late Promoter Trans Activation, and Cytotoxicity. *J. Virol.* **1990**, *64*, 4654–4660. [CrossRef]
11. Christensen, J.; Cotmore, S.F.; Tattersall, P. Minute Virus of Mice Transcriptional Activator Protein NS1 Binds Directly to the Transactivation Region of the Viral P38 Promoter in a Strictly ATP-Dependent Manner. *J. Virol.* **1995**, *69*, 5422–5430. [CrossRef]
12. Naeger, L.K.; Cater, J.; Pintel, D.J. The Small Nonstructural Protein (NS2) of the Parvovirus Minute Virus of Mice Is Required for Efficient DNA Replication and Infectious Virus Production in a Cell-Type-Specific Manner. *J. Virol.* **1990**, *64*, 6166–6175. [CrossRef] [PubMed]
13. Tullis, G.E.; Labieniec-Pintel, L.; Clemens, K.E.; Pintel, D. Generation and Characterization of a Temperature-Sensitive Mutation in the NS-1 Gene of the Autonomous Parvovirus Minute Virus of Mice. *J. Virol.* **1988**, *62*, 2736–2744. [CrossRef]
14. Cotmore, S.F.; Gottlieb, R.L.; Tattersall, P. Replication Initiator Protein NS1 of the Parvovirus Minute Virus of Mice Binds to Modular Divergent Sites Distributed throughout Duplex Viral DNA. *J. Virol.* **2007**, *81*, 13015–13027. [CrossRef]
15. McCarty, D.M.; Pereira, D.J.; Zolotukhin, I.; Zhou, X.; Ryan, J.H.; Muzyczka, N. Identification of Linear DNA Sequences That Specifically Bind the Adeno-Associated Virus Rep Protein. *J. Virol.* **1994**, *68*, 4988–4997. [CrossRef] [PubMed]
16. Niskanen, E.A.; Ihalainen, T.O.; Kalliolinna, O.; Häkkinen, M.M.; Vihinen-Ranta, M. Effect of ATP Binding and Hydrolysis on Dynamics of Canine Parvovirus NS1. *J. Virol.* **2010**, *84*, 5391–5403. [CrossRef]
17. Niskanen, E.A.; Kalliolinna, O.; Ihalainen, T.O.; Hakkinen, M.; Vihinen-Ranta, M. Mutations in DNA Binding and Transactivation Domains Affect the Dynamics of Parvovirus NS1 Protein. *J. Virol.* **2013**, *87*, 11762–11774. [CrossRef]
18. Brownstein, D.G.; Smith, A.L.; Johnson, E.A.; Pintel, D.J.; Naeger, L.K.; Tattersall, P. The Pathogenesis of Infection with Minute Virus of Mice Depends on Expression of the Small Nonstructural Protein NS2 and on the Genotype of the Allotropic Determinants VP1 and VP2. *J. Virol.* **1992**, *66*, 3118–3124. [CrossRef] [PubMed]
19. Ruiz, Z.; Mihaylov, I.S.; Cotmore, S.F.; Tattersall, P. Recruitment of DNA Replication and Damage Response Proteins to Viral Replication Centers during Infection with NS2 Mutants of Minute Virus of Mice (MVM). *Virology* **2011**, *410*, 375–384. [CrossRef] [PubMed]
20. Naeger, L.K.; Salomé, N.; Pintel, D.J. NS2 Is Required for Efficient Translation of Viral MRNA in Minute Virus of Mice-Infected Murine Cells. *J. Virol.* **1993**, *67*, 1034–1043. [CrossRef]
21. Cotmore, S.F.; D'Abramo, A.M.; Carbonell, L.F.; Bratton, J.; Tattersall, P. The NS2 Polypeptide of Parvovirus MVM Is Required for Capsid Assembly in Murine Cells. *Virology* **1997**, *231*, 267–280. [CrossRef]
22. Engelsma, D.; Valle, N.; Fish, A.; Salomé, N.; Almendral, J.M.; Fornerod, M. A Supraphysiological Nuclear Export Signal Is Required for Parvovirus Nuclear Export. *Mol. Biol. Cell* **2008**, *19*, 2544–2552. [CrossRef]
23. Eichwald, V.; Daeffler, L.; Klein, M.; Rommelaere, J.; Salomé, N. The NS2 Proteins of Parvovirus Minute Virus of Mice Are Required for Efficient Nuclear Egress of Progeny Virions in Mouse Cells. *J. Virol.* **2002**, *76*, 10307–10319. [CrossRef]
24. Miller, C.L.; Pintel, D.J. Interaction between Parvovirus NS2 Protein and Nuclear Export Factor Crm1 Is Important for Viral Egress from the Nucleus of Murine Cells. *J. Virol.* **2002**, *76*, 3257–3266. [CrossRef]
25. López-Bueno, A.; Valle, N.; Gallego, J.M.; Pérez, J.; Almendral, J.M. Enhanced Cytoplasmic Sequestration of the Nuclear Export Receptor CRM1 by NS2 Mutations Developed in the Host Regulates Parvovirus Fitness. *J. Virol.* **2004**, *78*, 10674–10684. [CrossRef]
26. Hashemi, H.; Condurat, A.L.; Stroh-Dege, A.; Weiss, N.; Geiss, C.; Pilet, J.; Bartolomé, C.C.; Rommelaere, J.; Salomé, N.; Dinsart, C. Mutations in the Non-Structural Protein-Coding Sequence of Protoparvovirus h-1pv Enhance the Fitness of the Virus and Show Key Benefits Regarding the Transduction Efficiency of Derived Vectors. *Viruses* **2018**, *10*, 150. [CrossRef]
27. Kelich, J.M.; Ma, J.; Dong, B.; Wang, Q.; Chin, M.; Magura, C.M.; Xiao, W.; Yang, W. Super-Resolution Imaging of Nuclear Import of Adeno-Associated Virus in Live Cells. *Mol. Ther. Methods Clin. Dev.* **2015**, *2*, 15047. [CrossRef]
28. Junod, S.L.; Saredy, J.; Yang, W. Nuclear Import of Adeno-Associated Viruses Imaged by High-Speed Single-Molecule Microscopy. *Viruses* **2021**, *13*, 167. [CrossRef]
29. Mäntylä, E.; Kann, M.; Vihinen-Ranta, M. Protoparvovirus Knocking at the Nuclear Door. *Viruses* **2017**, *9*, 286. [CrossRef]
30. Mäntylä, E.; Chacko, J.V.; Aho, V.; Parrish, C.R.; Shahin, V.; Kann, M.; Digman, M.A.; Gratton, E.; Vihinen-Ranta, M. Viral Highway to Nucleus Exposed by Image Correlation Analyses. *Sci. Rep.* **2018**, *8*, 1–11. [CrossRef]
31. Porwal, M.; Cohen, S.; Snoussi, K.; Popa-Wagner, R.; Anderson, F.; Dugot-Senant, N.; Wodrich, H.; Dinsart, C.; Kleinschmidt, J.A.; Panté, N.; et al. Parvoviruses Cause Nuclear Envelope Breakdown by Activating Key Enzymes of Mitosis. *PLoS Pathog.* **2013**, *9*, e1003671. [CrossRef]
32. Cohen, S.; Panté, N. Pushing the Envelope: Microinjection of Minute Virus of Mice into Xenopus Oocytes Causes Damage to the Nuclear Envelope. *J. Gen. Virol.* **2005**, *86*, 3243–3252. [CrossRef]
33. Cohen, S.; Behzad, A.R.; Carroll, J.B.; Panté, N. Parvoviral Nuclear Import: Bypassing the Host Nuclear-Transport Machinery. *J. Gen. Virol.* **2006**, *87*, 3209–3213. [CrossRef]
34. Ros, C.; Bayat, N.; Wolfisberg, R.; Almendral, J.M. Protoparvovirus Cell Entry. *Viruses* **2017**, *9*, 313. [CrossRef]

35. Lombardo, E.; Ramírez, J.C.; Garcia, J.; Almendral, J.M. Complementary Roles of Multiple Nuclear Targeting Signals in the Capsid Proteins of the Parvovirus Minute Virus of Mice during Assembly and Onset of Infection. *J. Virol.* **2002**, *76*, 7049–7059. [CrossRef]
36. Vihinen-Ranta, M.; Kakkola, L.; Kalela, A.; Vilja, P.; Vuento, M. Characterization of a Nuclear Localization Signal of Canine Parvovirus Capsid Proteins. *Eur. J. Biochem.* **1997**, *250*, 389–394. [CrossRef]
37. Vihinen-Ranta, M.; Wang, D.; Weichert, W.S.; Parrish, C.R. The VP1 N-Terminal Sequence of Canine Parvovirus Affects Nuclear Transport of Capsids and Efficient Cell Infection. *J. Virol.* **2002**, *76*, 1884–1891. [CrossRef]
38. Popa-Wagner, R.; Sonntag, F.; Schmidt, K.; King, J.; Kleinschmidt, J.A. Nuclear Translocation of Adeno-Associated Virus Type 2 Capsid Proteins for Virion Assembly. *J. Gen. Virol.* **2012**, *93*, 1887–1898. [CrossRef]
39. Liu, P.; Chen, S.; Wang, M.; Cheng, A. The Role of Nuclear Localization Signal in Parvovirus Life Cycle. *Virol. J.* **2017**, *14*, 1–6. [CrossRef]
40. Grieger, J.C.; Snowdy, S.; Samulski, R.J. Separate Basic Region Motifs within the Adeno-Associated Virus Capsid Proteins Are Essential for Infectivity and Assembly. *J. Virol.* **2006**, *80*, 5199–5210. [CrossRef] [PubMed]
41. Sonntag, F.; Bleker, S.; Leuchs, B.; Fischer, R.; Kleinschmidt, J.A. Adeno-Associated Virus Type 2 Capsids with Externalized VP1/VP2 Trafficking Domains Are Generated Prior to Passage through the Cytoplasm and Are Maintained until Uncoating Occurs in the Nucleus. *J. Virol.* **2006**, *80*, 11040–11054. [CrossRef]
42. Mäntylä, E.; Aho, V.; Kann, M.; Vihinen-Ranta, M. Cytoplasmic Parvovirus Capsids Recruit Importin Beta for Nuclear Delivery. *J. Virol.* **2019**, *94*. [CrossRef]
43. Fay, N.; Panté, N. Old Foes, New Understandings: Nuclear Entry of Small Non-Enveloped DNA Viruses. *Curr. Opin. Virol.* **2015**, *12*, 59–65. [CrossRef] [PubMed]
44. Tattersall, P.; Ward, D.C. Rolling Hairpin Model for Replication of Parvovirus and Linear Chromosomal DNA. *Nature* **1976**, *263*, 106–109. [CrossRef] [PubMed]
45. Adeyemi, R.O.; Landry, S.; Davis, M.E.; Weitzman, M.D.; Pintel, D.J. Parvovirus Minute Virus of Mice Induces a DNA Damage Response That Facilitates Viral Replication. *PLoS Pathog.* **2010**, *6*, 1001141. [CrossRef] [PubMed]
46. Bashir, T.; Rommelaere, J.; Cziepluch, C. In Vivo Accumulation of Cyclin A and Cellular Replication Factors in Autonomous Parvovirus Minute Virus of Mice-Associated Replication Bodies. *J. Virol.* **2001**, *75*, 4394–4398. [CrossRef] [PubMed]
47. Ihalainen, T.O.; Niskanen, E.A.; Jylhävä, J.; Paloheimo, O.; Dross, N.; Smolander, H.; Langowski, J.; Timonen, J.; Vihinen-Ranta, M. Parvovirus Induced Alterations in Nuclear Architecture and Dynamics. *PLoS ONE* **2009**, *4*. [CrossRef]
48. Adeyemi, R.O.; Pintel, D.J. Replication of Minute Virus of Mice in Murine Cells Is Facilitated by Virally Induced Depletion of P21. *J. Virol.* **2012**, *86*, 8328–8332. [CrossRef]
49. Adeyemi, R.O.; Pintel, D.J. Parvovirus-Induced Depletion of Cyclin B1 Prevents Mitotic Entry of Infected Cells. *PLoS Pathog.* **2014**, *10*, 1003891. [CrossRef]
50. Lou, S.; Luo, Y.; Cheng, F.; Huang, Q.; Shen, W.; Kleiboeker, S.; Tisdale, J.F.; Liu, Z.; Qiu, J. Human Parvovirus B19 DNA Replication Induces a DNA Damage Response That Is Dispensable for Cell Cycle Arrest at Phase G2/M. *J. Virol.* **2012**, *86*, 10748–10758. [CrossRef]
51. Majumder, K.; Etingov, I.; Pintel, D.J. Protoparvovirus Interactions with the Cellular DNA Damage Response. *Viruses* **2017**, *9*, 323. [CrossRef]
52. Majumder, K.; Wang, J.; Boftsi, M.; Fuller, M.S.; Rede, J.E.; Joshi, T.; Pintel, D.J. Parvovirus Minute Virus of Mice Interacts with Sites of Cellular DNA Damage to Establish and Amplify Its Lytic Infection. *eLife* **2018**, *7*. [CrossRef]
53. Cotmore, S.F.; Tattersall, P. Parvoviruses: Small Does Not Mean Simple. *Annu. Rev. Virol.* **2014**, *1*, 517–537. [CrossRef]
54. Op De Beeck, A.; Caillet-Fauquet, P. The NS1 Protein of the Autonomous Parvovirus Minute Virus of Mice Blocks Cellular DNA Replication: A Consequence of Lesions to the Chromatin? *J. Virol.* **1997**, *71*, 5323–5329. [CrossRef]
55. Moffatt, S.; Yaegashi, N.; Tada, K.; Tanaka, N.; Sugamura, K. Human Parvovirus B19 Nonstructural (NS1) Protein Induces Apoptosis in Erythroid Lineage Cells. *J. Virol.* **1998**, *72*, 3018–3028. [CrossRef] [PubMed]
56. Chen, A.Y.; Qiu, J. Parvovirus Infection-Induced Cell Death and Cell Cycle Arrest. *Future Virol.* **2010**, *5*, 731–743. [CrossRef]
57. Cotmore, S.F.; Tattersall, P. Parvovirus Diversity and DNA Damage Responses. *Cold Spring Harb. Perspect. Biol.* **2013**, *5*. [CrossRef] [PubMed]
58. Lombardo, E.; Ramírez, J.C.; Agbandje-McKenna, M.; Almendral, J.M. A Beta-Stranded Motif Drives Capsid Protein Oligomers of the Parvovirus Minute Virus of Mice into the Nucleus for Viral Assembly. *J. Virol.* **2000**, *74*, 3804–3814. [CrossRef]
59. Riolobos, L.; Reguera, J.; Mateu, M.G.; Almendral, J.M. Nuclear Transport of Trimeric Assembly Intermediates Exerts a Morphogenetic Control on the Icosahedral Parvovirus Capsid. *J. Mol. Biol.* **2006**, *357*, 1026–1038. [CrossRef]
60. Gil-Ranedo, J.; Hernando, E.; Riolobos, L.; Domínguez, C.; Kann, M.; Almendral, J.M. The Mammalian Cell Cycle Regulates Parvovirus Nuclear Capsid Assembly. *PLoS Pathog.* **2015**, *11*, e1004920. [CrossRef] [PubMed]
61. Russell, D.W.; Miller, A.D.; Alexander, I.E. Adeno-Associated Virus Vectors Preferentially Transduce Cells in S Phase *Proc. Natl. Acad. Sci. USA* **1994**, *91*, 8915–8919. [CrossRef] [PubMed]
62. Bartlett, J.S.; Wilcher, R.; Samulski, R.J. Infectious Entry Pathway of Adeno-Associated Virus and Adeno-Associated Virus Vectors *J. Virol.* **2000**, *74*, 2777–2785. [CrossRef] [PubMed]
63. Vihinen-Ranta, M.; Yuan, W.; Parrish, C.R. Cytoplasmic Trafficking of the Canine Parvovirus Capsid and Its Role in Infection and Nuclear Transport. *J. Virol.* **2000**, *74*, 4853–4859. [CrossRef] [PubMed]

64. Seisenberger, G.; Ried, M.U.; Endreß, T.; Büning, H.; Hallek, M.; Bräuchle, C. Real-Time Single-Molecule Imaging of the Infection Pathway of Anadeno-Associated Virus. *Science* **2001**, *294*, 1929–1932. [CrossRef]
65. Lee, D.W.; Allison, A.B.; Bacon, K.B.; Parrish, C.R.; Daniel, S. Single-Particle Tracking Shows That a Point Mutation in the Carnivore Parvovirus Capsid Switches Binding between Host-Specific Transferrin Receptors. *J. Virol.* **2016**, *90*, 4849–4853. [CrossRef]
66. Cureton, D.K.; Harbison, C.E.; Cocucci, E.; Parrish, C.R.; Kirchhausen, T. Limited Transferrin Receptor Clustering Allows Rapid Diffusion of Canine Parvovirus into Clathrin Endocytic Structures. *J. Virol.* **2012**, *86*, 5330–5340. [CrossRef]
67. Xiao, P.-J.; Samulski, R.J. Cytoplasmic Trafficking, Endosomal Escape, and Perinuclear Accumulation of Adeno-Associated Virus Type 2 Particles Are Facilitated by Microtubule Network. *J. Virol.* **2012**, *86*, 10462–10473. [CrossRef]
68. Saxton, M.J. A Biological Interpretation of Transient Anomalous Subdiffusion. I. Qualitative Model. *Biophys. J.* **2007**, *92*, 1178–1191. [CrossRef]
69. Warrington, K.H.; Gorbatyuk, O.S.; Harrison, J.K.; Opie, S.R.; Zolotukhin, S.; Muzyczka, N. Adeno-Associated Virus Type 2 VP2 Capsid Protein Is Nonessential and Can Tolerate Large Peptide Insertions at Its N Terminus. *J. Virol.* **2004**, *78*, 6595–6609. [CrossRef]
70. Lux, K.; Goerlitz, N.; Schlemminger, S.; Perabo, L.; Goldnau, D.; Endell, J.; Leike, K.; Kofler, D.M.; Finke, S.; Hallek, M.; et al. Green Fluorescent Protein-Tagged Adeno-Associated Virus Particles Allow the Study of Cytosolic and Nuclear Trafficking. *J. Virol.* **2005**, *79*, 11776–11787. [CrossRef]
71. Chandran, J.S.; Sharp, P.S.; Karyka, E.; Aves-Cruzeiro, J.M.D.C.; Coldicott, I.; Castelli, L.; Hautbergue, G.; Collins, M.O.; Azzouz, M. Site Specific Modification of Adeno-Associated Virus Enables Both Fluorescent Imaging of Viral Particles and Characterization of the Capsid Interactome. *Sci. Rep.* **2017**, *7*, 1–17. [CrossRef]
72. Michelfelder, S.; Varadi, K.; Raupp, C.; Hunger, A.; Körbelin, J.; Pahrmann, C.; Schrepfer, S.; Müller, O.J.; Kleinschmidt, J.A.; Trepel, M. Peptide Ligands Incorporated into the Threefold Spike Capsid Domain to Re-Direct Gene Transduction of AAV8 and AAV9 in Vivo. *PLoS ONE* **2011**, *6*, e23101. [CrossRef]
73. Digman, M.A.; Gratton, E. Lessons in Fluctuation Correlation Spectroscopy. *Annu. Rev. Phys. Chem.* **2011**, *62*, 645–668. [CrossRef]
74. Hinde, E.; Cardarelli, F.; Digman, M.A.; Gratton, E. In Vivo Pair Correlation Analysis of EGFP Intranuclear Diffusion Reveals DNA-Dependent Molecular Flow. *Proc. Natl. Acad. Sci. USA* **2010**, *107*, 16560–16565. [CrossRef] [PubMed]
75. Digman, M.A.; Gratton, E. Imaging Barriers to Diffusion by Pair Correlation Functions. *Biophys. J.* **2009**, *97*, 665–673. [CrossRef] [PubMed]
76. Liang, L.; Wang, X.; Xing, D.; Chen, T.; Chen, W.R. Noninvasive Determination of Cell Nucleoplasmic Viscosity by Fluorescence Correlation Spectroscopy. *J. Biomed. Opt.* **2009**, *14*, 024013. [CrossRef]
77. Seksek, O.; Biwersi, J.; Verkman, A.S. Translational Diffusion of Macromolecule-Sized Solutes in Cytoplasm and Nucleus. *J. Cell Biol.* **1997**, *138*, 131–142. [CrossRef] [PubMed]
78. Ros, C.; Baltzer, C.; Mani, B.; Kempf, C. Parvovirus Uncoating in Vitro Reveals a Mechanism of DNA Release without Capsid Disassembly and Striking Differences in Encapsidated DNA Stability. *Virology* **2006**, *345*, 137–147. [CrossRef] [PubMed]
79. Mani, B.; Baltzer, C.; Valle, N.; Almendral, J.M.; Kempf, C.; Ros, C. Low PH-Dependent Endosomal Processing of the Incoming Parvovirus Minute Virus of Mice Virion Leads to Externalization of the VP1 N-Terminal Sequence (N-VP1), N-VP2 Cleavage, and Uncoating of the Full-Length Genome. *J. Virol.* **2006**, *80*, 1015–1024. [CrossRef]
80. Suikkanen, S.; Antila, M.; Jaatinen, A.; Vihinen-Ranta, M.; Vuento, M. Release of Canine Parvovirus from Endocytic Vesicles. *Virology* **2003**, *316*, 267–280. [CrossRef]
81. Caliaro, O.; Marti, A.; Ruprecht, N.; Leisi, R.; Subramanian, S.; Hafenstein, S.; Ros, C. Parvovirus B19 Uncoating Occurs in the Cytoplasm without Capsid Disassembly and It Is Facilitated by Depletion of Capsid-Associated Divalent Cations. *Viruses* **2019**, *11*, 430. [CrossRef]
82. Cotmore, S.F.; D'Abramo, A.M.; Ticknor, C.M.; Tattersall, P. Controlled Conformational Transitions in the MVM Virion Expose the VP1 N-Terminus and Viral Genome without Particle Disassembly. *Virology* **1999**, *254*, 169–181. [CrossRef] [PubMed]
83. Cotmore, S.F.; Hafenstein, S.; Tattersall, P. Depletion of Virion-Associated Divalent Cations Induces Parvovirus Minute Virus of Mice to Eject Its Genome in a 3′-to-5′ Direction from an Otherwise Intact Viral Particle. *J. Virol.* **2010**, *84*, 1945–1956. [CrossRef]
84. Cotmore, S.F.; Tattersall, P. Mutations at the Base of the Icosahedral Five-Fold Cylinders of Minute Virus of Mice Induce 3′-to-5′ Genome Uncoating and Critically Impair Entry Functions. *J. Virol.* **2012**, *86*, 69–80. [CrossRef] [PubMed]
85. Ros, C.; Kempf, C. The Ubiquitin-Proteasome Machinery Is Essential for Nuclear Translocation of Incoming Minute Virus of Mice. *Virology* **2004**, *324*, 350–360. [CrossRef]
86. Bernaud, J.; Rossi, A.; Fis, A.; Gardette, L.; Aillot, L.; Büning, H.; Castelnovo, M.; Salvetti, A.; Faivre-Moskalenko, C. Characterization of AAV Vector Particle Stability at the Single-Capsid Level. *J. Biol. Phys.* **2018**, *44*, 181–194. [CrossRef]
87. Cziepluch, C.; Lampel, S.; Grewenig, A.; Grund, C.; Lichter, P.; Rommelaere, J. H-1 Parvovirus-Associated Replication Bodies: A Distinct Virus-Induced Nuclear Structure. *J. Virol.* **2000**, *74*, 4807–4815. [CrossRef]
88. Mäntylä, E.; Salokas, K.; Oittinen, M.; Aho, V.; Mäntysaari, P.; Palmujoki, L.; Kalliolinna, O.; Ihalainen, T.O.; Niskanen, E.A.; Timonen, J.; et al. Promoter-Targeted Histone Acetylation of Chromatinized Parvoviral Genome Is Essential for the Progress of Infection. *J. Virol.* **2016**, *90*, 4059–4066. [CrossRef] [PubMed]

89. Ihalainen, T.O.; Willman, S.F.; Niskanen, E.A.; Paloheimo, O.; Smolander, H.; Laurila, J.P.; Kaikkonen, M.U.; Vihinen-Ranta, M. Distribution and Dynamics of Transcription-Associated Proteins during Parvovirus Infection. *J. Virol.* **2012**, *86*, 13779–13784. [CrossRef] [PubMed]
90. Ihalainen, T.O.; Niskanen, E.A.; Jylhävä, J.; Turpeinen, T.; Rinne, J.; Timonen, J.; Vihinen-Ranta, M. Dynamics and Interactions of Parvoviral NS1 Protein in the Nucleus. *Cell. Microbiol.* **2007**, *9*, 1946–1959. [CrossRef] [PubMed]
91. Riolobos, L.; Valle, N.; Hernando, E.; Maroto, B.; Kann, M.; Almendral, J.M. Viral Oncolysis That Targets Raf-1 Signaling Control of Nuclear Transport. *J. Virol.* **2010**, *84*, 2090–2099. [CrossRef]
92. Wistuba, A.; Kern, A.; Weger, S.; Grimm, D.; Kleinschmidt, J.A. Subcellular Compartmentalization of Adeno-Associated Virus Type 2 Assembly. *J. Virol.* **1997**, *71*, 1341–1352. [CrossRef] [PubMed]
93. Naumer, M.; Sonntag, F.; Schmidt, K.; Nieto, K.; Panke, C.; Davey, N.E.; Popa-Wagner, R.; Kleinschmidt, J.A. Properties of the Adeno-Associated Virus Assembly-Activating Protein. *J. Virol.* **2012**, *86*, 13038–13048. [CrossRef] [PubMed]
94. Plevka, P.; Hafenstein, S.; Li, L.; D'Abrgamo, A.; Cotmore, S.F.; Rossmann, M.G.; Tattersall, P. Structure of a Packaging-Defective Mutant of Minute Virus of Mice Indicates That the Genome Is Packaged via a Pore at a 5-Fold Axis. *J. Virol.* **2011**, *85*, 4822–4827. [CrossRef] [PubMed]
95. Subramanian, S.; Organtini, L.J.; Grossman, A.; Domeier, P.P.; Cifuente, J.O.; Makhov, A.M.; Conway, J.F.; D'Abramo, A.; Cotmore, S.F.; Tattersall, P.; et al. Cryo-EM Maps Reveal Five-Fold Channel Structures and Their Modification by Gatekeeper Mutations in the Parvovirus Minute Virus of Mice (MVM) Capsid. *Virology* **2017**, *510*, 216–223. [CrossRef]
96. Maroto, B.; Valle, N.; Saffrich, R.; Almendral, J.M. Nuclear Export of the Nonenveloped Parvovirus Virion Is Directed by an Unordered Protein Signal Exposed on the Capsid Surface. *J. Virol.* **2004**, *78*, 10685–10694. [CrossRef] [PubMed]
97. Köhler, A.; Hurt, E. Exporting RNA from the Nucleus to the Cytoplasm. *Nat. Rev. Mol. Cell Biol.* **2007**, *8*, 761–773. [CrossRef] [PubMed]
98. Panté, N.; Kann, M. Nuclear Pore Complex Is Able to Transport Macromolecules with Diameters of 39 Nm. *Mol. Biol. Cell* **2002**, *13*, 425–434. [CrossRef]
99. Nicolson, S.C.; Samulski, R.J. Recombinant Adeno-Associated Virus Utilizes Host Cell Nuclear Import Machinery to Enter the Nucleus. *J. Virol.* **2014**, *88*, 4132–4144. [CrossRef]
100. Söderberg, O.; Leuchowius, K.J.; Gullberg, M.; Jarvius, M.; Weibrecht, I.; Larsson, L.G.; Landegren, U. Characterizing Proteins and Their Interactions in Cells and Tissues Using the in Situ Proximity Ligation Assay. *Methods* **2008**, *45*, 227–232. [CrossRef]
101. Gerold, G.; Bruening, J.; Pietschmann, T. Decoding Protein Networks during Virus Entry by Quantitative Proteomics. *Virus Res.* **2016**, *218*, 25–39. [CrossRef] [PubMed]
102. Richards, A.L.; Eckhardt, M.; Krogan, N.J. Mass Spectrometry-based Protein–Protein Interaction Networks for the Study of Human Diseases. *Mol. Syst. Biol.* **2021**, *17*, e8792. [CrossRef] [PubMed]
103. Pankow, S.; Bamberger, C.; Calzolari, D.; Bamberger, A.; Yates, J.R. Deep Interactome Profiling of Membrane Proteins by Co-Interacting Protein Identification Technology. *Nat. Protoc.* **2016**, *11*, 2515–2528. [CrossRef] [PubMed]
104. Kovács, I.A.; Luck, K.; Spirohn, K.; Wang, Y.; Pollis, C.; Schlabach, S.; Bian, W.; Kim, D.K.; Kishore, N.; Hao, T.; et al. Network-Based Prediction of Protein Interactions. *Nat. Commun.* **2019**, *10*, 1–8. [CrossRef]
105. Silverbush, D.; Sharan, R. A Systematic Approach to Orient the Human Protein–Protein Interaction Network. *Nat. Commun.* **2019**, *10*, 1–9. [CrossRef] [PubMed]
106. Liu, Q.; Zheng, J.; Sun, W.; Huo, Y.; Zhang, L.; Hao, P.; Wang, H.; Zhuang, M. A Proximity-Tagging System to Identify Membrane Protein–Protein Interactions. *Nat. Methods* **2018**, *15*, 715–722. [CrossRef] [PubMed]
107. Roux, K.J.; Kim, D.I.; Raida, M.; Burke, B. A Promiscuous Biotin Ligase Fusion Protein Identifies Proximal and Interacting Proteins in Mammalian Cells. *J. Cell Biol.* **2012**, *196*, 801–810. [CrossRef]
108. Sears, R.M.; May, D.G.; Roux, K.J. BioID as a tool for protein-proximity labeling in living cells. *Methods Mol. Biol.* **2019**. [CrossRef]
109. Liu, X.; Salokas, K.; Tamene, F.; Jiu, Y.; Weldatsadik, R.G.; Öhman, T.; Varjosalo, M. An AP-MS- and BioID-Compatible MAC-Tag Enables Comprehensive Mapping of Protein Interactions and Subcellular Localizations. *Nat. Commun.* **2018**, *9*, 1–16. [CrossRef]
110. Wang, X.; Xu, P.; Cheng, F.; Li, Y.; Wang, Z.; Hao, S.; Wang, J.; Ning, K.; Ganaie, S.S.; Engelhardt, J.F.; et al. Cellular Cleavage and Polyadenylation Specificity Factor 6 (CPSF6) Mediates Nuclear Import of Human Bocavirus 1 NP1 Protein and Modulates Viral Capsid Protein Expression. *J. Virol.* **2020**, *94*. [CrossRef]
111. Smith-Moore, S.; Neil, S.J.D.; Fraefel, C.; Michael Linden, R.; Bollen, M.; Rowe, H.M.; Henckaerts, E. Adeno-Associated Virus Rep Proteins Antagonize Phosphatase PP1 to Counteract KAP1 Repression of the Latent Viral Genome. *Proc. Natl. Acad. Sci. USA* **2018**, *115*, E3529–E3538. [CrossRef]
112. Mattola, S.; Salokas, K.; Aho, V.; Hakanen, S.; Salminen, S.; Mäntylä, E.; Niskanen, E.A.; Svirskaite, J.; Ihalainen, T.O.; Parrish, C.R.; et al. Parvovirus Nonstructural Protein 2 Interacts with Proteins of Cellular Machinery Regulating Chromatin Functions (Manuscript).
113. Lilley, C.E.; Schwartz, R.A.; Weitzman, M.D. Using or Abusing: Viruses and the Cellular DNA Damage Response. *Trends Microbiol.* **2007**, *15*, 119–126. [CrossRef]
114. Trigg, B.J.; Ferguson, B.J. Functions of DNA Damage Machinery in the Innate Immune Response to DNA virus Infection. *Curr. Opin. Virol.* **2015**, *15*, 56–62. [CrossRef]
115. Kastan, M.B.; Bartek, J. Cell-Cycle Checkpoints and Cancer. *Nature* **2004**, *432*, 316–323. [CrossRef]

116. Shiotani, B.; Zou, L. Single-Stranded DNA Orchestrates an ATM-to-ATR Switch at DNA Breaks. *Mol. Cell* **2009**, *33*, 547–558. [CrossRef] [PubMed]
117. Deng, X.; Yan, Z.; Cheng, F.; Engelhardt, J.F.; Qiu, J. Replication of an Autonomous Human Parvovirus in Non-Dividing Human Airway Epithelium Is Facilitated through the DNA Damage and Repair Pathways. *PLoS Pathog.* **2016**, *12*. [CrossRef]
118. Luo, Y.; Lou, S.; Deng, X.; Liu, Z.; Li, Y.; Kleiboeker, S.; Qiu, J. Parvovirus B19 Infection of Human Primary Erythroid Progenitor Cells Triggers ATR-Chk1 Signaling, Which Promotes B19 Virus Replication. *J. Virol.* **2011**, *85*, 8046–8055. [CrossRef]
119. Schwartz, R.A.; Carson, C.T.; Schuberth, C.; Weitzman, M.D. Adeno-Associated Virus Replication Induces a DNA Damage Response Coordinated by DNA-Dependent Protein Kinase. *J. Virol.* **2009**, *83*, 6269–6278. [CrossRef] [PubMed]
120. Vassin, V.M.; Anantha, R.W.; Sokolova, E.; Kanner, S.; Borowiec, J.A. Human RPA Phosphorylation by ATR Stimulates DNA Synthesis and Prevents SsDNA Accumulation during DNA-Replication Stress. *J. Cell Sci.* **2009**, *122*, 4070–4080. [CrossRef] [PubMed]
121. Majumder, K.; Boftsi, M.; Whittle, F.B.; Wang, J.; Fuller, M.S.; Joshi, T.; Pintel, D.J. The NS1 Protein of the Parvovirus MVM Aids in the Localization of the Viral Genome to Cellular Sites of DNA Damage. *PLoS Pathog.* **2020**, *16*, e1009002. [CrossRef] [PubMed]
122. Dixon, J.R.; Gorkin, D.U.; Ren, B. Chromatin Domains: The Unit of Chromosome Organization. *Mol. Cell* **2016**, *62*, 668–680. [CrossRef] [PubMed]
123. Ostling, O.; Johanson, K.J. Microelectrophoretic Study of Radiation-Induced DNA Damages in Individual Mammalian Cells. *Biochem. Biophys. Res. Commun.* **1984**, *123*, 291–298. [CrossRef]
124. Olive, P.L.; Banáth, J.P. The Comet Assay: A Method to Measure DNA Damage in Individual Cells. *Nat. Protoc.* **2006**, *1*, 23–29. [CrossRef]
125. Collins, A.R. The Comet Assay: A Heavenly Method! *Mutagenesis* **2015**, *30*, 1–4. [CrossRef]
126. Alekseev, O.; Donegan, W.E.; Donovan, K.R.; Limonnik, V.; Azizkhan-Clifford, J. HSV-1 Hijacks the Host Dna Damage Response in Corneal Epithelial Cells through ICP4-Mediated Activation of ATM. *Invest. Ophthalmol. Vis. Sci.* **2020**, *61*, 39. [CrossRef] [PubMed]
127. Luo, Y.; Kleiboeker, S.; Deng, X.; Qiu, J. Human Parvovirus B19 Infection Causes Cell Cycle Arrest of Human Erythroid Progenitors at Late S Phase That Favors Viral DNA Replication. *J. Virol.* **2013**, *87*, 12766–12775. [CrossRef]
128. Goodwin, S.; McPherson, J.D.; McCombie, W.R. Coming of Age: Ten Years of next-Generation Sequencing Technologies. *Nat. Rev. Genet.* **2016**, *17*, 333–351. [CrossRef]
129. Metzker, M.L. Sequencing Technologies the next Generation. *Nat. Rev. Genet.* **2010**, *11*, 31–46. [CrossRef] [PubMed]
130. Marx, V. Method of the Year: Spatially Resolved Transcriptomics. *Nat. Methods* **2021**, *18*, 9–14. [CrossRef] [PubMed]
131. Stickels, R.R.; Murray, E.; Kumar, P.; Li, J.; Marshall, J.L.; Di Bella, D.J.; Arlotta, P.; Macosko, E.Z.; Chen, F. Highly Sensitive Spatial Transcriptomics at Near-Cellular Resolution with Slide-SeqV2. *Nat. Biotechnol.* **2021**, *39*, 313–319. [CrossRef] [PubMed]
132. Larsson, L.; Frisén, J.; Lundeberg, J. Spatially Resolved Transcriptomics Adds a New Dimension to Genomics. *Nat. Methods* **2021**, *18*, 15–18. [CrossRef]
133. Zhuang, X. Spatially Resolved Single-Cell Genomics and Transcriptomics by Imaging. *Nat. Methods* **2021**, *18*, 18–22. [CrossRef]
134. Wang, Z.; Gerstein, M.; Snyder, M. RNA-Seq: A Revolutionary Tool for Transcriptomics. *Nat. Rev. Genet.* **2009**, *10*, 57–63. [CrossRef] [PubMed]
135. Buenrostro, J.D.; Giresi, P.G.; Zaba, L.C.; Chang, H.Y.; Greenleaf, W.J. Transposition of Native Chromatin for Fast and Sensitive Epigenomic Profiling of Open Chromatin, DNA-Binding Proteins and Nucleosome Position. *Nat. Methods* **2013**, *10*, 1213–1218. [CrossRef]
136. Aughey, G.N.; Cheetham, S.W.; Southall, T.D. DamID as a Versatile Tool for Understanding Gene Regulation. *Dev. Camb.* **2019**, *146*. [CrossRef] [PubMed]
137. Kaya-Okur, H.S.; Wu, S.J.; Codomo, C.A.; Pledger, E.S.; Bryson, T.D.; Henikoff, J.G.; Ahmad, K.; Henikoff, S. CUT&Tag for Efficient Epigenomic Profiling of Small Samples and Single Cells. *Nat. Commun.* **2019**, *10*, 1–10. [CrossRef]
138. Furey, T.S. ChIP-Seq and beyond: New and Improved Methodologies to Detect and Characterize Protein-DNA Interactions. *Nat. Rev. Genet.* **2012**, *13*, 840–852. [CrossRef] [PubMed]
139. De Wit, E.; de Laat, W. A Decade of 3C Technologies: Insights into Nuclear Organization. *Genes Dev.* **2012**, *26*, 11–24. [CrossRef]
140. Kong, X.; Wei, G.; Chen, N.; Zhao, S.; Shen, Y.; Zhang, J.; Li, Y.; Zeng, X.; Wu, X. Dynamic Chromatin Accessibility Profiling Reveals Changes in Host Genome Organization in Response to Baculovirus Infection. *PLoS Pathog.* **2020**, *16*, e1008633. [CrossRef]
141. Lamontagne, R.J.; Soldan, S.S.; Su, C.; Wiedmer, A.; Won, K.J.; Lu, F.; Goldman, A.R.; Wickramasinghe, J.; Tang, H.Y.; Speicher, D.W.; et al. A Multi-Omics Approach to Epstein-Barr Virus Immortalization of B-Cells Reveals EBNA1 Chromatin Pioneering Activities Targeting Nucleotide Metabolism. *PLoS Pathog.* **2021**, *17*, e1009208. [CrossRef]
142. Van Steensel, B.; Henikoff, S. Identification of in Vivo DNA Targets of Chromatin Proteins Using Tethered Dam Methyltransferase. *Nat. Biotechnol.* **2000**, *18*, 424–428. [CrossRef]
143. Kind, J.; Pagie, L.; De Vries, S.S.; Nahidiazar, L.; Dey, S.S.; Bienko, M.; Zhan, Y.; Lajoie, B.; De Graaf, C.A.; Amendola, M.; et al. Genome-Wide Maps of Nuclear Lamina Interactions in Single Human Cells. *Cell* **2015**, *163*, 134–147. [CrossRef]
144. Chen, X.; Shen, Y.; Draper, W.; Buenrostro, J.D.; Litzenburger, U.; Cho, S.W.; Satpathy, A.T.; Carter, A.C.; Ghosh, R.P.; East-Seletsky, A.; et al. ATAC-See Reveals the Accessible Genome by Transposase-Mediated Imaging and Sequencing. *Nat. Methods* **2016**, *13*, 1013–1020. [CrossRef] [PubMed]

145. Kind, J.; Pagie, L.; Ortabozkoyun, H.; Boyle, S.; De Vries, S.S.; Janssen, H.; Amendola, M.; Nolen, L.D.; Bickmore, W.A.; Van Steensel, B. Single-Cell Dynamics of Genome-Nuclear Lamina Interactions. *Cell* **2013**, *153*, 178–192. [CrossRef] [PubMed]
146. Takei, Y.; Yun, J.; Zheng, S.; Ollikainen, N.; Pierson, N.; White, J.; Shah, S.; Thomassie, J.; Suo, S.; Eng, C.H.L.; et al. Integrated Spatial Genomics Reveals Global Architecture of Single Nuclei. *Nature* **2021**, *590*, 344–350. [CrossRef] [PubMed]

Review

The VP1u of Human Parvovirus B19: A Multifunctional Capsid Protein with Biotechnological Applications

Carlos Ros *, Jan Bieri and Remo Leisi

Department of Chemistry and Biochemistry, University of Bern, 3012 Bern, Switzerland; jan.bieri@dcb.unibe.ch (J.B.); remo.leisi@dcb.unibe.ch (R.L.)
* Correspondence: carlos.ros@dcb.unibe.ch

Academic Editor: Giorgio Gallinella
Received: 2 December 2020; Accepted: 16 December 2020; Published: 18 December 2020

Abstract: The viral protein 1 unique region (VP1u) of human parvovirus B19 (B19V) is a multifunctional capsid protein with essential roles in virus tropism, uptake, and subcellular trafficking. These functions reside on hidden protein domains, which become accessible upon interaction with cell membrane receptors. A receptor-binding domain (RBD) in VP1u is responsible for the specific targeting and uptake of the virus exclusively into cells of the erythroid lineage in the bone marrow. A phospholipase A_2 domain promotes the endosomal escape of the incoming virus. The VP1u is also the immunodominant region of the capsid as it is the target of neutralizing antibodies. For all these reasons, the VP1u has raised great interest in antiviral research and vaccinology. Besides the essential functions in B19V infection, the remarkable erythroid specificity of the VP1u makes it a unique erythroid cell surface biomarker. Moreover, the demonstrated capacity of the VP1u to deliver diverse cargo specifically to cells around the proerythroblast differentiation stage, including erythroleukemic cells, offers novel therapeutic opportunities for erythroid-specific drug delivery. In this review, we focus on the multifunctional role of the VP1u in B19V infection and explore its potential in diagnostics and erythroid-specific therapeutics.

Keywords: parvovirus B19; B19V; VP1u; receptor; PLA_2; virus entry; erythroid cells; biomarker; drug delivery; nanocarrier

1. Introduction

The *Parvoviridae* is a family of nonenveloped viruses that packages a linear, single-stranded DNA genome (~5 kb) within a small (~25 nm) icosahedral capsid. As a direct consequence of their limited coding potential, parvoviruses are particularly dependent on host cellular factors for their replication [1,2]. Parvoviruses are widely spread in nature and their host range might span the entire animal kingdom [3]. Depending on their host, members of the family *Parvoviridae* are subdivided into the subfamilies *Parvovirinae*, infecting vertebrates and *Densovirinae*, infecting insects and other arthropods. Viruses that infect vertebrates, including humans, are further divided into the dependoparvoviruses and the autonomous parvoviruses [4]. The dependoparvoviruses replicate only in the presence of a helper virus, such as adenovirus or herpesvirus. The adeno-associated viruses (AAVs) are not linked with any known pathology, have a wide tissue specificity, and replicate in dividing and nondividing cells. These properties make AAVs useful gene transfer vehicles for therapeutic applications [5]. Although autonomous parvoviruses use similar strategies for cell entry and replication, they differ substantially in their pathogenic potential, which ranges from subclinical to severe or even lethal infections [2]. As autonomous parvoviruses can only replicate in dividing cells, when the host cell DNA replication machinery becomes available, they tend to cause more severe infections in young than in adult hosts.

While most ssDNA viruses show a circular genome structure, parvoviruses have a linear genome, that is typically organized in two open reading frames (ORFs). The ORFs are flanked by palindromic sequences of variable length, which fold into hairpin structures and are essential for replication [6,7]. The 5′ ORF (ns or rep gene) encodes for the regulatory nonstructural protein(s) required for viral DNA replication and packaging. The 3′ ORF (cap gene) encodes two to four variants of a single capsid protein (VP). Following a principle of genetic economy, the different VPs are generated by alternative splicing or alternative codon usage, but also by post-translational proteolytic processing during entry, resulting in a common C-terminal sequence but different N-terminal extensions of variable length [8–10]. The T = 1 icosahedral parvovirus capsid is assembled from 60 VPs, however, the number of N-terminal VP variants used to assemble the infectious particles varies from two (VP1 and VP2) to four (VP1–VP4) depending on the genus. The VP variants are numbered in order of length, with VP1 being the largest variant. The common C-terminal region of the VPs forms the capsid shell, which consists of a conserved alpha-helix and a jelly roll motif containing eight antiparallel β-strands. The different configurations of the loops connecting the conserved β-strands delineate the surface topology, which is characteristic to each parvovirus genus and define the virus tropism and antigenicity [11]. Despite low sequence identity, the parvovirus capsids display structural features that are conserved across different genera, i.e., a narrow depression at the twofold axis of symmetry, protrusions of variable size and shape at the threefold axis and a canyon-like structure encircling a cylindrical pore at the fivefold axis connecting with the interior of the capsid.

The minor protein VP1 has an N-terminal extension of variable length, the so-called VP1-unique region (VP1u), and is present at 3 to 10 copies per virion depending on the parvovirus genus. VP1u is not required for virus assembly but contains several essential motifs required for the infection. Nuclear localization signals (NLSs) consisting of a stretch of basic amino acids have been identified in the VP1u from several parvoviruses. These motifs were shown to confer nuclear import potential to the incoming particles [12–19]. Another motif found in VP1u, except for amdoparvoviruses, is a phospholipase A_2 (PLA_2) enzyme domain, which enables viruses to escape from endosomal vesicles into the cytosol during cell entry [20–26]. Other motifs in VP1u were found to be essential for the infection. In AAV these motifs include signals that are known to be involved in protein interaction, endosomal sorting, and signal transduction in eukaryotic cells [27]. In B19V, a receptor-binding domain (RBD) required for virus uptake was identified at the N-terminal of the VP1u [28].

To infect the cell, parvoviruses follow an intricate path from the cell surface to the nucleus where they deliver the viral DNA for replication. During the process of entry, the incoming parvovirus capsids undergo a program of conformational rearrangements triggered by specific cellular factors that facilitate their intracellular transport [29,30]. A major capsid rearrangement that is largely conserved among parvoviruses involves the externalization of the VP1u region. Initially sequestered in mature virions, VP1u and its essential motifs become accessible at the particle surface during entry triggered by the acidic endosomal environment [10,16,31]. Besides low pH, AAVs may require additional cellular factors [31]. An exception is B19V, whose VP1u becomes accessible during the initial interactions with cellular receptors [32]. VP1u exposure occurs through the five-fold channel that connects with the interior of the capsid [33–36]. Structural and in vitro studies suggest that these channels serve not only as portals for the externalization of N-terminal capsid protein sequences but also for the packaging and release of the viral genome [16,31,37–41]. Mutations that perturb the functional structure of the channel result in defective genome encapsidation, uncoating and VP1u externalization [38,42,43]. B19V represents again an exception as the VP1u seems to be already exposed on the capsid surface, although in a conformation that is not accessible to antibodies [44].

This review focuses on the VP1u of B19V, which shares common aspects with other parvoviruses but has unique features, like its structural conformation relative to the virion, immunodominance, extraordinary length or the presence of a receptor-binding domain responsible for the restricted tropism of the virus. We present the current knowledge on the different VP1u motifs, their functions in the virus infection and the potential biotechnological applications of the B19V VP1u in human therapy and diagnostics.

2. Human Erythroparvovirus B19 (B19V)

B19V is the most prominent and well-characterized human pathogen within the *Parvoviridae* causing a mild childhood rash disease named *erythema infectiosum* or fifth disease [45]. The infection is often asymptomatic; however, in adults, B19V infection may induce a wide range of more severe pathological conditions, such as arthralgias and arthritis [46]. B19V infection may lead to aplastic crisis in patients with pre-existing bone marrow disorders and shortened red cell survival [47] and persistent infection in immunocompromised persons. Infection during pregnancy may result in *hydrops fetalis* and fetal death [48]. B19V was the first parvovirus known to cause disease in humans [49]. Since 2005, other human parvoviruses have been identified and include human bocavirus (HBoV1-4), parvovirus 4, bufavirus, tusavirus and cutavirus. Except for HBoV, which has been implicated in acute respiratory tract infections [50], the rest are emergent human parvoviruses with uncertain clinical significance [45,51].

B19V is transmitted via aerosol droplets that come into contact with the upper respiratory tract mucosa [47]. The virus crosses the mucosal epithelium through a yet unknown mechanism and disseminates with the bloodstream to the bone marrow, where it infects erythroid precursors at a particular erythropoietin (EPO)-dependent stage of differentiation [52–54]. The extraordinary narrow tropism of B19V is mediated at different levels of the viral life cycle. Crucial steps of the viral infection, such as uptake, genome replication, transcription, splicing and packaging, are restricted to the EPO-dependent erythroid differentiation around the proerythroblast stage [54–60]. The lytic replication cycle results in the destruction of the erythroid precursor cells [61,62], which accounts for the hematological syndromes observed during the infection [47]. Acute infection frequently results in high-titer viremia, which precedes the onset of clinical manifestations and has been associated with B19V transmission through transfusion and plasma-derived medicinal products [63].

3. B19V Capsid

The ssDNA genome of B19V is packaged into a small, nonenveloped, T = 1 icosahedral capsid. Similar to the genome of dependoparvoviruses, the B19V genome has two identical inverted terminal repeats (ITRs; ~383 nt), which serve as the origin of replication [64]. The capsid consists of 60 structural subunits of two N-terminal VP variants, VP1 and VP2. Approximately 95% are VP2 (major VP; 60 kDa) and 5% are VP1 (minor VP; 86 kDa) [65]. VP1 and VP2 are generated through alternative splicing, resulting in the same C-terminal sequence but VP1 contains 227 additional residues at the VP1 N-terminal region, the so-called VP1 "unique region" (VP1u). The 60 protomers form 20 trimeric capsomers in the cytoplasm of the infected cell, which are assembled to an icosahedral capsid structure in the host nucleus. Due to the T = 1 symmetry, all protein subunits can be assembled in the same orientation to each other. This perfect symmetry enables an optimal thermodynamic sink for each protomer interaction, forming a very stable capsid around the ssDNA genome.

Large-scale propagation of native B19V is not possible due to the lack of a fully permissive cell culture system. Accordingly, structural studies have been performed with recombinant B19V-like particles, which are similar, although not identical, to infectious native capsids. The structure of the VP2 recombinant particle has been determined to ~3.5 Å resolution [36]. Similar to other parvoviruses and many icosahedral viruses, the major capsid protein VP2 is structured as a "jelly roll" with a β-barrel motif. The loops connecting the strands of the β-barrel define the capsid surface topology that differentiates B19V from other parvoviruses. B19V lacks the prominent protrusions at the icosahedral threefold axes characteristic in other parvoviruses. The channel at the fivefold icosahedral axis is surrounded by a large canyon-like depression. Different from other parvoviruses, the channel in B19 VP2 capsids is constricted at its outside end. However, a cluster of glycine residues at this position may confer sufficient flexibility to open the channel upon specific cellular triggers during the infection. A striking difference between B19V and other parvoviruses is the external position of the N-VP2 and probably also VP1u [36,66]. Accordingly, the role of the fivefold channel in B19V would be limited to the externalization and packaging of the viral genome.

4. VP1u Is the Immunodominant Region of the Capsid but It Is Not Accessible in Native Virions

Although VP1u may occupy a surface position in the B19V capsid, different regions of the protein were shown to be inaccessible to antibodies. However, exposure of native capsids to heat or low pH rendered these regions accessible without capsid disassembly. In contrast to native virions, VP1u is always accessible in recombinant B19V-like particles [44]. The inaccessibility of VP1u in native virions is not well understood and may be explained by a compact structural conformation, or by the presence of a masking structure hiding the essential protein domains to the immune system. Despite the non-accessible conformation of VP1u in native particles, this protein is the immunodominant part of the capsid and contains clusters of critical neutralizing epitopes [67,68] (Figure 1).

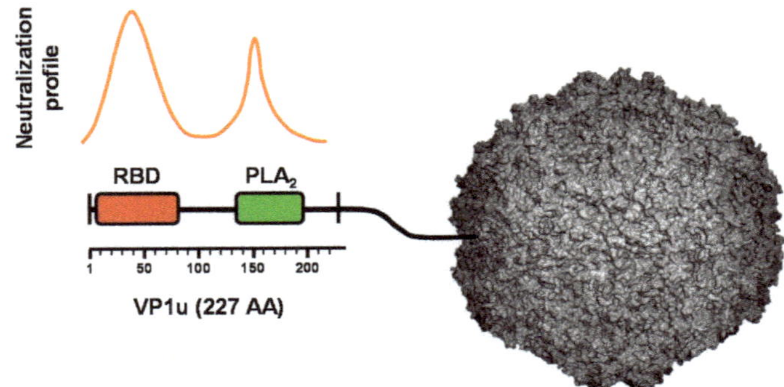

Figure 1. Schematic depiction of the neutralization profile and functional domains in the VP1u of B19V. The neutralizing profile revealed a cluster of important epitopes in the N-terminal region of the VP1 corresponding to functional domains. RBD; receptor-binding domain. PLA_2; phospholipase A_2 domain.

Typically, neutralizing antibodies prevent the viral infection by interfering with early steps of the viral life cycle, i.e., attachment to cellular receptors, uptake, fusion, or conformational changes required for entry [69,70]. Importantly, the inhibition by neutralizing antibodies should be distinguished from the opsonization of viruses by antibodies, which can hamper the viral infection by immobilization of the virus and subsequent degradation of the immune complex by the complement system, immune cells or also by the cytoplasmic TRIM21/proteasome mechanism [71]. However, the specific targeting of essential capsid protein domains by neutralizing antibodies is required to efficiently interfere with the viral infection. Upon B19 viremia, the humoral immune response first generates IgM antibodies, which predominantly target the major capsid protein VP2. With the class-switch and long-term immunity, an increasing percentage of B lymphocytes secrete neutralizing antibodies against the VP1u region [72]. In this regard, a deficient immune response to VP1u has been associated with persistent infections, emphasizing the important role of the immune response against VP1u in clearing the virus [73,74].

Immunization experiments with vaccine candidates based on virus-like particles (VLPs) demonstrated that VP1u is essential to raise a strong neutralizing response against B19V [75,76]. However, the neutralization mechanism of antibodies targeting VP1u has remained largely elusive. Being the immunodominant region of the capsid, the originally inaccessible VP1u should become exposed in the extracellular milieu, and not inside endosomes, as shown for other parvoviruses. In line with this assumption, it has been shown that a neutralizing antibody against the N-terminal part of the VP1u was unable to bind native cell-free virions but was able to block virus entry into susceptible cells. Moreover, capsids without VP1u were unable to internalize into susceptible cells, demonstrating the involvement of the VP1u in B19V uptake [32,54]. These findings explain the high neutralization potential of VP1u antibodies, which target exclusively capsids during the initial interaction with cell receptors and block virus uptake, and further emphasize the importance of VP1u as an essential component of prospective B19V vaccines.

5. Role of VP1u in the Restricted Tropism of B19V

B19V has a remarkable narrow tropism. The virus shows productive infection exclusively in erythroid precursor cells at EPO-dependent intermediate erythroid differentiation stages, with increasing permissiveness from BFU-E to erythroblasts [53]. Viral tropism can be determined already at the cell surface by the expression of specific cell receptors required for virus entry and/or intracellularly by receptor-independent post-entry replication steps. The marked erythroid tropism of B19V is determined at multiple steps, i.e., the receptor-mediated uptake, genome replication, transcription, splicing and packaging [56,57,77]. A virus requiring such strict intracellular conditions for replication would also require a selective mechanism of cell entry to target exclusively the few cells where the virus can replicate. This strategy would allow the virus to avoid internalizing non-permissive cells, which would lead to abortive infections and inefficient viral propagation. Accordingly, it would be expected that B19V uses an erythroid-specific surface molecule as an entry receptor.

5.1. VP1u Contains a Receptor-Binding Domain That Is Essential for Virus Entry into Permissive Cells

The neutral glycosphingolipid globoside (Gb4), also known as P antigen, has long been considered the primary receptor of B19V [78]. A large body of evidence suggests that B19V recognizes Gb4 and that the interaction is required for the infection [79–82]. However, the wide-range Gb4 expression does not correlate well with B19V binding and uptake and cannot either explain the pathogenesis or the remarkable narrow tissue tropism of the virus [83]. By using a knockout cell line, we demonstrated that Gb4 does not have the expected function as the primary cell surface receptor required for B19V entry. Instead, Gb4 has an essential role at a post-entry step after virus uptake and before the delivery of the viral genome into the nucleus for replication [84]. Other receptor molecules, such as $\alpha5\beta1$ integrin [85] and Ku80 autoantigen [86] have been proposed as potential coreceptors for B19V infection. However, the restricted uptake of B19V does not correspond with their wide expression profiles.

In an earlier study, we showed that the VP1u harbors a receptor-binding domain (RBD), which enables the uptake of the virus. Purified recombinant VP1u (recVP1u) was able to bind and to internalize exclusively into B19V permissive cells. Moreover, incorporation of VP1u subunits on bacteriophage VLPs by chemical coupling enabled their internalization into B19V permissive cells (Figure 2) [59]. The VP1u cognate receptor has not yet been identified, but its expression profile corresponds with the restricted tropism of B19V, being expressed exclusively in cells at erythropoietin-dependent erythroid differentiation stages [54,59].

5.2. Mapping and Structural Characterization of the Receptor-Binding Domain in the VP1u

The receptor-binding domain (RBD) in the VP1u was identified by using recVP1u variants with increasing N- and C-terminal truncations. The VP1u variants internalized normally when they were truncated less than 5 AA at the N-terminus or less than 147 AA at the C-terminus. Longer truncations at both ends decreased or blocked VP1u uptake [28]. According to these results, the RBD spans the region between AA 5–80 of VP1u, which explains the detectable exposure of this domain on the surface of susceptible cells before uptake [32,87], as well as the presence of a cluster of neutralizing epitopes [67,68].

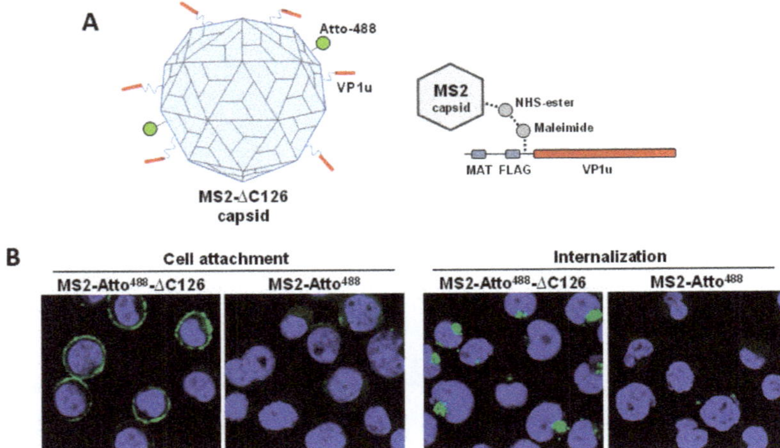

Figure 2. Binding and internalization of recombinant MS2-VP1u labeled with Atto-488. (**A**) Schematic depiction of MS2-VP1u particles. (**B**) Confocal fluorescence microscopy of MS2-VP1u bound to UT7/Epo cells at 4 °C and internalized at 37 °C. MS2-Atto^{488}-Δ126 (100 N-terminal AA of VP1u); MS2-Atto^{488} (without VP1u).

The secondary structure analysis of the N-terminal of VP1u (AA 1–80) from natural B19V isolates, predicted a cluster of three α-helices with high confidence: helix 1 (AA 14–31), helix 2 (35–45), and helix 3 (59–68) (Figure 3A). However, only helix 1 was conserved among other erythroparvoviruses (Figure 3B) and displayed a prominent amphiphilic character. The marked segregation of polar and hydrophobic amino acids between the two opposite flanks of the α-helix is well suited for receptor binding. Compared with the residues of the hydrophilic side, the amino acids of the hydrophobic side were highly conserved (Figure 3C). Point mutations on the hydrophobic side blocked VP1u binding and internalization, suggesting a critical role of these residues in the interaction of VP1u with its cognate cellular receptor [28].

The sequence analysis of the first 80 amino acids of VP1u predicted two additional helices (Figure 3A). Disruption of the tertiary conformation of these domains by the introduction of flexible sequences strongly impaired VP1u internalization. This observation suggests that the spatial configuration of the three helices is crucial for VP1u binding to its cognate receptor and subsequent uptake. An ab initio modeling of the RBD by the QUARK algorithm [88] predicted a helix-like spatial configuration of the three helices (Figure 4A,B), where a cluster of conserved and internalization-relevant amino acids was modeled in close proximity (Figure 4C,D) [28]. The spatial proximity of function-relevant residues may correspond to a critical receptor-interacting site.

5.3. *VP1u Cognate Receptor Facilitates B19V Targeting and Uptake Exclusively into Permissive Cells*

B19V requires a strict intracellular environment for replication that can only be found in the erythroid progenitor cells (EPCs) in the bone marrow. The essential intracellular factors appear to be simultaneously upregulated in EPCs during the EPO-dependent differentiation stages. A study has shown that the internalization and the replication of B19V are considerably enhanced when CD34+ hematopoietic stem cells were stimulated with EPO [57]. Not surprisingly, the two cell lines that are most frequently used to study B19V infection, the megakaryocyte-erythroid UT7/Epo cells [89] and the erythroleukemic KU812Ep6 cells [90], are both derived from an EPO-dependent subclone. EPO signaling maintains the survival of cells that entered the intermediate erythroid differentiation stages [91–93]. Besides EPO signaling, B19V infection requires hypoxic conditions, which characterize the bone marrow microenvironment where the virus replicates. Hypoxia upregulates the signal transducer and activator of transcription 5 (STAT5) pathway, which facilitates viral DNA replication [61,94,95]. During the EPO-dependent differentiation stages, a cluster of erythroid-specific genes is upregulated [96], including the VP1u cognate receptor [59], which jointly are essential for B19V replication. In this regard, the main role of the VP1u receptor would be to facilitate the targeting and the uptake of B19V exclusively into cells providing a permissive intracellular environment for the infection. This strategy prevents the internalization into non-permissive cells, which would result in abortive infection.

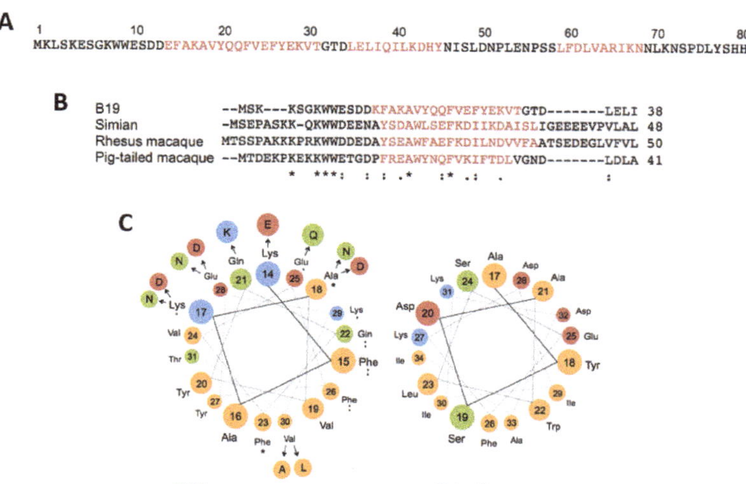

Figure 3. Structural motifs within the VP1u RBD. (**A**) Amino acid sequence of the N-terminal of VP1u (AA 1–80) and the three predicted alpha helices (underlined red). (**B**) Conservation of alpha helix 1 among erythroparvoviruses. (**C**) Modeled helical wheel of the conserved helix 1 (AA 14–31) shows the spatial arrangement of hydrophobic and polar amino acids within helix 1. Amino acid differences found in B19V isolates are shown in a wider radius. Hydrophobic = orange; polar = green; basic = blue; acid = red. The helical wheel of the simian parvovirus helix 1 is shown.

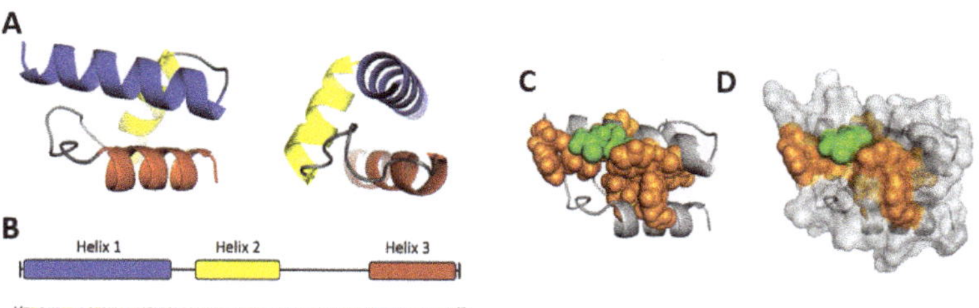

Figure 4. Ab initio modeling of the RBD in the VP1u. (**A**) Front and side views. Helix 1 appears in blue (AA 14–31), helix 2 in yellow (AA 35–45), helix 3 in red (AA 57–68). (**B**) Helix distribution and sequence of the modeled AA 14–68. The amino acids required for VP1u internalization are colored in orange (hydrophobic) and green (polar). (**C**) The spatial distribution of essential amino acids is shown as spheres in the helical structure (**D**) and in the surface model of the RBD.

5.4. Evolutionary Aspects of B19V Restricted Tropism and the Origin of the RBD in the VP1u

The origin of the marked tropism of B19V for erythroid precursors in the bone marrow is not known. The erythroparvovirus and dependoparvovirus genomes show striking similarities, both having identical hairpin telomeres at both sides, and related replication mechanism [97]. It is conceivable that the erythroparvovirus ancestor was dependent on helper virus co-infections. In line with this hypothesis, several studies observed the enhancement of B19V replication and gene expression in non-permissive cells in the presence of helper virus genes [98,99]. Besides the enhanced genome replication, adenovirus genes transactivated the B19V promoters, including the p44 promoter in the middle of the genome (nt 2247), which is normally silenced during B19V infection [100,101]. The p44 promoter is homologous to the promoters that initiate the expression of the structural capsid proteins in other parvoviruses. Interestingly, the expression of the structural proteins represents a limiting factor in B19V infection in non-erythroid cells. The transcription of the structural genes from the p44 promoter might have played an important role in the helper-dependent ancestors of erythroparvoviruses but could have been replaced during evolution by an alternative helper-independent replication in erythroid cells. In contrast to most other parvoviruses, B19V shows alternative splicing in the transcript from the p6 promoter that also enables the expression of the distal genes [102]. This exceptional splicing mechanism of B19V, which strikingly occurs only in EPCs, makes the internal and helper-dependent p44 promoter dispensable. Interestingly, there is another putative internal promoter (p55) at nt 2308 that might have similar properties.

According to this evolutionary model (Figure 5), the erythroparvovirus ancestor would have generally exhibited a helper virus-dependent replication in different tissues, and sporadically, a helper-independent replication in EPCs. However, without a specific targeting and internalization into the erythroid progenitor cells, the overall infection still depended on the helper virus co-infection. The erythroid-specific transcription of the structural genes from the p6 promoter generates a transcript with a longer 5′-UTR that possibly allows the displacement of the start codon of the capsid proteins and consequently, a longer VP1u region (Figure 6). The additional N-terminal amino acid sequence, expressed only during the helper-independent infection in erythroid cells, might have evolved to the RBD in the VP1u. The erythroid-specific targeting boosted the infection in the EPCs and thus represented a positive feedback loop that promoted the autonomous replication in the erythroid tissue. Vice versa, the helper-independent replication was the driving force for the positive feedback mechanism. The positive feedback enhanced the reliance on additional erythroid-specific factors and thus finally led to the extreme tropism of erythroparvoviruses.

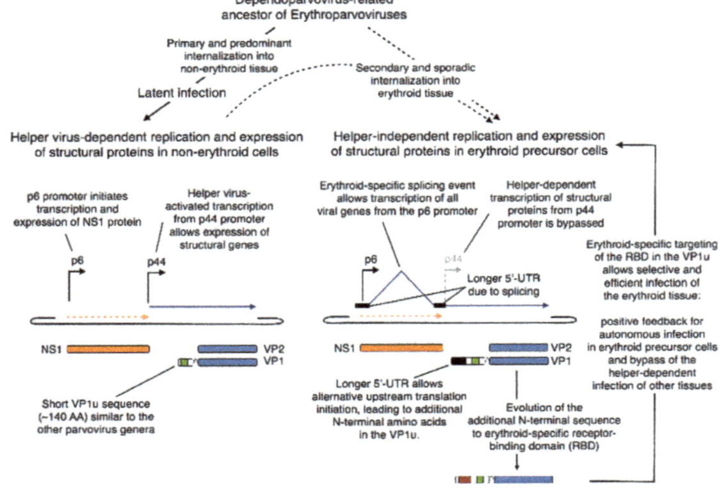

Figure 5. Proposed origin of the marked erythroid tropism of B19V (see text for details).

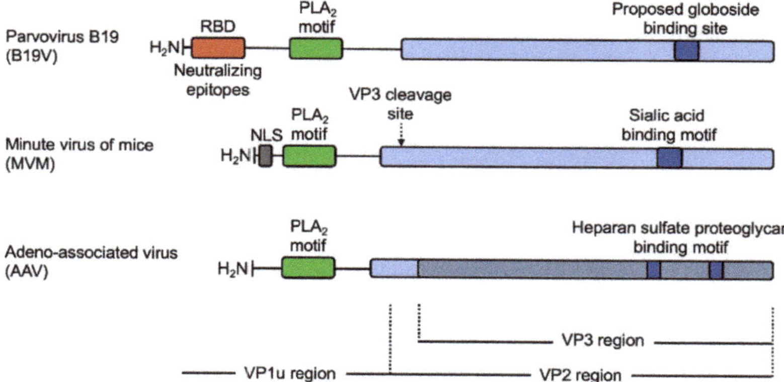

Figure 6. Infection-relevant functional domains in the structural proteins of three representative parvoviruses. B19V exhibits a longer VP1u region compared to other parvoviruses. The additional N-terminal stretch of 80–90 amino acids contain the functional RBD [28] and a cluster of neutralizing epitopes [67,68]. NLS, nuclear localization signal.

The helper-dependent replication and expression of B19V genes in non-erythroid cells might still represent a significant aspect of the pathogenesis of B19V infection. The unspecific entry of the virus into non-erythroid tissues would not necessarily end in an abortive infection. The internalization of B19V during the late viremic phase by the antibody-dependent enhancement (ADE) provides a basis for latent infections in diverse tissues. These latent viruses might be sporadically reactivated by helper virus infections, which would explain many of the B19V-associated diseases as well as the recurrent detection of B19V DNA in the serum and different tissues [99,103].

6. Role of VP1u in the Subcellular Trafficking of Incoming B19V

To infect the cell, parvoviruses follow a complex route from the plasma membrane to the nucleus where they replicate. Various domains in the VP1u of parvoviruses have been shown to play a critical role in the process by assisting the transport of the incoming capsids throughout the different membrane-enclosed organelles and the highly crowded cytosol and by promoting their translocation through the nuclear pore complex (NPC) into the nucleus (Figure 6).

6.1. The Phospholipase A_2 (PLA_2) Domain

Following the interaction of the VP1u RBD with its cognate receptor, B19V is internalized by clathrin-mediated endocytosis and enters the endosomal pathway [104]. Endosomes provide cues that trigger capsid conformational rearrangements required for subsequent trafficking steps and contribute to the transport of incoming viruses to the nuclear vicinity. However, the mechanism followed by parvoviruses to escape from endosomal vesicles into the cytosol remains unclear. Phospholipase A_2 enzymes (PLA_2s) catalyze the hydrolysis of phospholipids and the release of lipid mediator precursors. Accordingly, PLA_2s are key enzymes in many cellular processes such as lipid membrane metabolism, inflammation, membrane remodeling, host defense, and signal transduction [105]. PLA_2s are found in mammalian tissues as well as in arachnids, insects, mollusks, reptiles, plants, and bacteria. A PLA_2 domain containing the typical catalytic motif HDXXY and the calcium-binding site GXG is conserved in the VP1u of parvoviruses, including B19V (except for amdoparvoviruses) [21–23,25,106]. Mutations in either of these motifs disturbed both, the enzymatic activity and viral infectivity [20,21,23]. The pharmacological disruption of endosomal membranes or co-infection with endosomolytically active adenovirus, but not with inactive variants, partially rescued the infectivity of the PLA_2 mutants, suggesting a role of the VP1u PLA_2 in altering the endosomal membrane integrity to enable endosomal escape of viruses into the cytosol [24,26]. In B19V, VP1u mutations not related to the critical PLA_2 motifs were also shown to reduce the enzymatic activity, probably by disrupting the three-dimensional rearrangement surrounding the PLA_2 domain [107]. Although the PLA_2 may facilitate the endosomal escape of incoming B19V, the mechanism involved is poorly understood. Interestingly, B19V endosomal escape was shown to occur without detectable endosomal membrane permeabilization or damage [104] and the enzymatic requirements for the PLA_2 activity, i.e., pH and calcium concentration are not optimal in the endocytic compartment [106]. Accordingly, it remains unclear how the PLA_2 activity of VP1u supports the escape of the endosomal capsids.

B19V PLA_2 has been shown to up-regulate Ca^{2+} entry [108], to inhibit Na^+/K^+ ATPase activity and K^+ channels [109,110], and to up-regulate ENaC [111]. These activities may contribute to the pathophysiology of B19V infections. Moreover, due to the inflammatory-like effects exerted by recombinant VP1u in cultured fibroblast [112] and in UT7/Epo cells [107], it has been hypothesized that the PLA_2 domain of VP1u may contribute to B19V-associated syndromes, such as arthropathy and autoimmunity.

6.2. Nuclear Localization Signals (NLSs)

Following endosomal escape, parvovirus capsids are imported into the nucleus. Nuclear import of most proteins involves classical nuclear localization signals (NLSs) consisting of a stretch of basic amino acids, which interact with importin-α/importin-β to mediate transport through the nuclear pore [113]. The size of the parvovirus capsid is below the diameter limit of the nuclear pore complex (NPC). Therefore, capsids can theoretically be translocated through the NPC intact or without major disassembly [114]. NLSs have been identified in the VP1u from several parvoviruses [12–19] and confer nuclear import potential to the incoming particles via interaction with importin-β [115]. The VP1u of B19V does not display a motif resembling a classical NLS, however, when expressed in eukaryotic cells, VP1u accumulates in the cytoplasm and in the nucleus [107]. The stretch of basic amino acids found in VP2 occupies an internal position in the capsid and has been implicated in the nuclear translocation of assembly intermediates [116]. Accordingly, the mechanism of nuclear import of B19V remains uncertain and might differ fundamentally from that of other parvoviruses.

7. Biotechnological Applications of the VP1u of B19V

Nanocarriers are designed to efficiently deliver therapeutic molecules to specific tissues minimizing adverse effects [117]. Despite important progress, the drug delivery technology based on synthetic nanocarriers remains highly inefficient. One meta-analysis revealed that over 99% of the drugs do not reach the diseased cells and accumulate instead in non-target tissues or are cleared from the body [118]. Ideally, nanocarriers must specifically internalize into the target cells, escape from the endocytic compartment, and release their payload into the cytosol. These processes resemble the early infection steps of viruses, which operate as powerful natural nanocarriers to efficiently deliver genetic material into target cells by complex mechanisms shaped by evolution. The targeting machinery that is engaged in the early viral infection steps can be utilized to generate virus-inspired nanocarriers as efficient drug or gene delivery vehicles [119,120]. In this regard, the VP1u of B19V includes many interesting features that can potentially be exploited for drug delivery and diagnostics, i.e., specific cell targeting, efficient cell entry, and endosomal escape.

7.1. Specific Biomarker for EPO-Dependent Erythroid Differentiation Stages

Diverse hematological conditions (e.g., leukemia, thalassemic and myelodysplastic syndromes, bone marrow metastases of solid tumors, septicemia, or severe health conditions after surgery) are typically associated with the presence of erythroblasts outside the bone marrow [121–126]. Accordingly, the screening of peripheral blood for nucleated red blood cells (NRBCs) is used to recognize hematological disorders or severe health conditions. Assays to detect NRBCs must be very sensitive because the presence of only a few NRBCs can indicate serious underlying disorders.

Unfortunately, automated hematology analyzers may not detect low levels of NRBCs. Besides, they generate suspect flags, which should be examined manually [127]. The currently used automated detection of NRBCs in peripheral blood has a detection limit of 1-2 erythroblasts per 100 white blood cells [123,128]. In comparison, VP1u decorated MS2 capsids were able to detect as few as one erythroleukemic UT7/Epo cell in 100,000 isolated white blood cells (unpublished observations). The sensitive identification of erythroblasts in the peripheral blood by fluorescent VP1u bioconjugates has the potential to improve the detection of diverse hematological disorders or severe health conditions and to facilitate an early diagnosis without the systematic need of an invasive technique such as bone marrow biopsy.

The precise identification and isolation of erythroid progenitor cells is important in hematological research and in diagnostics to characterize and treat bone marrow disorders. However, the technique remains rather complex and laborious, since the currently used markers are not lineage-specific (CD36, CD38, CD44, CD45, CD71, CD105, EPOR) or are broadly expressed during the erythroid development (glycophorin A). Therefore, the combination of several antibodies is necessary to achieve the correct identification [124,129–134]. In contrast, the fluorescent VP1u bioconjugate appeared as a unique and highly sensitive marker for the EPO-dependent erythroid differentiation stages and readily detected these cells in heterogeneous cell populations from different tissues [54]. The findings show the potential of the VP1u as a biomarker to identify and sort erythroid differentiation stages in a simpler procedure than it has been practiced so far.

It is expected that the future biotechnological applications of the VP1u will be spurred by the identification of its cognate receptor. However, the identity of the VP1u receptor will not necessarily be determinant for the applicability of the VP1u as a specific cellular marker. Historically, it is not uncommon to use cell surface markers to identify cell populations based on empirical evidence without knowing the identity and/or the function of the targeted receptors.

7.2. Specific Drug Delivery and Chemotherapy

7.2.1. β-Hemoglobin Disorders

β-hemoglobin disorders are a group of highly prevalent hereditary diseases caused by mutations in the gene encoding for the β-chain of hemoglobin, resulting in qualitative and quantitative defects in β-globin production. β-thalassemias are a heterogeneous group of genetic disorders characterized by the partial or complete absence of β-globin chain production, leading to anemia and iron overload. The disease is highly prevalent with 80–90 million carriers worldwide. Without diagnosis and appropriate treatment, the severe forms of β-thalassemia lead to death before age 20 [135]. Sickle cell disease (SCD) is the most common and severe hemoglobinopathy. In SCD, a single mutation in the β-globin gene results in the production of an aberrant hemoglobin molecule, which causes the rigid sickle-like shape of erythrocytes. Without treatment, SCD is lethal before age five [136].

Patients with severe β-hemoglobin disorders require regular blood transfusions, which lead to iron overload and related complications. Accordingly, iron chelation therapies are also required [137,138]. The most severe forms of the disease have been successfully treated by allogeneic hematopoietic stem cell transplantation from a matched related donor. However, major drawbacks are the difficulty to find a histocompatible donor and the need for extensive immunosuppressive regimens, with the risk of immunological complication. Besides, this approach is not accessible for many affected individuals [139,140]. Gene therapy and gene editing strategies to restore the globin genes have generated promising results. However, these approaches lack cell-specific vectors, resulting in poor efficiency and the risk of insertional oncogenesis [141–143].

Due to the numerous drawbacks associated with the current therapeutic strategies, there is a great interest in developing novel therapeutic options. The therapeutic targeting of RNA by double-stranded RNA-mediated interference (RNAi) or by antisense oligonucleotides (ASOs) allows specific inhibition of the target of interest and a very rapid transferability to the clinics [144]. However, the delivery of nucleic acid molecules to the bone marrow remains highly inefficient. The MS2 capsid is a well-studied vector for drug delivery and can be easily loaded with therapeutic ASOs or small interfering RNAs (siRNA) [145,146]. This strategy provides protection of the therapeutic nucleic acid molecules in the extracellular milieu, avoids solubility problems, and thus allows more the options to improve the modifications of the oligonucleotides. In a previous study, we showed that anchoring of VP1u subunits to the surface of MS2 capsids retargets the particles to erythroid cells. This finding offers the opportunity to deliver encapsidated genetic material specifically to this cell population [59]. Potential targets of therapeutic ASOs or siRNA might be different factors involved in the regulation of erythropoiesis, such as transferrin receptor 2, or regulatory elements of fetal hemoglobin, such as B-cell lymphoma/leukemia 11A and erythroid Kruppel-like factor. Specific downregulation of such factors in erythroid progenitor cells would significantly alleviate symptoms of β-hemoglobin disorders [147–150].

7.2.2. Erythroleukemia

Acute erythroleukemia is a rare disorder associated with a poor prognosis. A study reported a median overall survival of 8 months [151]. The treatment of erythroleukemia is compromised due to the systemic distribution and resistance of the malignant cells to chemotherapeutics [152,153]. Therefore, the successful elimination of erythroleukemic cells by a cytotoxin requires a "magic bullet" strategy—an efficient and specific targeting of the toxin to cancer cells—minimizing adverse effects to the surrounding healthy cells [154]. Erythroleukemias exhibit proliferating cancer cells in the early and intermediate erythroid differentiation stages [155], which are the target cells of the VP1u. Accordingly, the VP1u-mediated toxin delivery represents a possible strategy to overcome the resistance of erythroleukemia to chemotherapeutics. In previous studies, VP1u successfully targeted a toxin specifically to malignant erythroid precursors and thus selectively eliminated these cells from a mixed cell culture [156].

The immunity of many individuals against B19V would represent a serious obstacle for the application of the VP1u-targeted delivery. About half of the human population is seropositive for anti-B19V antibodies. Similar problems are faced in the application of AAV vectors for gene therapy, where many individuals have antibodies against serotypes 2 and 3 [157]. Following the natural mechanism of viruses to evade the immune system, the AAV researchers are searching for AAV isolates and isotypes, which are not neutralized by the common pool of antibodies, but still offer the beneficial properties of the original virus [158–160]. In the case of a short protein with a single function as with the RBD of the VP1u, an immune escape by antigenic drift is easier to achieve without disturbing the receptor binding and internalization capacity. The natural mutations observed in various B19V isolates (Figure 3C) together with the mutational studies already performed [28], provide an excellent basis to mimic an antigenic drift of the VP1u RBD without decreasing the targeting function of the protein. Furthermore, there exist different options to reduce the antigenicity of a therapeutic protein, such as a fusion with an abundant endogenous protein as serum albumin or the immunoglobulin constant fragments [161–163]. The coupling to these endogenous proteins does not only circumvent the immune response, but also considerably increases the solubility, stability, and serum half-life of the therapeutic proteins. In line with this concept, bovine serum albumin (BSA) was used as an adaptor molecule for the attachment of the toxins to a VP1u-NeutrAvidin complex. The results showed that the modified BSA remained soluble after the attachment of 20–30 fluorescein or toxin molecules to the protein and was targeted exclusively to VP1u-expressing cells. The stability of the drug attachment might be increased by packing the effector molecule into a capsid, as shown with the MS2 bacteriophage in previous studies [145,146,164]. The specific delivery of an encapsidated effector allows a higher dose per delivered particle without increasing toxicity. Besides, the capsid can be engineered to incorporate multiple residues to improve the targeting efficiency.

8. Concluding Remarks

The VP1u is a key component of the capsid of human parvovirus B19 with essential functions in multiple steps of the infection, such as tissue tropism, uptake, intracellular trafficking, and entry. The VP1u is also the immunodominant region of the capsid and a crucial component for prospective vaccines. In the future, efforts will be focused to better understand the essential functions of VP1u in B19V infection and to identify the VP1u interactome, notably its cognate cell receptor.

Recent innovations in protein engineering and nanomaterials science have the potential to revolutionize the conventional methods of diagnosis and treatment, bringing new hopes to patients. However, to date, a major barrier in their clinical application remains their poor selective targeting. Only a few clinically approved nanoscale delivery vehicles integrate molecules to selectively target the cargo to the tissue of interest. In this regard, the remarkable erythroid specificity of the VP1u offers novel opportunities to generate virus-inspired biomarkers and nanocarriers to specifically target erythroid cells. This approach may contribute to a better understanding of the mechanisms governing erythroid development and to treat disorders of the erythroid lineage. Efforts to circumvent the VP1u immune response and to optimize the stability and density of cargo delivery will facilitate its transferability to human diagnostics and therapies.

Author Contributions: Conceptualization, R.L. and C.R.; writing—original draft preparation, C.R.; review and editing, R.L., J.B. and C.R.; Supervision, funding acquisition, C.R. All authors have read and agreed to the published version of the manuscript.

Funding: This work was supported by the Swiss National Science Foundation (grant number 31003A_179384) to J.B.

Conflicts of Interest: The authors declare no conflict of interest.

References

1. Cotmore, S.F.; Tattersall, P. Parvoviruses: Small does not mean simple. *Annu. Rev. Virol.* **2014**, *1*, 517–537. [CrossRef] [PubMed]
2. Kailasan, S.; Agbandje-Mckenna, M.; Parrish, C.R. Parvovirus Family Conundrum: What Makes a Killer? *Annu. Rev. Virol.* **2015**, *2*, 425–450. [CrossRef] [PubMed]
3. François, S.; Filloux, D.; Roumagnac, P.; Bigot, D.; Gayral, P.; Martin, D.P.; Froissart, R.; Ogliastro, M. Discovery of parvovirus-related sequences in an unexpected broad range of animals. *Sci. Rep.* **2016**, *6*, 1–13. [CrossRef] [PubMed]
4. Cotmore, S.F.; Agbandje-McKenna, M.; Canuti, M.; Chiorini, J.A.; Eis-Hubinger, A.M.; Hughes, J.; Mietzsch, M.; Modha, S.; Ogliastro, M.; Pénzes, J.J.; et al. ICTV virus taxonomy profile: Parvoviridae. *J. Gen. Virol.* **2019**, *100*, 367–368. [CrossRef] [PubMed]
5. Samulski, R.J.; Muzyczka, N. AAV-mediated gene therapy for research and therapeutic purposes. *Annu. Rev. Virol.* **2014**, *1*, 427–451. [CrossRef] [PubMed]
6. Cotmore, S.F.; Tattersall, P. Characterization and molecular cloning of a human parvovirus genome. *Science* **1984**, *226*, 1161–1165. [CrossRef]
7. Cotmore, S.F.; Tattersall, P. DNA replication in the autonomous parvoviruses. *Semin. Virol.* **1995**, *6*, 271–281. [CrossRef]
8. Tattersall, P.; Cawte, P.J.; Shatkin, A.J.; Ward, D.C. Three structural polypeptides coded for by minite virus of mice, a parvovirus. *J. Virol.* **1976**, *20*, 273–289. [CrossRef]
9. Becerra, S.P.; Koczot, F.; Fabisch, P.; Rose, J.A. Synthesis of adeno-associated virus structural proteins requires both alternative mRNA splicing and alternative initiations from a single transcript. *J. Virol.* **1988**, *62*, 2745–2754. [CrossRef]
10. Mani, B.; Baltzer, C.; Valle, N.; Almendral, J.M.; Kempf, C.; Ros, C. Low pH-dependent endosomal processing of the incoming parvovirus minute virus of mice virion leads to externalization of the VP1 N-terminal sequence (N-VP1), N-VP2 cleavage, and uncoating of the full-length genome. *J. Virol.* **2006**, *80*, 1015–1024. [CrossRef]

11. Mietzsch, M.; Pénzes, J.J.; Agbandje-Mckenna, M. Twenty-five years of structural parvovirology. *Viruses* **2019**, *11*, 362. [CrossRef] [PubMed]
12. Vihinen-Ranta, M.; Kakkola, L.; Kalela, A.; Vilja, P.; Vuento, M. Characterization of a nuclear localization signal of canine parvovirus capsid proteins. *Eur. J. Biochem.* **1997**, *250*, 389–394. [CrossRef] [PubMed]
13. Lombardo, E.; Ramírez, J.C.; Garcia, J.; Almendral, J.M. Complementary Roles of Multiple Nuclear Targeting Signals in the Capsid Proteins of the Parvovirus Minute Virus of Mice during Assembly and Onset of Infection. *J. Virol.* **2002**, *76*, 7049–7059. [CrossRef] [PubMed]
14. Vihinen-Ranta, M.; Wang, D.; Weichert, W.S.; Parrish, C.R. The VP1 N-Terminal Sequence of Canine Parvovirus Affects Nuclear Transport of Capsids and Efficient Cell Infection. *J. Virol.* **2002**, *76*, 1884–1891. [CrossRef] [PubMed]
15. Grieger, J.C.; Snowdy, S.; Samulski, R.J. Separate Basic Region Motifs within the Adeno-Associated Virus Capsid Proteins Are Essential for Infectivity and Assembly. *J. Virol.* **2006**, *80*, 5199–5210. [CrossRef] [PubMed]
16. Sonntag, F.; Bleker, S.; Leuchs, B.; Fischer, R.; Kleinschmidt, J.A. Adeno-Associated Virus Type 2 Capsids with Externalized VP1/VP2 Trafficking Domains Are Generated prior to Passage through the Cytoplasm and Are Maintained until Uncoating Occurs in the Nucleus. *J. Virol.* **2006**, *80*, 11040–11054. [CrossRef] [PubMed]
17. Li, Q.; Zhang, Z.; Zheng, Z.; Ke, X.; Luo, H.; Hu, Q.; Wang, H. Identification and characterization of complex dual nuclear localization signals in human bocavirus NP1. *J. Gen. Virol.* **2013**, *94*, 1335–1342. [CrossRef]
18. Boisvert, M.; Bouchard-Levesque, V.; Fernandes, S.; Tijssen, P. Classic Nuclear Localization Signals and a Novel Nuclear Localization Motif Are Required for Nuclear Transport of Porcine Parvovirus Capsid Proteins. *J. Virol.* **2014**, *88*, 11748–11759. [CrossRef]
19. Liu, P.; Chen, S.; Wang, M.; Cheng, A. The role of nuclear localization signal in parvovirus life cycle. *Virol. J.* **2017**, *14*, 17–22. [CrossRef]
20. Li, Y.; Zádori, Z.; Bando, H.; Dubuc, R.; Fédière, G.; Szelei, J.; Tijssen, P. Genome organization of the densovirus from Bombyx mori (BmDNV-1) and enzyme activity of its capsid. *J. Gen. Virol.* **2001**, *82*, 2821–2825. [CrossRef]
21. Zádori, Z.; Szelei, J.; Lacoste, M.C.; Li, Y.; Gariépy, S.; Raymond, P.; Allaire, M.; Nabi, I.R.; Tijssen, P. A Viral Phospholipase A2 Is Required for Parvovirus Infectivity. *Dev. Cell* **2001**, *1*, 291–302. [CrossRef]
22. Dorsch, S.; Liebisch, G.; Kaufmann, B.; von Landenberg, P.; Hoffmann, J.H.; Drobnik, W.; Modrow, S. The VP1 Unique Region of Parvovirus B19 and Its Constituent Phospholipase A2-Like Activity. *J. Virol.* **2002**, *76*, 2014–2018. [CrossRef] [PubMed]
23. Girod, A.; Wobus, C.E.; Zádori, Z.; Ried, M.; Leike, K.; Tijssen, P.; Kleinschmidt, J.A.; Hallek, M. The VP1 capsid protein of adeno-associated virus type 2 is carrying a phospholipase A2 domain required for virus infectivity. *J. Gen. Virol.* **2002**, *83*, 973–978. [CrossRef] [PubMed]
24. Farr, G.A.; Zhang, L.G.; Tattersall, P. Parvoviral virions deploy a capsid-tethered lipolytic enzyme to breach the endosomal membrane during cell entry. *Proc. Natl. Acad. Sci. USA* **2005**, *102*, 17148–17153. [CrossRef] [PubMed]
25. Qu, X.W.; Liu, W.P.; Qi, Z.Y.; Duan, Z.J.; Zheng, L.S.; Kuang, Z.Z.; Zhang, W.J.; Hou, Y. De Phospholipase A2-like activity of human bocavirus VP1 unique region. *Biochem. Biophys. Res. Commun.* **2008**, *365*, 158–163. [CrossRef] [PubMed]
26. Stahnke, S.; Lux, K.; Uhrig, S.; Kreppel, F.; Hösel, M.; Coutelle, O.; Ogris, M.; Hallek, M.; Büning, H. Intrinsic phospholipase A2 activity of adeno-associated virus is involved in endosomal escape of incoming particles. *Virology* **2011**, *409*, 77–83. [CrossRef]
27. Popa-Wagner, R.; Porwal, M.; Kann, M.; Reuss, M.; Weimer, M.; Florin, L.; Kleinschmidt, J.A. Impact of VP1-Specific Protein Sequence Motifs on Adeno-Associated Virus Type 2 Intracellular Trafficking and Nuclear Entry. *J. Virol.* **2012**, *86*, 9163–9174. [CrossRef]
28. Leisi, R.; Di Tommaso, C.; Kempf, C.; Ros, C. The receptor-binding domain in the VP1u region of parvovirus B19. *Viruses* **2016**, *8*, 61. [CrossRef]
29. Harbison, C.E.; Chiorini, J.A.; Parrish, C.R. The parvovirus capsid odyssey: From the cell surface to the nucleus. *Trends Microbiol.* **2008**, *16*, 208–214. [CrossRef]
30. Ros, C.; Bayat, N.; Wolfisberg, R.; Almendral, J.M. Protoparvovirus cell entry. *Viruses* **2017**, *9*, 313. [CrossRef]
31. Kronenberg, S.; Böttcher, B.; von der Lieth, C.W.; Bleker, S.; Kleinschmidt, J.A. A Conformational Change in the Adeno-Associated Virus Type 2 Capsid Leads to the Exposure of Hidden VP1 N Termini. *J. Virol.* **2005**, *79*, 5296–5303. [CrossRef] [PubMed]

32. Bönsch, C.; Zuercher, C.; Lieby, P.; Kempf, C.; Ros, C. The Globoside Receptor Triggers Structural Changes in the B19 Virus Capsid That Facilitate Virus Internalization. *J. Virol.* **2010**, *84*, 11737–11746. [CrossRef] [PubMed]
33. Tsao, J.; Chapman, M.S.; Agbandje, M.; Keller, W.; Smith, K.; Wu, H.; Luo, M.; Smith, T.J.; Rossmann, M.G.; Compans, R.W.; et al. The three-dimensional structure of canine parvovirus and its functional implications. *Science* **1991**, *251*, 1456–1464. [CrossRef] [PubMed]
34. Chapman, M.S.; Rossmann, M.G. Structure, sequence, and function correlations among parvoviruses. *Virology* **1993**, *194*, 491–508. [CrossRef]
35. Llamas-Saiz, A.L.; Agbandje-McKenna, M.; Wikoff, W.R.; Bratton, J.; Tattersall, P.; Rossmann, M.G. Structure determination of minute virus of mice. *Acta Crystallogr. Sect. D Biol. Crystallogr.* **1997**, *53*, 93–102. [CrossRef] [PubMed]
36. Kaufmann, B.; Simpson, A.A.; Rossmann, M.G. The structure of human parvovirus B19. *Proc. Natl. Acad. Sci. USA* **2004**, *101*, 11628–11633. [CrossRef]
37. Agbandje-McKenna, M.; Llamas-Saiz, A.L.; Wang, F.; Tattersall, P.; Rossmann, M.G. Functional implications of the structure of the murine parvovirus, minute virus of mice. *Structure* **1998**, *6*, 1369–1381. [CrossRef]
38. Bleker, S.; Sonntag, F.; Kleinschmidt, J.A. Mutational Analysis of Narrow Pores at the Fivefold Symmetry Axes of Adeno-Associated Virus Type 2 Capsids Reveals a Dual Role in Genome Packaging and Activation of Phospholipase A2 Activity. *J. Virol.* **2005**, *79*, 2528–2540. [CrossRef]
39. Farr, G.A.; Cotmore, S.F.; Tattersall, P. VP2 Cleavage and the Leucine Ring at the Base of the Fivefold Cylinder Control pH-Dependent Externalization of both the VP1 N Terminus and the Genome of Minute Virus of Mice. *J. Virol.* **2006**, *80*, 161–171. [CrossRef]
40. Plevka, P.; Hafenstein, S.; Li, L.; D'Abrgamo, A.; Cotmore, S.F.; Rossmann, M.G.; Tattersall, P. Structure of a Packaging-Defective Mutant of Minute Virus of Mice Indicates that the Genome Is Packaged via a Pore at a 5-Fold Axis. *J. Virol.* **2011**, *85*, 4822–4827. [CrossRef]
41. Castellanos, M.; Pérez, R.; Rodríguez-Huete, A.; Grueso, E.; Almendral, J.M.; Mateu, M.G. A slender tract of glycine residues is required for translocation of the VP2 protein N-terminal domain through the parvovirus MVM capsid channel to initiate infection. *Biochem. J.* **2013**, *455*, 87–94. [CrossRef] [PubMed]
42. Farr, G.A.; Tattersall, P. A conserved leucine that constricts the pore through the capsid fivefold cylinder plays a central role in parvoviral infection. *Virology* **2004**, *323*, 243–256. [CrossRef] [PubMed]
43. Cotmore, S.F.; Tattersall, P. Mutations at the Base of the Icosahedral Five-Fold Cylinders of Minute Virus of Mice Induce 3′-to-5′ Genome Uncoating and Critically Impair Entry Functions. *J. Virol.* **2012**, *86*, 69–80. [CrossRef] [PubMed]
44. Ros, C.; Gerber, M.; Kempf, C. Conformational changes in the VP1-unique region of native human parvovirus B19 lead to exposure of internal sequences that play a role in virus neutralization and infectivity. *J. Virol.* **2006**, *80*, 12017–12024. [CrossRef] [PubMed]
45. Qiu, J.; Söderlund-Venermo, M.; Young, N.S. Human Parvoviruses. *Clin. Microbiol. Rev.* **2017**, *30*, 43–113. [CrossRef] [PubMed]
46. Vassilopoulos, D.; Calabrese, L.H. Virally associated arthritis 2008: Clinical, epidemiologic, and pathophysiologic considerations. *Arthritis Res. Ther.* **2008**, *10*, 1–8. [CrossRef]
47. Heegaard, E.D.; Brown, K.E. Human Parvovirus B19. *Clin. Microbiol. Rev.* **2002**, *15*, 485–505. [CrossRef]
48. Bonvicini, F.; Bua, G.; Gallinella, G. Parvovirus B19 infection in pregnancy—Awareness and opportunities. *Curr. Opin. Virol.* **2017**, *27*, 8–14. [CrossRef]
49. Cossart, Y.E.; Cant, B.; Field, A.M.; Widdows, D. Parvovirus-Like Particles in Human Sera. *Lancet* **1975**, *305*, 72–73. [CrossRef]
50. Christensen, A.; Kesti, O.; Elenius, V.; Eskola, A.L.; Døllner, H.; Altunbulakli, C.; Akdis, C.A.; Söderlund-Venermo, M.; Jartti, T. Human bocaviruses and paediatric infections. *Lancet Child Adolesc. Health* **2019**, *3*, 418–426. [CrossRef]
51. Söderlund-Venermo, M. Emerging Human Parvoviruses: The Rocky Road to Fame. *Annu. Rev. Virol.* **2019**, *6*, 71–91. [CrossRef] [PubMed]
52. Ozawa, K.; Kurtzman, G.; Young, N. Replication of the B19 parvovirus in human bone marrow cell cultures. *Science* **1986**, *233*, 883–886. [CrossRef] [PubMed]
53. Takahashi, T.; Ozawa, K.; Takahashi, K.; Asano, S.; Takaku, F. Susceptibility of human erythropoietic cells to B19 parvovirus in vitro increases with differentiation. *Blood* **1990**, *75*, 603–610. [CrossRef] [PubMed]

54. Leisi, R.; Ruprecht, N.; Kempf, C.; Ros, C. Parvovirus B19 uptake is a highly selective process controlled by VP1u, a novel determinant of viral tropism. *J. Virol.* **2013**, *87*, 13161–13167. [CrossRef]
55. Guan, W.; Cheng, F.; Yoto, Y.; Kleiboeker, S.; Wong, S.; Zhi, N.; Pintel, D.J.; Qiu, J. Block to the Production of Full-Length B19 Virus Transcripts by Internal Polyadenylation Is Overcome by Replication of the Viral Genome. *J. Virol.* **2008**, *82*, 9951–9963. [CrossRef]
56. Wong, S.; Zhi, N.; Filippone, C.; Keyvanfar, K.; Kajigaya, S.; Brown, K.E.; Young, N.S. Ex Vivo-Generated CD36+ Erythroid Progenitors Are Highly Permissive to Human Parvovirus B19 Replication. *J. Virol.* **2008**, *82*, 2470–2476. [CrossRef]
57. Chen, A.Y.; Guan, W.; Lou, S.; Liu, Z.; Kleiboeker, S.; Qiu, J. Role of Erythropoietin Receptor Signaling in Parvovirus B19 Replication in Human Erythroid Progenitor Cells. *J. Virol.* **2010**, *84*, 12385–12396. [CrossRef]
58. Wolfisberg, R.; Ruprecht, N.; Kempf, C.; Ros, C. Impaired genome encapsidation restricts the in vitro propagation of human parvovirus B19. *J. Virol. Methods* **2013**, *193*, 215–225. [CrossRef]
59. Leisi, R.; Nordheim, M.V.; Ros, C.; Kempf, C. The VP1u receptor restricts parvovirus B19 uptake to permissive erythroid cells. *Viruses* **2016**, *8*, 265. [CrossRef]
60. Gallinella, G.; Manaresi, E.; Zuffi, E.; Venturoli, S.; Bonsi, L.; Bagnara, G.P.; Musiani, M.; Zerbini, M. Different patterns of restriction to B19 parvovirus replication in human blast cell lines. *Virology* **2000**, *278*, 361–367. [CrossRef]
61. Ganaie, S.S.; Qiu, J. Recent advances in replication and infection of human parvovirus B19. *Front. Cell. Infect. Microbiol.* **2018**, *8*, 1–12. [CrossRef] [PubMed]
62. Luo, Y.; Qiu, J. Human parvovirus B19: A mechanistic overview of infection and DNA replication. *Future Virol.* **2015**, *10*, 155–167. [CrossRef] [PubMed]
63. Stramer, S.L.; Dodd, R.Y. Transfusion-transmitted emerging infectious diseases: 30 years of challenges and progress. *Transfusion* **2013**, *53*, 2375–2383. [CrossRef] [PubMed]
64. Deiss, V.; Tratschin, J.D.; Weitz, M.; Siegl, G. Cloning of the human parvovirus B19 genome and structural analysis of its palindromic termini. *Virology* **1990**, *175*, 247–254. [CrossRef]
65. Cotmore, S.F.; McKie, V.C.; Anderson, L.J.; Astell, C.R.; Tattersall, P. Identification of the major structural and nonstructural proteins encoded by human parvovirus B19 and mapping of their genes by procaryotic expression of isolated genomic fragments. *J. Virol.* **1986**, *60*, 548–557. [CrossRef] [PubMed]
66. Kaufmann, B.; Chipman, P.R.; Kostyuchenko, V.A.; Modrow, S.; Rossmann, M.G. Visualization of the Externalized VP2 N Termini of Infectious Human Parvovirus B19. *J. Virol.* **2008**, *82*, 7306–7312. [CrossRef]
67. Saikawa, T.; Anderson, S.; Momoeda, M.; Kajigaya, S.; Young, N.S. Neutralizing linear epitopes of B19 parvovirus cluster in the VP1 unique and VP1-VP2 junction regions. *J. Virol.* **1993**, *67*, 3004–3009. [CrossRef]
68. Anderson, S.; Momoeda, M.; Kawase, M.; Kajigaya, S.; Young, N.S. Peptides derived from the unique region of B19 parvovirus minor capsid protein elicitneutralizing antibodies in rabbits. *Virology* **1995**, *206*, 626–632. [CrossRef]
69. Klasse, P.J.; Sattentau, Q.J. Occupancy and mechanism in antibody-mediated neutralization of animal viruses. *J. Gen. Virol.* **2002**, *83*, 2091–2108. [CrossRef]
70. Klasse, P.J. Neutralization of Virus Infectivity by Antibodies: Old Problems in New Perspectives. *Adv. Biol.* **2014**, *2014*, 157895. [CrossRef]
71. Burton, D.R. Antibodies, viruses and vaccines. *Nat. Rev. Immunol.* **2002**, *2*, 706–713. [CrossRef] [PubMed]
72. Modrow, S.; Dorsch, S. Antibody responses in parvovirus B19 infected patients. *Pathol. Biol.* **2002**, *50*, 326–331. [CrossRef]
73. Kurtzman, G.J.; Cohen, B.J.; Field, A.M.; Oseas, R.; Blaese, R.M.; Young, N.S. Immune response to B19 parvovirus and an antibody defect in persistent viral infection. *J. Clin. Investig.* **1989**, *84*, 1114–1123. [CrossRef] [PubMed]
74. Kurtzman, G.; Frickhofen, N.; Kimball, J.; Jenkins, D.W.; Nienhuis, A.W.; Young, N.S. Pure Red-Cell Aplasia of 10 Years' Duration Due to Persistent Parvovirus B19 Infection and Its Cure with Immunoglobulin Therapy. *N. Engl. J. Med.* **1989**, *321*, 519–523. [CrossRef] [PubMed]
75. Bernstein, D.I.; El Sahly, H.M.; Keitel, W.A.; Wolff, M.; Simone, G.; Segawa, C.; Wong, S.; Shelly, D.; Young, N.S.; Dempsey, W. Safety and immunogenicity of a candidate parvovirus B19 vaccine. *Vaccine* **2011**, *29*, 7357–7363. [CrossRef]

76. Chandramouli, S.; Medina-Selby, A.; Coit, D.; Schaefer, M.; Spencer, T.; Brito, L.A.; Zhang, P.; Otten, G.; Mandl, C.W.; Mason, P.W.; et al. Generation of a parvovirus B19 vaccine candidate. *Vaccine* **2013**, *31*, 3872–3878. [CrossRef]
77. Bua, G.; Manaresi, E.; Bonvicini, F.; Gallinella, G. Parvovirus B19 Replication and Expression in Differentiating Erythroid Progenitor Cells. *PLoS ONE* **2016**, *11*, 1–19. [CrossRef]
78. Brown, K.E.; Anderson, S.M.; Young, N.S. Erythrocyte P antigen: Cellular receptor for B19 parvovirus. *Science* **1993**, *262*, 114–117. [CrossRef]
79. Brown, K.E.; Cohen, B.J. Haemagglutination by parvovirus B19. *J. Gen. Virol.* **1992**, *73*, 2147–2149. [CrossRef]
80. Brown, K.E.; Hibbs, J.R.; Gallinella, G.; Anderson, S.M.; Lehman, E.D.; McCarthy, P.; Young, N.S. Resistance to parvovirus B19 infection due to lack of virus receptor (erythrocyte P antigen). *N. Engl. J. Med.* **1994**, *330*, 1192–1196. [CrossRef]
81. Chipman, P.R.; Agbandje-Mckenna, M.; Kajigaya, S.; Brown, K.E.; Young, N.S.; Baker, T.S.; Rossmann, M.G. Cryo-electron microscopy studies of empty capsids of human parvovirus B19 complexed with its cellular receptor. *Proc. Natl. Acad. Sci. USA* **1996**, *93*, 7502–7506. [CrossRef] [PubMed]
82. Nasir, W.; Nilsson, J.; Olofsson, S.; Bally, M.; Rydell, G.E. Parvovirus B19 VLP recognizes globoside in supported lipid bilayers. *Virology* **2014**, *456–457*, 364–369. [CrossRef] [PubMed]
83. Weigel-Kelley, K.A.; Yoder, M.C.; Srivastava, A. Recombinant Human Parvovirus B19 Vectors: Erythrocyte P Antigen Is Necessary but Not Sufficient for Successful Transduction of Human Hematopoietic Cells. *J. Virol.* **2001**, *75*, 4110–4116. [CrossRef] [PubMed]
84. Bieri, J.; Ros, C. Globoside Is Dispensable for Parvovirus B19 Entry but Essential at a Postentry Step for Productive Infection. *J. Virol.* **2019**, *93*, 1–15. [CrossRef]
85. Weigel-Kelley, K.A.; Yoder, M.C.; Srivastava, A. $\alpha5\beta1$ integrin as a cellular coreceptor for human parvovirus B19: Requirement of functional activation of $\beta1$ integrin for viral entry. *Blood* **2003**, *102*, 3927–3933. [CrossRef] [PubMed]
86. Munakata, Y.; Saito-Ito, T.; Kumura-Ishii, K.; Huang, J.; Kodera, T.; Ishii, T.; Hirabayashi, Y.; Koyanagi, Y.; Sasaki, T. Ku80 autoantigen as a cellular coreceptor for human parvovirus B19 infection. *Blood* **2005**, *106*, 3449–3456. [CrossRef]
87. Bönsch, C.; Kempf, C.; Ros, C. Interaction of parvovirus B19 with human erythrocytes alters virus structure and cell membrane integrity. *J. Virol.* **2008**, *82*, 11784–11791. [CrossRef]
88. Xu, D.; Zhang, Y. Ab initio protein structure assembly using continuous structure fragments and optimized knowledge-based force field. *Proteins Struct. Funct. Bioinform.* **2012**, *80*, 1715–1735. [CrossRef]
89. Shimomura, S.; Komatsu, N.; Frickhofen, N.; Anderson, S.; Kajigaya, S.; Young, N.S. First continuous propagation of B19 parvovirus in a cell line. *Blood* **1992**, *79*, 18–24. [CrossRef]
90. Miyagawa, E.; Yoshida, T.; Takahashi, H.; Yamaguchi, K.; Nagano, T.; Kiriyama, Y.; Okochi, K.; Sato, H. Infection of the erythroid cell line, KU812Ep6 with human parvovirus B19 and its application to titration of B19 infectivity. *J. Virol. Methods* **1999**, *83*, 45–54. [CrossRef]
91. Koury, M.J.; Bondurant, M.C. Maintenance by erythropoietin of viability and maturation of murine erythroid precursor cells. *J. Cell. Physiol.* **1988**, *137*, 65–74. [CrossRef] [PubMed]
92. Koury, M.J.; Bondurant, M.C. Erythropoietin retards DNA breakdown and prevents programmed death in erythroid progenitor cells. *Science* **1990**, *248*, 378–381. [CrossRef] [PubMed]
93. Pop, R.; Shearstone, J.R.; Shen, Q.; Liu, Y.; Hallstrom, K.; Koulnis, M.; Gribnau, J.; Socolovsky, M. A key commitment step in erythropoiesis is synchronized with the cell cycle clock through mutual inhibition between PU.1 and S-phase progression. *PLoS Biol.* **2010**, *8*, e1000484. [CrossRef] [PubMed]
94. Chen, A.Y.; Kleiboeker, S.; Qiu, J. Productive parvovirus B19 infection of primary human erythroid progenitor cells at hypoxia is regulated by STAT5A and MEK signaling but not HIFα. *PLoS Pathog.* **2011**, *7*, e1002088. [CrossRef]
95. Ganaie, S.S.; Zou, W.; Xu, P.; Deng, X.; Kleiboeker, S.; Qiu, J. Phosphorylated STAT5 directly facilitates parvovirus B19 DNA replication in human erythroid progenitors through interaction with the MCM complex. *PLoS Pathog.* **2017**, *13*, 1–27. [CrossRef]
96. An, X.; Schulz, V.P.; Li, J.; Wu, K.; Liu, J.; Xue, F.; Hu, J.; Mohandas, N.; Gallagher, P.G. Global transcriptome analyses of human and murine terminal erythroid differentiation. *Blood* **2014**, *123*, 3466–3477. [CrossRef]
97. Kerr, J.R.; Linden, M.R. Human dependovirus infection. In *Parvoviruses*; Hodder Arnold, Oxford University Press: London, UK, 2006; pp. 381–384.

98. Guan, W.; Wong, S.; Zhi, N.; Qiu, J. The Genome of Human Parvovirus B19 Can Replicate in Nonpermissive Cells with the Help of Adenovirus Genes and Produces Infectious Virus. *J. Virol.* **2009**, *83*, 9541–9553. [CrossRef]
99. Winter, K.; von Kietzell, K.; Heilbronn, R.; Pozzuto, T.; Fechner, H.; Weger, S. Roles of E4orf6 and VA I RNA in Adenovirus-Mediated Stimulation of Human Parvovirus B19 DNA Replication and Structural Gene Expression. *J. Virol.* **2012**, *86*, 5099–5109. [CrossRef]
100. Doerig, C.; Hirt, B.; Antonietti, J.P.; Beard, P. Nonstructural protein of parvoviruses B19 and minute virus of mice controls transcription. *J. Virol.* **1990**, *64*, 387–396. [CrossRef]
101. Ponnazhagan, S.; Woody, M.J.; Wang, X.S.; Zhou, S.Z.; Srivastava, A. Transcriptional transactivation of parvovirus B19 promoters in nonpermissive human cells by adenovirus type 2. *J. Virol.* **1995**, *69*, 8096–8101. [CrossRef]
102. Ozawa, K.; Ayub, J.; Hao, Y.S.; Kurtzman, G.; Shimada, T.; Young, N. Novel transcription map for the B19 (human) pathogenic parvovirus. *J. Virol.* **1987**, *61*, 2395–2406. [CrossRef] [PubMed]
103. Von Kietzell, K.; Pozzuto, T.; Heilbronn, R.; Grössl, T.; Fechner, H.; Weger, S. Antibody-Mediated Enhancement of Parvovirus B19 Uptake into Endothelial Cells Mediated by a Receptor for Complement Factor C1q. *J. Virol.* **2014**, *88*, 8102–8115. [CrossRef] [PubMed]
104. Quattrocchi, S.; Ruprecht, N.; Bnsch, C.; Bieli, S.; Zürcher, C.; Boller, K.; Kempf, C.; Ros, C. Characterization of the early steps of human parvovirus B19 infection. *J. Virol.* **2012**, *86*, 9274–9284. [CrossRef]
105. Dennis, E.A.; Cao, J.; Hsu, Y.H.; Magrioti, V.; Kokotos, G. Phospholipase A2 enzymes: Physical structure, biological function, disease implication, chemical inhibition, and therapeutic intervention. *Chem. Rev.* **2011**, *111*, 6130–6185. [CrossRef] [PubMed]
106. Canaan, S.; Zádori, Z.; Ghomashchi, F.; Bollinger, J.; Sadilek, M.; Moreau, M.E.; Tijssen, P.; Gelb, M.H. Interfacial Enzymology of Parvovirus Phospholipases A2. *J. Biol. Chem.* **2004**, *279*, 14502–14508. [CrossRef]
107. Deng, X.; Dong, Y.; Yi, Q.; Huang, Y.; Zhao, D.; Yang, Y.; Tijssen, P.; Qiu, J.; Liu, K.; Li, Y. The Determinants for the Enzyme Activity of Human Parvovirus B19 Phospholipase A2 (PLA2) and Its Influence on Cultured Cells. *PLoS ONE* **2013**, *8*, e61440. [CrossRef]
108. Lupescu, A.; Bock, C.-T.; Lang, P.A.; Aberle, S.; Kaiser, H.; Kandolf, R.; Lang, F. Phospholipase A2 Activity-Dependent Stimulation of Ca2+ Entry by Human Parvovirus B19 Capsid Protein VP1. *J. Virol.* **2006**, *80*, 11370–11380. [CrossRef]
109. Almilaji, A.; Szteyn, K.; Fein, E.; Pakladok, T.; Munoz, C.; Elvira, B.; Towhid, S.T.; Alesutan, I.; Shumilina, E.; Bock, C.T.; et al. Down-regulation of Na/K + ATPase activity by human parvovirus B19 capsid protein VP1. *Cell. Physiol. Biochem.* **2013**, *31*, 638–648. [CrossRef]
110. Ahmed, M.; Almilaji, A.; Munoz, C.; Elvira, B.; Shumilina, E.; Bock, C.T.; Kandolf, R.; Lang, F. Down-regulation of K+ channels by human parvovirus B19 capsid protein VP1. *Biochem. Biophys. Res. Commun.* **2014**, *450*, 1396–1401. [CrossRef]
111. Ahmed, M.; Honisch, S.; Pelzl, L.; Fezai, M.; Hosseinzadeh, Z.; Bock, C.T.; Kandolf, R.; Lang, F. Up-regulation of epithelial Na+ channel ENaC by human parvovirus B19 capsid protein VP1. *Biochem. Biophys. Res. Commun.* **2015**, *468*, 179–184. [CrossRef]
112. Lu, J.; Zhi, N.; Wong, S.; Brown, K.E. Activation of Synoviocytes by the Secreted Phospholipase A_2 Motif in the VP1-Unique Region of Parvovirus B19 Minor Capsid Protein. *J. Infect. Dis.* **2006**, *193*, 582–590. [CrossRef] [PubMed]
113. Lange, A.; Mills, R.E.; Lange, C.J.; Stewart, M.; Devine, S.E.; Corbett, A.H. Classical nuclear localization signals: Definition, function, and interaction with importin α. *J. Biol. Chem.* **2007**, *282*, 5101–5105. [CrossRef]
114. Panté, N.; Kann, M. Nuclear pore complex is able to transport macromolecules with diameters of ~39 nm. *Mol. Biol. Cell* **2002**, *13*, 425–434. [CrossRef] [PubMed]
115. Mäntylä, E.; Aho, V.; Kann, M.; Vihinen-Ranta, M. Cytoplasmic Parvovirus Capsids Recruit Importin Beta for Nuclear Delivery. *J. Virol.* **2020**, *94*, e01532-19. [CrossRef]
116. Pillet, S.; Annan, Z.; Fichelson, S.; Morinet, F. Identification of a nonconventional motif necessary for the nuclear import of the human parvovirus B19 major capsid protein (VP2). *Virology* **2003**, *306*, 25–32. [CrossRef]
117. Peer, D.; Karp, J.M.; Hong, S.; Farokhzad, O.C.; Margalit, R.; Langer, R. 84 Nat nanotech 2007 R Langer Nanocarriers as an emerging platform for cancer therapy.pdf. *Nat. Nanotechnol.* **2007**, *2*, 751–760. [CrossRef]
118. Wilhelm, S.; Tavares, A.; Dai, Q.; Ohta, S.; Audet, J.; Dvorak, H.F.; Chan, W.C.W. Analysis of nanoparticle delivery to tumours. *Nat. Rev. Mater.* **2016**, *1*, 16014. [CrossRef]

119. Ma, Y.; Nolte, R.J.M.; Cornelissen, J.J.L.M. Virus-based nanocarriers for drug delivery. *Adv. Drug Deliv. Rev.* **2012**, *64*, 811–825. [CrossRef]
120. Somiya, M.; Liu, Q.; Kuroda, S. Current progress of virus-mimicking nanocarriers for drug delivery. *Nanotheranostics* **2017**, *1*, 415–429. [CrossRef]
121. Stachon, A.; Holland-Letz, T.; Krieg, M. High in-hospital mortality of intensive care patients with nucleated red blood cells in blood. *Clin. Chem. Lab. Med.* **2004**, *42*, 933–938. [CrossRef]
122. Stachon, A.; Kempf, R.; Holland-Letz, T.; Friese, J.; Becker, A.; Krieg, M. Daily monitoring of nucleated red blood cells in the blood of surgical intensive care patients. *Clin. Chim. Acta* **2006**, *366*, 329–335. [CrossRef] [PubMed]
123. Buttarello, M.; Plebani, M. Automated blood cell counts: State of the art. *Am. J. Clin. Pathol.* **2008**, *130*, 104–116. [CrossRef] [PubMed]
124. Koury, M.J. Abnormal erythropoiesis and the pathophysiology of chronic anemia. *Blood Rev.* **2014**, *28*, 49–66. [CrossRef] [PubMed]
125. Danise, P.; Maconi, M.; Barrella, F.; Di Palma, A.; Avino, D.; Rovetti, A.; Gioia, M.; Amendola, G. Evaluation of nucleated red blood cells in the peripheral blood of hematological diseases. *Clin. Chem. Lab. Med.* **2011**, *50*, 357–360. [CrossRef]
126. Buoro, S.; Vavassori, M.; Pipitone, S.; Benegiamo, A.; Lochis, E.; Fumagalli, S.; Falanga, A.; Marchetti, M.; Crippa, A.; Ottomano, C.; et al. Evaluation of nucleated red blood cell count by Sysmex XE-2100 in patients with thalassaemia or sickle cell anaemia and in neonates. *Blood Transfus.* **2015**, *13*, 588–594. [CrossRef]
127. Constantino, B.T.; Cogionis, B. Nucleated RBCs—Significance in the peripheral blood film. *Lab. Med.* **2000**, *31*, 223–229. [CrossRef]
128. Da Rin, G.; Vidali, M.; Balboni, F.; Benegiamo, A.; Borin, M.; Ciardelli, M.L.; Dima, F.; Di Fabio, A.; Fanelli, A.; Fiorini, F.; et al. Performance evaluation of the automated nucleated red blood cell count of five commercial hematological analyzers. *Int. J. Lab. Hematol.* **2017**, *39*, 663–670. [CrossRef]
129. Anagnostou, A.; Liu, Z.; Steiner, M.; Chin, K.; Lee, E.S.; Kessimian, N.; Noguchi, C.T. Erythropoietin receptor mRNA expression in human endothelial cells. *Proc. Natl. Acad. Sci. USA* **1994**, *91*, 3974–3978. [CrossRef]
130. Spivak, J.L. The anaemia of cancer: Death by a thousand cuts. *Nat. Rev. Cancer* **2005**, *5*, 543–555. [CrossRef]
131. Jelkmann, W.; Bohlius, J.; Hallek, M.; Sytkowski, A.J. The erythropoietin receptor in normal and cancer tissues. *Crit. Rev. Oncol. Hematol.* **2008**, *67*, 39–61. [CrossRef]
132. Chen, K.; Liu, J.; Heck, S.; Chasis, J.A.; An, X.; Mohandas, N. Resolving the distinct stages in erythroid differentiation based on dynamic changes in membrane protein expression during erythropoiesis. *Proc. Natl. Acad. Sci. USA* **2009**, *106*, 17413–17418. [CrossRef]
133. Hu, J.; Liu, J.; Xue, F.; Halverson, G.; Reid, M.; Guo, A.; Chen, L.; Raza, A.; Galili, N.; Jaffray, J.; et al. Isolation and functional characterization of human erythroblasts at distinct stages: Implications for understanding of normal and disordered erythropoiesis in vivo. *Blood* **2013**, *121*, 3246–3253. [CrossRef] [PubMed]
134. MacHherndl-Spandl, S.; Suessner, S.; Danzer, M.; Proell, J.; Gabriel, C.; Lauf, J.; Sylie, R.; Klein, H.U.; Béné, M.C.; Weltermann, A.; et al. Molecular pathways of early CD105-positive erythroid cells as compared with CD34-positive common precursor cells by flow cytometric cell-sorting and gene expression profiling. *Blood Cancer J.* **2013**, *3*, 1–13. [CrossRef]
135. Origa, R. β-Thalassemia. *Genet. Med.* **2017**, *19*, 609–619. [CrossRef] [PubMed]
136. Kato, G.J.; Piel, F.B.; Reid, C.D.; Gaston, M.H.; Ohene-Frempong, K.; Krishnamurti, L.; Smith, W.R.; Panepinto, J.A.; Weatherall, D.J.; Costa, F.F.; et al. Sickle cell disease. *Nat. Rev. Dis. Prim.* **2018**, *4*, 1–22. [CrossRef] [PubMed]
137. Bayanzay, K.; Alzoebie, L. Reducing the iron burden and improving survival in transfusion-dependent thalassemia patients: Current perspectives. *J. Blood Med.* **2016**, *7*, 159–169. [CrossRef] [PubMed]
138. Coates, T.D.; Wood, J.C. How we manage iron overload in sickle cell patients. *Br. J. Haematol.* **2017**, *177*, 703–716. [CrossRef]
139. Peters, C. Allogeneic Hematopoietic Stem Cell Transplantation to Cure Transfusion-Dependent Thalassemia: Timing Matters! *Biol. Blood Marrow Transpl.* **2018**, *24*, 1107–1108. [CrossRef]
140. Gluckman, G.; Cappelli, B.; Bernaudin, F.; Labopin, M.; Volt, F.; Carreras, J.; Pinto Simoes, B.; Ferster, A.; Dupont, S.; de la Fuente, J.; et al. Sickle cell disease: An international survey of results of HLA-identical sibling hematopoietic stem cell transplantation. *Blood* **2017**, *129*, 1548–1556. [CrossRef]

141. Karponi, G.; Zogas, N. Gene therapy for beta-thalassemia: Updated perspectives. *Appl. Clin. Genet.* **2019**, *12*, 167–180. [CrossRef]
142. Lidonnici, M.R.; Ferrari, G. Gene therapy and gene editing strategies for hemoglobinopathies. *Blood Cells Mol. Dis.* **2018**, *70*, 87–101. [CrossRef] [PubMed]
143. Lino, C.A.; Harper, J.C.; Carney, J.P.; Timlin, J.A. Delivering crispr: A review of the challenges and approaches. *Drug Deliv.* **2018**, *25*, 1234–1257. [CrossRef] [PubMed]
144. Chery, J. RNA therapeutics: RNAi and antisense mechanisms and clinical applications. *Postdr. J.* **2016**, *4*, 35–50. [CrossRef] [PubMed]
145. Galaway, F.A.; Stockley, P.G. MS2 viruslike particles: A robust, semisynthetic targeted drug delivery platform. *Mol. Pharm.* **2013**, *10*, 59–68. [CrossRef] [PubMed]
146. Ashley, C.E.; Carnes, E.C.; Phillips, G.K.; Durfee, P.N.; Buley, M.D.; Lino, C.A.; Padilla, D.P.; Phillips, B.; Carter, M.B.; Willman, C.L.; et al. Cell-specific delivery of diverse cargos by bacteriophage MS2 virus-like particles. *ACS Nano* **2011**, *5*, 5729–5745. [CrossRef] [PubMed]
147. Nai, A.; Lidonnici, M.R.; Rausa, M.; Mandelli, G.; Pagani, A.; Silvestri, L.; Ferrari, G.; Camaschella, C. The second transferrin receptor regulates red blood cell production in mice. *Blood* **2015**, *125*, 1170–1179. [CrossRef]
148. Artuso, I.; Lidonnici, M.R.; Altamura, S.; Mandelli, G.; Pettinato, M.; Muckenthaler, M.U.; Silvestri, L.; Ferrari, G.; Camaschella, C.; Nai, A. Transferrin receptor 2 is a potential novel therapeutic target for β-thalassemia: Evidence from a murine model. *Blood* **2018**, *132*, 2286–2297. [CrossRef]
149. Casu, C.; Pettinato, M.; Liu, A.; Aghajan, M.; Lo Presti, V.; Lidonnici, M.R.; Munoz, K.A.; O'Hara, E.; Olivari, V.; Di Modica, S.M.; et al. Correcting β-thalassemia by combined therapies that restrict iron and modulate erythropoietin activity. *Blood* **2020**, *136*, 1968–1979. [CrossRef]
150. Lohani, N.; Bhargava, N.; Munshi, A.; Ramalingam, S. Pharmacological and molecular approaches for the treatment of β-hemoglobin disorders. *J. Cell. Physiol.* **2018**, *233*, 4563–4577. [CrossRef]
151. Hasserjian, R.P.; Zuo, Z.; Garcia, C.; Tang, G.; Kasyan, A.; Luthra, R.; Abruzzo, L.V.; Kantarjian, H.M.; Medeiros, L.J.; Wang, S.A. Acute erythroid leukemia: A reassessment using criteria refined in the 2008 WHO classification. *Blood* **2010**, *115*, 1985–1992. [CrossRef]
152. Wickrema, A.; Crispino, J.D. Erythroid and megakaryocytic transformation. *Oncogene* **2007**, *26*, 6803–6815. [CrossRef] [PubMed]
153. Mazzella, F.M.; Kowal-Vern, A.; Atef Shrit, M.; Rector, J.T.; Cotelingam, J.D.; Schumacher, H.R. Effects of multidrug resistance gene expression in acute erythroleukemia. *Mod. Pathol.* **2000**, *13*, 407–413. [CrossRef] [PubMed]
154. Schwartz, R.S. Paul Ehrlich's magic bullets. *N. Engl. J. Med.* **2004**, *350*, 1079–1080. [CrossRef]
155. Boddu, P.; Benton, C.B.; Wang, W.; Borthakur, G.; Khoury, J.D.; Pemmaraju, N. Erythroleukemia-historical perspectives and recent advances in diagnosis and management. *Blood Rev.* **2018**, *32*, 96–105. [CrossRef] [PubMed]
156. Leisi, R.; Von Nordheim, M.; Kempf, C.; Ros, C. Specific targeting of proerythroblasts and erythroleukemic cells by the VP1u region of parvovirus B19. *Bioconjug. Chem.* **2015**, *26*, 1923–1930. [CrossRef]
157. Rapti, K.; Louis-Jeune, V.; Kohlbrenner, E.; Ishikawa, K.; Ladage, D.; Zolotukhin, S.; Hajjar, R.J.; Weber, T. Neutralizing antibodies against AAV serotypes 1, 2, 6, and 9 in sera of commonly used animal models. *Mol. Ther.* **2012**, *20*, 73–83. [CrossRef]
158. Rutledge, E.A.; Halbert, C.L.; Russell, D.W. Infectious Clones and Vectors Derived from Adeno-Associated Virus (AAV) Serotypes Other Than AAV Type 2. *J. Virol.* **1998**, *72*, 309–319. [CrossRef]
159. Gao, G.; Vandenberghe, L.H.; Alvira, M.R.; Lu, Y.; Calcedo, R.; Zhou, X.; Wilson, J.M. Clades of Adeno-Associated Viruses Are Widely Disseminated in Human Tissues. *J. Virol.* **2004**, *78*, 6381–6388. [CrossRef]
160. Zinn, E.; Pacouret, S.; Khaychuk, V.; Turunen, H.T.; Carvalho, L.S.; Andres-Mateos, E.; Shah, S.; Shelke, R.; Maurer, A.C.; Maurer, E.; et al. In silico reconstruction of the viral evolutionary lineage yields a potent gene therapy vector. *Cell Rep.* **2015**, *12*, 1056–1068. [CrossRef]
161. Yeh, P.; Landais, D.; Lemaître, M.; Maury, I.; Crenne, J.Y.; Becquart, J.; Murry-Brelier, A.; Boucher, F.; Montay, G.; Fleer, R.; et al. Design of yeast-secreted albumin derivatives for human therapy: Biological and antiviral properties of a serum albumin-CD4 genetic conjugate. *Proc. Natl. Acad. Sci. USA* **1992**, *89*, 1904–1908. [CrossRef]

162. Czajkowsky, D.M.; Hu, J.; Shao, Z.; Pleass, R.J. Fc-fusion proteins: New developments and future perspectives. *EMBO Mol. Med.* **2012**, *4*, 1015–1028. [CrossRef] [PubMed]
163. Shen, B.Q.; Xu, K.; Liu, L.; Raab, H.; Bhakta, S.; Kenrick, M.; Parsons-Reponte, K.L.; Tien, J.; Yu, S.F.; Mai, E.; et al. Conjugation site modulates the in vivo stability and therapeutic activity of antibody-drug conjugates. *Nat. Biotechnol.* **2012**, *30*, 184–189. [CrossRef] [PubMed]
164. Wu, M.; Brown, W.L.; Stockley, P.G. Cell-Specific Delivery of Bacteriophage-Encapsidated Ricin a Chain. *Bioconjug. Chem.* **1995**, *6*, 587–595. [CrossRef] [PubMed]

Publisher's Note: MDPI stays neutral with regard to jurisdictional claims in published maps and institutional affiliations.

Article

Oncolytic H-1 Parvovirus Enters Cancer Cells through Clathrin-Mediated Endocytosis

Tiago Ferreira [1], Amit Kulkarni [2], Clemens Bretscher [1], Karsten Richter [3], Marcelo Ehrlich [4] and Antonio Marchini [1,2,*]

1 Laboratory of Oncolytic Virus Immuno-Therapeutics, German Cancer Research Centre, Im Neuenheimer Feld 242, 69120 Heidelberg, Germany; t.ferreira@dkfz.de (T.F.); c.bretscher@dkfz.de (C.B.)
2 Laboratory of Oncolytic Virus Immuno-Therapeutics, Luxembourg Institute of Health, 84 Val Fleuri, L-1526 Luxembourg, Luxembourg; amit.kulkarni@lih.lu
3 Core Facility Electron Microscopy, German Cancer Research Centre, Im Neuenheimer Feld 280, 69120 Heidelberg, Germany; k.richter@dkfz-heidelberg.de
4 Laboratory of Signal Transduction and Membrane Biology, The Shumins School for Biomedicine and Cancer Research, George S. Wise Faculty of Life Sciences, Tel Aviv University, 69978 Tel Aviv, Israel; marceloe@post.tau.ac.il
* Correspondence: a.marchini@dkfz.de or antonio.marchini@lih.lu; Tel.: +49-6221-424969 or +352-26-970-856

Received: 8 October 2020; Accepted: 19 October 2020; Published: 21 October 2020

Abstract: H-1 protoparvovirus (H-1PV) is a self-propagating virus that is non-pathogenic in humans and has oncolytic and oncosuppressive activities. H-1PV is the first member of the *Parvoviridae* family to undergo clinical testing as an anticancer agent. Results from clinical trials in patients with glioblastoma or pancreatic carcinoma show that virus treatment is safe, well-tolerated and associated with first signs of efficacy. Characterisation of the H-1PV life cycle may help to improve its efficacy and clinical outcome. In this study, we investigated the entry route of H-1PV in cervical carcinoma HeLa and glioma NCH125 cell lines. Using electron and confocal microscopy, we detected H-1PV particles within clathrin-coated pits and vesicles, providing evidence that the virus uses clathrin-mediated endocytosis for cell entry. In agreement with these results, we found that blocking clathrin-mediated endocytosis using specific inhibitors or small interfering RNA-mediated knockdown of its key regulator, AP2M1, markedly reduced H-1PV entry. By contrast, we found no evidence of viral entry through caveolae-mediated endocytosis. We also show that H-1PV entry is dependent on dynamin, while viral trafficking occurs from early to late endosomes, with acidic pH necessary for a productive infection. This is the first study that characterises the cell entry pathways of oncolytic H-1PV.

Keywords: oncolytic viruses; rodent protoparvovirus H-1PV; virus entry; clathrin-mediated endocytosis

1. Introduction

The rodent H-1 protoparvovirus (H-1PV) belongs to the *Parvoviridae* family, genus *Protoparvovirus* [1]. This genus also includes *Rodent protoparvovirus 1* (H-1PV, Kilham rat virus, LuIII virus, minute virus of mice (MVM), mouse parvovirus, tumour virus X, rat minute virus), *Rodent protoparvovirus 2* (rat parvovirus 1), *Carnivore protoparvovirus 1* (canine parvovirus (CPV) and feline panleukopenia parvovirus (FPV)), *Primate protoparvovirus 1* (bufavirus) and *Ungulate parvovirus 1* (porcine parvovirus (PPV)) [2,3]. Protoparvoviruses (PtPVs) are single-stranded DNA viruses with an icosahedral capsid of about 25 nm diameter. Their genomes encompass the non-structural (NS) and the viral particle (VP) transcriptional units, whose expressions are regulated by the P4 and P38 promoters, respectively. The NS transcriptional unit encodes the NS1 and NS2 proteins, whereas the VP transcriptional unit encodes the VP1 and VP2 capsid proteins and the small alternatively translated protein [4].

Owing to their ability to specifically infect, replicate and kill human cancer cells, rodent PtPVs are under investigation as potential anticancer therapies. Pre-clinical studies have revealed that H-1PV in particular has remarkable oncolytic and oncosuppressive activity in a number of cell culture and animal models of cancers from different origins [5]. Notably, H-1PV-induced cancer cell death and lysis are immunogenic and stimulate the immune system to participate in the elimination of cancer cells [6]. NS1 is the major effector of H-1PV oncotoxicity [7].

Although viral oncolytic activity is shared between rodent PtPVs, H-1PV is the only member of the genus to have reached the clinic as an anticancer therapy. In a phase I/IIa clinical trial in patients with recurrent glioblastoma (ParvOryx01), H-1PV treatment was safe, well-tolerated and associated with first evidence of anticancer efficacy. This evidence included the ability of H-1PV to cross the blood–brain barrier after intravenous administration, its wide distribution in the tumour bed, the induction of tumour necrosis and immuno-conversion of the tumour microenvironment. As a result, virus treatment led to an improved progression-free survival and median overall survival of patients in comparison with historical controls [8]. A dose-escalation phase I/IIa pilot study in patients with metastatic pancreatic cancer recently confirmed the excellent safety and tolerability of H-1PV treatment. In accordance with the results of ParvOryx01, patients who responded to the treatment showed evident changes in the tumour microenvironment and induction of specific immune responses [9].

The PtPV life cycle is strictly dependent on host cellular factors for a productive infection, from cell surface attachment and entry to virus DNA replication, gene expression, multiplication and egress. Some of these factors are frequently overexpressed or dysregulated in cancer cells. The list includes cell cycle regulators, transcription factors, modulators of the DNA damage response, kinases and cytoskeleton components (reviewed in Reference [10]). However, unlike for other PtPVs, the early steps of H-1PV infection remain to be characterised.

The first interaction between PtPVs and the target cell occurs through binding to a specific surface receptor exposed on the host plasma membrane. Cellular receptors for some PtPVs have been described, such as the transferrin receptor for CPV and FPV. H-1PV, like MVM and PPV, uses sialic acid (SA) for cell surface attachment and entry. However, it is unclear whether SA itself acts as a functional viral receptor for the virus or is a component of an as yet unidentified receptor(s) or receptor complex [3,11,12].

After docking to the cellular membrane, viruses are internalised through different pathways [13]. Clathrin- and caveolae-mediated endocytosis are two dynamin-dependent pathways, whereas macropinocytosis, lipid-raft-mediated endocytosis and caveolae/clathrin-independent endocytosis are dynamin-independent pathways [14,15]. Clathrin-mediated endocytosis is the pathway commonly used by small viruses, including PtPVs [16–20]. The mechanism begins with the recruitment of adaptor protein 2 (AP-2) complexes on the plasma membrane, followed by the assembly of a three-dimensional clathrin coat that leads to a progressive invagination of the membrane. Dynamin self-assembles around the vesicle neck and mediates its scission, and the vesicle is subsequently released into the interior of the cell [21].

PtPVs also use alternative endocytic pathways. For instance, MVM prototype strain takes at least three different endocytic routes: clathrin-, caveolae- and clathrin-independent carrier-mediated endocytosis [22]. Even though endocytosis seems to be the default entry pathway for PtPVs, differences between members of the family may contribute to the tropism of these viruses.

As the PtPV is trafficked within the cellular endosome, its capsid undergoes slow structural changes. In particular, the acidic environment exposes the catalytic phospholipase 2 domain of VP1. This conformational change promotes the digestion of the endosomal membrane, resulting in the release of viral particles from the late endosome to the cytosol [3]. Thereafter, incoming PtPV particles are transported to the nucleus in a process that is dependent on the cytoskeleton and associated motor proteins [23].

In this study, we used electron microscopy (EM) and immunofluorescence (IF), together with a number of chemical inhibitors and siRNA-mediated knockdown, to identify which of these pathways H-1PV uses to enter cancer cells. We found that H-1PV cell uptake occurs preferentially through clathrin-mediated, but not caveolae-mediated, endocytosis in cervical carcinoma HeLa and glioma NCH125 cell lines. Entry was also dependent on dynamin activity. We show that after its internalisation, H-1PV, like other PtPVs, passes through early endosomes to late endosomes/lysosomes during its cytosolic trafficking. Productive infection relies heavily on the acidic pH in the endosomes.

2. Materials and Methods

2.1. Cells and Viruses

The cervical carcinoma-derived HeLa [7] and the glioblastoma-derived NCH125 cell lines [24] were cultured in Dulbecco's modified Eagle's medium (DMEM) supplemented with 10% FBS, 100 units/mL penicillin, 100 µg/mL streptomycin and 2 mM L-glutamine (all from Gibco, Thermo Fischer Scientific, Darmstadt, Germany) in a humidified incubator at 37 °C. Both cell lines were tested for mycoplasma contamination by PCR in a regular base.

Both wild-type H-1PV and recombinant H-1PV harbouring the green fluorescent protein-encoding gene (recH-1PV-EGFP) were produced, purified and titrated as previously described [25,26].

2.2. Electron Microscopy

HeLa cells were seeded on punched sheets of ACLAR-Fluoropolymer films (Electron Microscopy Sciences) at a density of 8×10^4 cells/well in 24-well plates. On the following day, cells were infected with H-1PV at a multiplicity of infection (MOI) of 2000 plaque forming units (pfu) per cell in DMEM 5% FBS at 4 °C for 1 h to allow virus attachment to the cell surface and promote a synchronised infection. In order to catch the internalisation event, cells were shifted to 37 °C for 0, 5, 10, 20 and 30 min. After incubation, ACLAR-Fluoropolymer films were embedded in epoxy resin for ultrathin sectioning according to standard procedures. Briefly, chemical fixation was carried out in buffered aldehyde (4% formaldehyde, 2% glutaraldehyde, 1 mM $CaCl_2$, 1 mM $MgCl_2$ in 100 mM Ca-cacodylate, pH 7.2), followed by post-fixation in buffered 1% osmium tetroxide and en bloc staining in 1% uranylacetate. Following dehydration in graded steps of ethanol, the adherent cells were flat-embedded in epoxy resin (mixture of glycid ether, methylnadic anhydride and dodecenyl-succinic-anhydride; Serva). Ultrathin sections of nominal thickness 60 nm and contrast-stained with lead-citrate and uranylacetate were analysed using a Zeiss EM 910 (Carl Zeiss, Oberkochen, Germany) at 120 kV and micrographs taken using a slow scan charge-coupled device camera (TRS, Olympus, Moorenweis, Germany).

2.3. Co-Localisation of H-1PV and Cellular Proteins by Confocal Microscopy

HeLa cells were seeded at a density of 3.5×10^3 cells/spot on spot slides and grown in 50 µL of complete cellular medium. On the following day, cells were placed on ice for 15 min and then infected with wild-type H-1PV at a MOI of 500 (pfu/cell) in a total of 70 µL of 5% FCS-containing medium. At 1 h post-infection, cells were shifted to 37 °C for varied times depending on the experiment, before being fixed with 3.7% paraformaldehyde on ice for 15 min and permeabilised with 1% Triton X-100 for 10 min. Immunostaining was carried out with the following antibodies, all used at dilution 1:500 for 1 h: mouse monoclonal anti-H-1PV capsid (a conformational antibody kindly provided by Barbara Leuchs; DKFZ Virus Production and Development Unit, Heidelberg, Germany) [27], rabbit monoclonal anti-clathrin heavy chain (D3C6; Cell Signalling Technology, Leiden, Netherlands), rabbit monoclonal anti-EEA1 (3288; Cell Signalling Technology), rabbit monoclonal anti-Rab7 (9367T; Cell Signalling Technology) and rabbit polyclonal anti-LAMP-1 (CD107a) (AB2971; Merck, Darmstadt, Germany). Anti-mouse Alexa Fluor 594 IgG (A11005; Thermo Fisher Scientific, Bleiswijk, Netherlands) or anti-rabbit Alexa Fluor 488 IgG (A11008; Thermo Fisher Scientific) were used as secondary antibodies. Nuclei were stained by 4′,6-diamidin-2-phenylindol (DAPI). Images in the green channel (H-1PV), red channel (varied cellular proteins) or blue channel (DAPI) were acquired with a confocal microscope (Leica TCS SP5 II, Wetzlar, Germany). Picture analysis was carried out using the Leica LAS X Software.

2.4. Treatment with Inhibitors of Endocytosis Pathways

Hypertonic sucrose (Carl Roth), 0.40 M, chlorpromazine (Sigma-Aldrich Chemie GmbH, Steinheim, Germany), 2.5 μg/mL for HeLa or 5 μg/mL for NCH125, and pitstop 2 (Sigma-Aldrich Chemie GmbH), 30 μM, were used to inhibit clathrin-mediated endocytosis. Dynole™ Series Kit containing Dynole 31–2 (active drug) and 34–2 (negative control) (ab120474; Abcam, Cambridge, UK), 5 μM, were used to inhibit dynamin activity. Bafilomycin A1 (BafA1; Cell Signalling Technology), 10 nM, and ammonium chloride (NH4CL; 1145, Merck), 25 mM, were used to prevent pH acidification. The concentrations of the aforementioned inhibitors were used at the highest doses before affecting the cellular proliferation. Nystatin (Sigma-Aldrich), 10 μg/mL, and methyl-β-cyclodextrin (MβCD; Sigma-Aldrich Chemie GmbH), 10 mM, were used to inhibit caveolae-mediated endocytosis. These concentrations were selected according to the literature [28,29] and also did not affect the proliferation of the cell lines used for the experiments.

Briefly, HeLa cells were seeded at a density of 8×10^4 cells/well in 24-well plates and, on the following day, pre-treated for 45 min with the various inhibitors. Cells were then infected with recH-1PV-EGFP at a MOI of 0.2–0.3 transduction units (TU)/cell for 4 h, washed twice with PBS and grown in culture medium for an additional 20 h. At 24 h post-infection, cells were washed once with PBS, fixed with 3.7% paraformaldehyde on ice for 15 min, permeabilised with 1% Triton X-100 for 10 min and stained with DAPI. Fluorescence images of EGFP-positive cells were acquired with a BZ-9000 fluorescence microscope (Keyence Corporation, Osaka, Japan) with 4X or 10X objective. DAPI staining was used to visualise the nuclei (cells). At least two independent experiments, each performed in duplicate, were performed for every condition tested.

2.5. Cell Proliferation Assay

Cell proliferation was monitored in real time through the xCELLigence system (ACEA Biosciences Inc., San Diego, CA, USA) according to the manufacturer's instructions. Briefly, 8×10^4 HeLa or NCH125 cells per well were seeded in a 96-well E-plate (Roche) in a total volume of 100 μL of complete DMEM medium. On the following day, cells were treated with different inhibitors for 45 min, and subsequently washed with PBS. Cell proliferation was monitored every 30 min in real time over a period of 72 h. Data is expressed as "Normalised cell index", where all curves were normalised to an arbitrary value of 1.0 at the timepoint before treatment. Average values of each experimental condition assessed in triplicate are presented with the respective standard deviation (SD).

2.6. siRNA-Mediated Knockdown

Cells were seeded at a density of 4×10^4 cells/well in a 24-well plate and grown in 500 μL of normal growth medium. After 24 h, cells were transfected with 10 nM siRNA using Lipofectamine RNAimax (Thermo Fisher Scientific, Carlsbad, CA, USA) according to the manufacturer's instructions. For *AP2M1*, we used the *AP2M1* ON-TARGET plus Human siRNA SMARTpool (L-008170–00-0005) and, as a negative control, the plus Non-targeting pool (D-001810–10-05) (Dharmacon, Thermo Fisher Scientific). For CAV-1, we used two Silencer Select Validated siRNAs (s2446 and s2448; Life Technologies, Paisley, Scotland) and Silencer Select Negative Control #2 siRNA (4390846, Life Technologies) as a control. After 24 h, the medium was replaced, and cells were grown for an additional 24 h to allow efficient gene silencing. The cells were then infected for 24 h with recH-1PV-EGFP at 0.2–0.3 TU/cell. Cells were then washed once with PBS and processed as described above for fluorescence microscopy. At least two independent experiments, each performed in duplicate, were performed for every condition tested.

2.7. Western Blotting

Cells were harvested, washed in PBS, and then lysed on ice for 30 min in RIPA buffer (10 mM Tris-HCl pH 7.5, 150 mM NaCl, 1 mM EDTA pH 8, 1% NP-40, 0.5% Na-deoxylcholate and 0.5% SDS supplemented with complete EDTA-free protease inhibitor (11697498001; Roche, Mannheim, Germany). Cellular debris was removed by centrifugation, and protein concentration in cell lysates was measured by bicinchoninic acid (BCA) assay (Thermo Fisher Scientific), according to manufacturer's instructions. SDS-PAGE analysis was performed on 50 µg of total protein extract. After separation, proteins were transferred to Hybond-P membrane (GE Healthcare, Freiburg, Germany). Immunoblotting was carried out with the following antibodies: mouse monoclonal anti-vinculin (sc-25336; Santa Cruz Biotechnology, Heidelberg, Germany) and rabbit monoclonal CAV-1 (D46G3; Cell Signalling Technology) at 1:1000 dilution. After incubation with horseradish peroxidase conjugated secondary antibodies (Santa Cruz Biotechnology) at 1:1000 dilution, proteins were revealed with Western Blot Chemiluminescence Reagent *Plus* (Perkin Elmer Life Sciences) and exposed to Hyperfilm™ ECL radiographic films (GE Healthcare).

3. Results

3.1. Electron Microscopy Analysis Reveals H-1PV within Clathrin-Coated Pits

In order to investigate the H-1PV internalisation pathway, we performed EM analysis of HeLa cells infected with H-1PV. Infection was carried out at 4 °C for 1 h to allow virus cell surface binding. Cells were then shifted back to 37 °C for various intervals to allow virus internalisation. EM analysis of infected cells at 4 °C showed the virus at the cell surface bound to thickened regions resembling clathrin-associated plasma membrane (Figure 1A) [30]. In the first 5 min after cells were shifted back to 37 °C, the invagination of clathrin-rich regions started with H-1PV particles remaining associated with these regions (Figure 1B). From 10 to 30 min, viral particles were detected both in clathrin-coated pits (Figure 1C) or in the cytosol inside completely invaginated vesicles (we found up to nine particles in a single vesicle) (Figure 1D). Furthermore, in the course of the experiment, no virus internalisation was found in vesicles with the small flask-shaped invaginations that are characteristic of caveolae-mediated endocytosis [31], which suggests that H-1PV is internalised mainly (if not exclusively) via clathrin-mediated endocytosis.

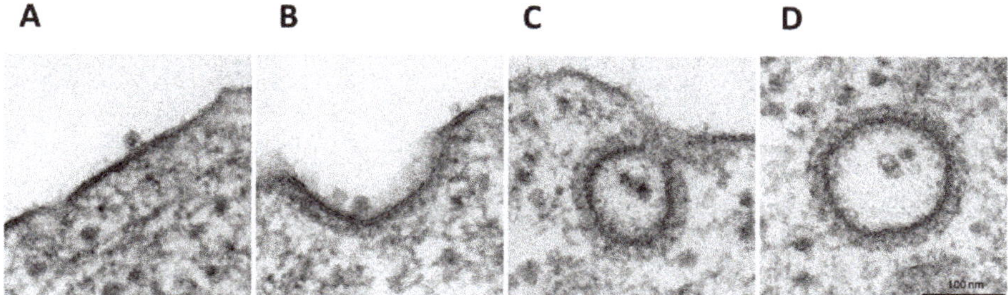

Figure 1. Endocytosis of H-1PV is clathrin-dependent. HeLa cells were infected with H-1PV for 1 h at 4 °C to allow H-1PV cell surface attachment but not entry. Cells were then shifted to 37 °C to allow H-1PV cell internalisation. Cells were collected every 5 min for a total of 30 min and processed for EM analysis. (**A**) At 4 °C, H-1PV particles are found attached to electro-dense (clathrin-rich) regions on the plasma membrane. (**B**) In the first 5 min after release at 37 °C, H-1PV particles are detected in early-forming clathrin-coated pits. (**C**) From 10 to 30 min, H-1PV particles moved into the cells within deeply invaginated clathrin-coated pits that were still connected to the plasma membrane, forming an hourglass-like membrane neck. (**D**) Later in the infection (10–30 min at 37 °C), H-1PV particles are seen being trafficked within the cell inside clathrin-coated vesicles.

3.2. H-1PV Co-Localises with Clathrin Upon Entry

The EM analysis provided first evidence that H-1PV uses clathrin-mediated endocytosis to enter cells. To confirm these results independently, we checked for possible co-localisation of H-1PV and clathrin-heavy chain (CHC). To this end, HeLa cells were infected with H-1PV for 1 h at 4 °C and then shifted back to 37 °C. After 30 min, a fraction of internalised H-1PV was clearly detected in association with CHC by confocal microscopy (Figure 2), providing further evidence that H-1PV is internalised through a clathrin-dependent pathway.

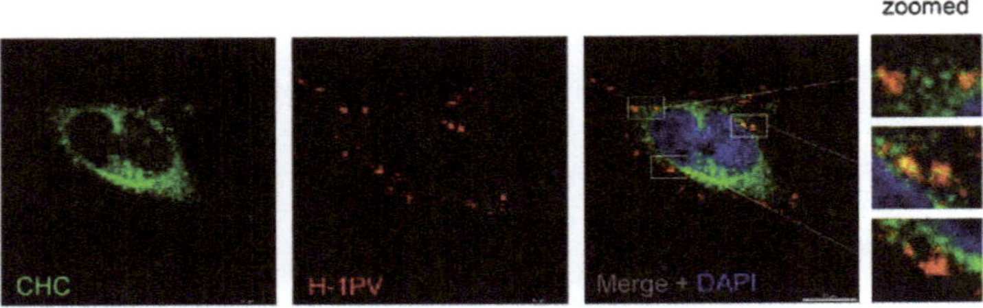

Figure 2. H-1PV co-localises with clathrin upon entry. HeLa cells were infected with H-1PV for 1 h at 4 °C and then shifted to 37 °C for 30 min before being processed for immunostaining using anti-H-1PV full capsid and anti-clathrin-heavy chain (CHC) antibodies. Cell nuclei were visualised by DAPI staining. Confocal microscopy analysis showed that H-1PV particles (Alexa Fluor 594, red) co-localised with CHC (Alexa Fluor 488, green) early upon infection. Three examples of regions where co-localisation is observed are framed by white boxes and shown zoomed in.

3.3. H-1PV Enters Cells Preferentially via Clathrin-Mediated Endocytosis

Next, we investigated whether targeting regulators of clathrin-mediated endocytosis would affect H-1PV infection. A recombinant H-1PV expressing the EGFP reporter gene (recH-1PV-EGFP) was used for the experiments. This non-replicative parvovirus shares the same capsid of the wild type but harbours the EGFP gene under the control of the natural P38 late promoter, whose activity is regulated by NS1 viral protein [25]. Therefore, the EGFP signal directly correlates to the ability of the virus to reach the nucleus and initiate its own gene transcription.

The effects of various inhibitors on H-1PV entry were assessed in HeLa and in glioma-derived NCH125 cell lines. The latter, like HeLa, is highly permissive to H-1PV infection. Pharmacological inhibitors included hypertonic sucrose, chlorpromazine (CPZ) and pitstop 2. Hypertonic sucrose is a classical inhibitor that traps clathrin in microcages [32]. CPZ is a cationic, amphiphilic drug that induces the misassembly of clathrin lattices at the cell surface and on endosomes [33]. Pitstop 2 interferes with the binding of proteins to the N-terminal domain of clathrin [34]. The internalisation of TexasRed-labelled transferrin, a protein known to be exclusively internalised through clathrin-mediated endocytosis, was monitored to check the effectiveness of each treatment [35]. At the concentrations used, the three inhibitors blocked transferrin uptake efficiently but did not affect cell proliferation (Supplementary Figure S1).

Pre-treatment with hypertonic sucrose decreased H-1PV transduction by more than 90% in HeLa and 80% in NCH125 cells compared to untreated cells (Figure 3A,B). When the compound was applied 3 h post-infection (by which time the virus is already internalised), no significant changes in H-1PV transduction were observed, indicating that the drug interferes with H-1PV transduction at the level of virus entry (Supplementary Figure S2). Strong inhibition of H-1PV transduction was also achieved by pre-treating cells with CPZ (approximately 90% reduction in both cell lines) or with pitstop 2 (60% reduction in Hela and over 70% in NCH125 cells) compared to untreated cells.

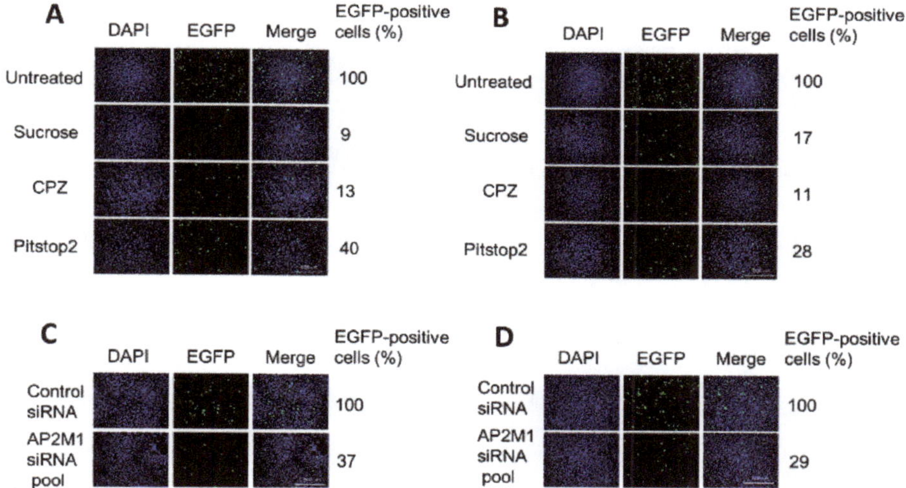

Figure 3. Blocking clathrin-mediated endocytosis results in a significant reduction of H-1PV transduction. (**A**) HeLa and (**B**) NCH125 cells were pre-treated with different clathrin-mediated endocytosis inhibitors (hypertonic sucrose, chlorpromazine (CPZ) or pitstop 2) or left untreated. Cells were then infected with recH-1PV-EGFP for 4 h in the presence of the inhibitor. At 20 h post-infection, cells were processed as described in the Materials and Methods (M&M) section. (**C**) HeLa and (**D**) NCH125 cells were transfected with a pool of siRNAs targeting either *AP2M1* or negative control. At 48 h post-transfection, cells were infected with recH-1PV-EGFP for 4 h and grown for an additional 20 h. Cells were then processed as described in the M&M section. Numbers represent the average percentage of EGFP-positive cells relative to the number of EGFP-positive cells observed in untreated or scrambled siRNA-transfected cells, which was arbitrarily set as 100%.

The AP-2 complex is a heterotetramer that plays an essential role in clathrin-mediated endocytosis [21]. To confirm the involvement of clathrin-mediated endocytosis in H-1PV cell entry, we silenced the expression of *AP2M1*, the gene encoding subunit μ1 of AP-2 [36]. To this end, HeLa and NCH125 cells were transfected with either a siRNA pool targeting *AP2M1* or scrambled siRNA (negative control), and subsequently infected with recH-1PV-EGFP. Under conditions in which the silencing of *AP2M1* successfully reduced transferrin uptake, we observed a strong decrease in H-1PV transduction (over 60% compared to the scrambled siRNA-treated cells) in both cell lines (Figure 3C,D). Taken together, these results show that H-1PV uses clathrin-mediated endocytosis to enter HeLa and NCH125 cells.

3.4. H-1PV Does Not Enter Cells via Caveolae-Dependent Endocytosis

To investigate whether H-1PV uses pathways other than clathrin-mediated endocytosis to enter HeLa and NCH125 cancer cells, we inhibited clathrin-independent endocytosis using nystatin and methyl-β-cyclodextrin (MβCD). Both drugs selectively disrupt lipid rafts (e.g., cholesterol), including those required for caveolae-dependent entry [28,29,37]. Yet, neither nystatin nor MβCD decreased H-1PV transduction, providing evidence that the virus does not use this endocytic route to enter these cells (Figure 4A,B). Similar results were also obtained using the two drugs at lower or higher concentrations (0.1–100 μg/mL for nystatin, 0.1–100 μM for MβCD). We also carried out siRNA-mediated knockdown of *CAV1* (which encodes for caveolin-1) by using two different siRNAs. Knock-down of *CAV1* gene expression did not decrease H-1PV transduction activity compared with scrambled siRNA-transfected cells, but instead increased it (Figure 4C,D). Together, these results indicate that caveolae-dependent endocytosis is not involved in H-1PV entry of HeLa and NCH125 cells.

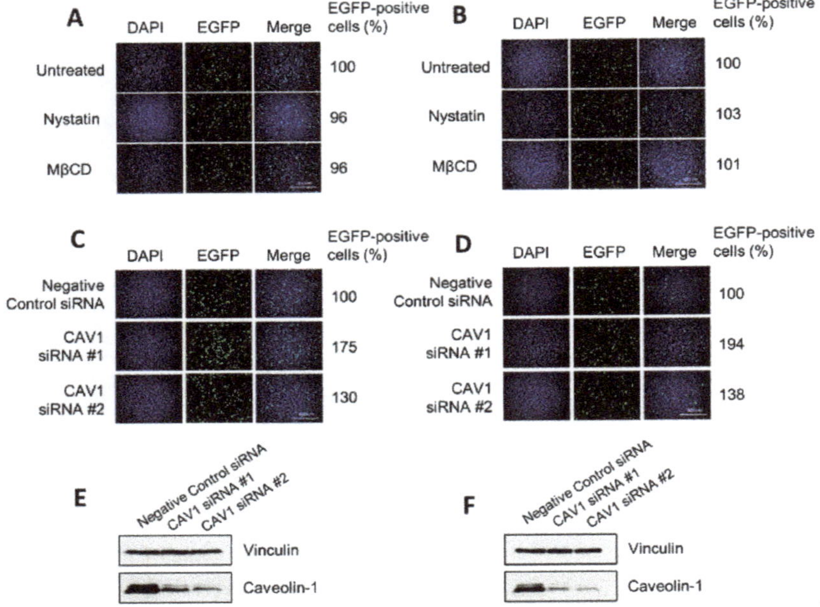

Figure 4. Disruption of clathrin-independent endocytosis does not decrease H-1PV transduction. (**A**) HeLa and (**B**) NCH125 cells were either pre-treated with cholesterol-sequestering drugs (nystatin or methyl-β cyclodextrin (MβCD)) for 45 min or left untreated. Cells were then infected with recH-1PV-EGFP for 4 h in the presence (or absence) of the inhibitor. At 20 h post-infection, cells were processed as described in the M&M section for immunofluorescence analysis. (**C**) HeLa and (**D**) NCH125 cells were transfected with siRNAs targeting *CAV1* or a negative control siRNA. At 48 h post-transfection, cells were infected with recH-1PV-EGFP for 4 h and grown for an additional 20 h. Cells were then processed as described in panel A. Numbers represent the average percentage of EGFP-positive cells relative to the number of EGFP-positive cells observed in untreated cells, which was arbitrarily set as 100%. The steady protein levels of caveolin-1 on lysates derived from (**E**) HeLa or (**F**) NCH125 siRNA-transfected cells were analysed by Western blotting. Vinculin was used as a loading control.

3.5. H-1PV Internalisation Is Dependent on Dynamin

Dynamin is a large GTPase with an essential role in cellular membrane fission for newly formed vesicles. It is therefore required for clathrin- and caveolae-mediated endocytosis but not for macropinocytosis [38]. We used the highly selective Dynole 34–2 to inhibit dynamin activity [39,40], and Dynole 31–2, its inactive form, as a negative control. At a concentration that blocked transferrin uptake, Dynole 34–2 drastically reduced virus transduction to just 8% in HeLa and 13% in NCH125 cells (Figure 5) compared to untreated cells. As expected, Dynole 31–2 did not have any significant effect on H-1PV transduction. These results demonstrate that dynamin plays an essential role in H-1PV infection.

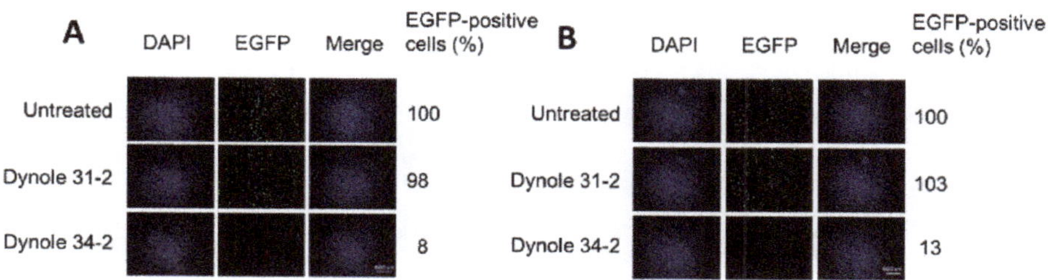

Figure 5. H-1PV requires dynamin activity for entry. (**A**) HeLa and (**B**) NCH125 cells were pre-treated with either Dynole 34–2 or its inactive form, Dynole 31–2. Cells were subsequently infected with recH-1PV-EGFP for 4 h in the presence of the inhibitor. At 20 h post-infection, cells were processed as described in Figure 4A. Numbers represent the average percentage of EGFP-positive cells relative to the number of EGFP-positive cells observed in untreated cells that was arbitrarily set as 100%.

3.6. H-1PV Hijacks Endosomes for Trafficking into the Cytosol and Acidic pH Is Required for Productive H-1PV Infection

The Rab family of proteins and their effectors play a key role in the formation, maintenance and trafficking of endosomes [41]. Early endosome antigen 1 (EEA1) is a Rab5 effector protein that is involved in sorting endocytic vesicles at the early endosome level [42]. Rab7 is considered to be the key regulator of late endosome trafficking. Lysosomal-associated membrane protein 1 (LAMP-1) is enriched in late endosomes and lysosomes, where, among other functions, it maintains lysosomal integrity and pH [43,44].

To provide direct evidence that H-1PV hijacks endocytosis for its intracellular trafficking, we infected HeLa cells with H-1PV for 1 h, fixed the cells, and then stained viral capsids as well as EEA1, Rab7 and LAMP-1 proteins with antibodies. Confocal microscopy analysis showed that H-1PV co-localised with EEA1, Rab7 and LAMP-1 markers during infection (Figure 6A).

Next, we hypothesised that the acidic pH in the endocytic compartments would provide the environmental conditions required for a productive H-1PV infection [3] (as it does for other PtPV infections), possibly triggering the conformational changes necessary for uncoating and nuclear translocation. To test whether low endosomal pH is required for H-1PV infection, we pre-treated HeLa and NCH125 cells with ammonium chloride (NH_4Cl), a lysosomotropic weak base [45], or bafilomycin A1 (BafA1), a blocker of vacuolar H^+-ATPases [46]. Treatment with NH_4Cl resulted in a strong decrease of virus transduction in both HeLa and NCH125 cells (39% and 30%, respectively) (Figure 6B,C), while treatment with BafA1 completely abolished H-1PV transduction (Figure 6D,E), in comparison to untreated cells. Together, these results provide evidence that H-1PV, like other PtPVs, requires acidic endosomal pH for a productive infection.

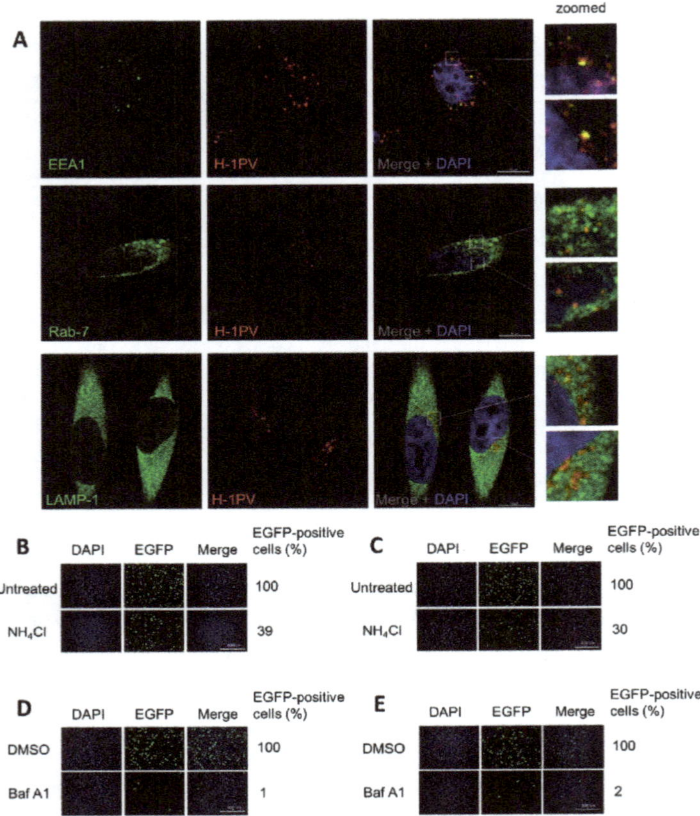

Figure 6. H-1PV trafficking occurs via the endosomal system with acidic pH being crucial for a productive infection. (**A**) HeLa cells were infected with H-1PV at MOI of 500 pfu/cell for 1 h at 4 °C and shifted to 37 °C for 30 min. Cells were then fixed, permeabilised and stained with DAPI and H-1PV capsid antibody, together with one of the endosomal markers (EEA1, Rab-7 or LAMP-1). Confocal microscopy analysis indicates that H-1PV particles (Alexa Fluor 594) co-localise with all three markers, EEA1, Rab-7 and LAMP-1 (Alexa Fluor 488), upon infection. Two examples of regions where co-localisation is observed are framed in white boxes and shown enlarged. (**B**,**D**) HeLa and (**C**,**E**) NCH125 cells were incubated with ammonium chloride (NH_4CL) or bafilomycin A1 (BafA1) respectively, for 45 min, or left untreated. Cells were subsequently infected with recH-1PV-EGFP for 4 h in the presence of the inhibitor. At 20 h post-infection, cells were processed as described in the M&M section. Numbers represent the average percentage of EGFP-positive cells relative to the number of EGFP-positive cells observed in untreated cells, which was arbitrarily set as 100%.

4. Discussion

H-1PV is a promising oncolytic virus. Early-phase clinical studies show that virus treatment is safe, well-tolerated and associated with first signs of anticancer efficacy. The present study aimed to fill a gap in knowledge of H-1PV biology regarding the pathways used by this virus to enter cancer cells. We chose two cancer cell lines as models to investigate the early steps of H-1PV infection: the cervical carcinoma-derived HeLa cell line and the glioma-derived NCH125 cell line. Previous laboratory studies have shown that these two cell lines are highly permissive to H-1PV infection and susceptible to its oncolytic activity [7,24,47].

In our study, for the first time, we show that H-1PV exploits clathrin-mediated endocytosis to enter cancer cells. EM analysis of H-1PV-infected HeLa cells revealed virus particles associated with clathrin-coated pits and then internalised inside clathrin-coated vesicles (Figure 1). In agreement with this finding, confocal microscopy analysis showed co-localisation of H-1PV and clathrin (Figure 2). Furthermore, pharmacological inhibition of clathrin-mediated endocytosis using hypertonic sucrose, CPZ and pitstop 2, as well as siRNA-mediated silencing of *AP2M1*, a key regulator of clathrin-mediated endocytosis, confirmed the heavy dependence of H-1PV on this pathway for successful entry into both HeLa and NCH125 cancer cells (Figure 3).

Previous research has reported that other PtPVs use clathrin-mediated endocytosis to gain access to cells and progress the infection (reviewed in Reference [3]). Our results are therefore in agreement with the idea that clathrin-mediated endocytosis is the default entry pathway for PtPVs. However, a number of PtPVs have also been shown to hijack alternative endocytic pathways. For instance, MVMp uses at least three different endocytic routes, such as clathrin-, caveolae- and clathrin-independent carrier-mediated endocytosis [22]. Moreover, PPV uses both clathrin-dependent endocytosis and macropinocytosis [16]. FPV and CPV are taken up by cells via binding to the transferrin receptor, which is typically endocytosed by clathrin-mediated endocytosis. However, FPV may also enter cells through alternative internalisation mechanisms, as deletions or mutations in the internalisation motif of the transferrin receptor, while decreasing FPV cellular uptake, did not completely arrest viral infection [48]. Although we cannot completely rule out the possibility that H-1PV, like other PtPVs, can also use different pathways to enter HeLa and NCH125 cancer cells, our results seem to exclude caveolae-mediated endocytosis as a major entry pathway. Indeed, pre-treatment of cells with nystatin and MβCD, two inhibitors of caveolae-mediated endocytosis, did not decrease H-1PV transduction levels, suggesting that these drugs did not modify H-1PV entry. These results are in agreement with previous studies showing that bovine parvovirus (BPV) [17] and PPV [16] do not use caveolae-mediated endocytosis for their cell internalisation. Surprisingly, *CAV1* siRNA-mediated knockdown increased H-1PV transduction (Figure 4), suggesting that caveolin-1 (the protein encoded by *CAV1*) interferes with H-1PV infection. Human immunodeficiency virus (HIV) infection is restricted by caveolin-1 in various ways, e.g., at the transcriptional level by suppressing NF-kB p65 acetylation in macrophages, or by interacting with HIV viral proteins to impair viral infectivity [49,50]. Similarly, abundant caveolin-1 levels prevent Influenza A virus from infecting mouse embryo fibroblasts, which is reversed by depleting caveolin-1 [51]. Further experiments are required to find out whether and at what level caveolin-1 represents a negative modulator of H-1PV infection. Along these lines, caveolin-1 antagonists or inhibitors could offer an interesting strategy to improve H-1PV efficacy in cancer with elevated levels of caveolin-1.

Our study also provides important evidence that dynamin is involved in H-1PV entry (Figure 5). Dynamin is also required for MVMp entry in murine A9 fibroblasts, a process that occurs through both clathrin- and caveolae-mediated endocytosis [22]. However, the same study showed that dynamin is not involved in MVMp entry into mouse mammary cells transformed with polyomavirus middle T antigen, which instead occurs via clathrin-independent carrier-mediated endocytosis [22].

Another cell entry route used by viruses is macropinocytosis, generally described as a dynamin-independent process [52,53]. Among PtPVs, PPV has been reported to use this route of entry [16]. However, we showed that inhibition of dynamin almost completely abolished H-1PV infection, which makes it unlikely that H-1PV uses macropinocytosis as an alternative pathway to enter HeLa and NCH125 cancer cells.

Numerous viruses are known to hijack Rab-dependent pathways to enter cells. Of these, Rab5 and Rab7 GTPases are the key regulators of transport to early and late endosomes, while LAMP-1 is present mainly in late endosomes/lysosomes [41,54]. In the present study, we show that H-1PV particles co-localise with EEA1, an early endosome marker, and with Rab7 and LAMP-1, two late endosome markers (Figure 6A), indicating that H-1PV, like other PtPVs [3], uses endosomes for its cytosolic trafficking into the cells. However, a fraction of H-1PV particles is most likely trapped in LAMP1-positive lysosomes. This has been observed for other PtPVs such as MVM, for which sequestration in LAMP1-positive lysosomes limits the efficiency of its nuclear translocation [55], and CPV, which accumulates in perinuclear LAMP2-positive lysosomes [56].

Previous studies have shown that the acidic environment inside the endosomes changes in redox conditions, and acid proteases and phosphatases drive the conformational rearrangements in the catalytic phospholipase 2 domain of the VP1 protein [55,57,58]. These changes are required first for the release of PtPV particles from the late endosome to the cytosol, and then for translocation into the nucleus [23,57,59,60]. For instance, the endosomal acidic environment is required for a productive infection of B19V [20], CPV [18,61,62] and MVM [55,63], among other parvoviruses. In agreement with these observations, we also found that NH_4Cl and BafA1 strongly hampered H-1PV transduction (Figure 6).

In summary, our study shows, for the first time, that H-1PV internalisation occurs via clathrin-mediated and dynamin-dependent endocytosis, while requiring endosomal acidification, with EEA1 and Rab7 involved in the infection process. However, we cannot rule out the possibility of H-1PV taking other pathways in other cell types. Important questions remain, namely which receptor H-1PV uses to initiate uptake and whether other co-receptors are involved. In addition, the exact mechanisms of endosomal escape and subsequent nuclear entry, and finally the site where the H-1PV genome becomes accessible for replication, need to be investigated [3]. A better understanding of the early steps of H-1PV (and, more generally, PtPV) infection is crucial not only to decipher viral tropism and inherent oncolytic properties, but also to improve the clinical potential of H-1PV in cancer virotherapy.

Supplementary Materials: The following are available online at http://www.mdpi.com/1999-4915/12/10/1199/s1, Figure S1: Entry pathway inhibitors did not alter cell proliferation. Figure S2: Sucrose, a classical inhibitor of clathrin-mediated endocytosis, has no impact on virus transduction if added 4 h after virus infection.

Author Contributions: Conceptualisation, T.F. and A.M.; methodologies, T.F., A.K., M.E. and K.R.; investigation, T.F., C.B. (confocal microscopy analysis) and K.R. (EM analysis); data curation, T.F. and A.M.; writing—original draft preparation, T.F. and A.M.; writing—review and editing, A.M.; visualisation, T.F., C.B. and A.M.; supervision, A.M.; funding acquisition, A.M. All authors have read and agreed to the published version of the manuscript.

Funding: This study has been partially funded by a research grant from the Cooperational Research Program of the Deutsches Krebsforschungszentrum, Heidelberg, and the Ministry of Science and Technology (MOST), Israel, to A.M. and M.E.

Acknowledgments: We are grateful to Barbara Leuchs (DKFZ, Heidelberg, Germany) for kindly providing the H-1PV capsid monoclonal antibody. We would also like to extend our acknowledgements to Tiina Marttila for producing both the H-1PV wild-type and the recH-1PV-EGFP viruses. We thank Caroline Hadley (INLEXIO) for critically reading the manuscript.

Conflicts of Interest: A.M. is co-inventor in several H-1PV-related patents. No other conflicts of interest are declared by the authors.

References

1. Cotmore, S.F.; Tattersall, P. Parvoviral host range and cell entry mechanisms. *Adv. Virus Res.* **2007**, *70*, 183–232. [PubMed]
2. Cotmore, S.F.; Agbandje-McKenna, M.; Chiorini, J.A.; Mukha, D.V.; Pintel, D.J.; Qiu, J.; Soderlund-Venermo, M.; Tattersall, P.; Tijssen, P.; Gatherer, D. The family parvoviridae. *Arch. Virol.* **2014**, *159*, 1239–1247. [CrossRef] [PubMed]
3. Ros, C.; Bayat, N.; Wolfisberg, R.; Almendral, J.M.J.V. Protoparvovirus cell entry. *Viruses* **2017**, *9*, 313. [CrossRef] [PubMed]
4. Bretscher, C.; Marchini, A. H-1 parvovirus as a cancer-killing agent: Past, present, and future. *Viruses* **2019**, *11*, 562. [CrossRef] [PubMed]
5. Hartley, A.; Kavishwar, G.; Salvato, I.; Marchini, A. A roadmap for the success of oncolytic parvovirus-based anticancer therapies. *Annu. Rev. Virol.* **2020**, *7*, 537–557. [CrossRef]

6. Marchini, A.; Daeffler, L.; Pozdeev, V.I.; Angelova, A.; Rommelaere, J. Immune conversion of tumor microenvironment by oncolytic viruses: The protoparvovirus H-1PV case study. *Front. Immunol.* **2019**, *10*, 1848. [CrossRef]
7. Hristov, G.; Kramer, M.; Li, J.; El-Andaloussi, N.; Mora, R.; Daeffler, L.; Zentgraf, H.; Rommelaere, J.; Marchini, A. through its nonstructural protein NS1, parvovirus H-1 induces apoptosis via accumulation of reactive oxygen species. *J. Virol.* **2010**, *84*, 5909–5922. [CrossRef]
8. Geletneky, K.; Huesing, J.; Rommelaere, J.; Schlehofer, J.R.; Leuchs, B.; Dahm, M.; Krebs, O.; von Knebel Doeberitz, M.; Huber, B.; Hajda, J. Phase I/IIa study of intratumoral/intracerebral or intravenous/intracerebral administration of Parvovirus H-1 (ParvOryx) in patients with progressive primary or recurrent glioblastoma multiforme: ParvOryx01 protocol. *BMC Cancer* **2012**, *12*, 99. [CrossRef]
9. Hajda, J.; Lehmann, M.; Krebs, O.; Kieser, M.; Geletneky, K.; Jäger, D.; Dahm, M.; Huber, B.; Schöning, T.; Sedlaczek, O. A non-controlled, single arm, open label, phase II study of intravenous and intratumoral administration of ParvOryx in patients with metastatic, inoperable pancreatic cancer: ParvOryx02 protocol. *BMC Cancer* **2017**, *17*, 1–11. [CrossRef]
10. Nüesch, J.P.; Lacroix, J.; Marchini, A.; Rommelaere, J. Molecular pathways: Rodent parvoviruses—Mechanisms of oncolysis and prospects for clinical cancer treatment. *Clin. Cancer Res.* **2012**, *18*, 3516–3523. [CrossRef]
11. López-Bueno, A.; Rubio, M.-P.; Bryant, N.; McKenna, R.; Agbandje-McKenna, M.; Almendral, J.M. Host-selected amino acid changes at the sialic acid binding pocket of the parvovirus capsid modulate cell binding affinity and determine virulence. *J. Virol.* **2006**, *80*, 1563–1573. [CrossRef] [PubMed]
12. Allaume, X.; El-Andaloussi, N.; Leuchs, B.; Bonifati, S.; Kulkarni, A.; Marttila, T.; Kaufmann, J.K.; Nettelbeck, D.M.; Kleinschmidt, J.; Rommelaere, J. Retargeting of rat parvovirus H-1PV to cancer cells through genetic engineering of the viral capsid. *J. Virol.* **2012**, *86*, 3452–3465. [CrossRef] [PubMed]
13. Harbison, C.E.; Chiorini, J.A.; Parrish, C.R. The parvovirus capsid odyssey: From the cell surface to the nucleus. *Trends Microbiol.* **2008**, *16*, 208–214. [CrossRef]
14. Doherty, G.J.; McMahon, H.T. Mechanisms of endocytosis. *Annu. Rev. Biochem.* **2009**, *78*, 857–902. [CrossRef] [PubMed]
15. Mercer, J.; Schelhaas, M.; Helenius, A. Virus entry by endocytosis. *Annu. Rev. Biochem.* **2010**, *79*, 803–833. [CrossRef]
16. Boisvert, M.; Fernandes, S.; Tijssen, P. Multiple pathways involved in porcine parvovirus cellular entry and trafficking toward the nucleus. *J. Virol.* **2010**, *84*, 7782–7792. [CrossRef]
17. Dudleenamjil, E.; Lin, C.-Y.; Dredge, D.; Murray, B.K.; Robison, R.A.; Johnson, F.B. Bovine parvovirus uses clathrin-mediated endocytosis for cell entry. *J. Gen. Virol.* **2010**, *91*, 3032–3041. [CrossRef]
18. Parker, J.S.; Parrish, C.R. Cellular uptake and infection by canine parvovirus involves rapid dynamin-regulated clathrin-mediated endocytosis, followed by slower intracellular trafficking. *J. Virol.* **2000**, *74*, 1919–1930. [CrossRef]
19. Vendeville, A.; Ravallec, M.; Jousset, F.-X.; Devise, M.; Mutuel, D.; López-Ferber, M.; Fournier, P.; Dupressoir, T.; Ogliastro, M. Densovirus infectious pathway requires clathrin-mediated endocytosis followed by trafficking to the nucleus. *J. Virol.* **2009**, *83*, 4678–4689. [CrossRef]
20. Quattrocchi, S.; Ruprecht, N.; Bönsch, C.; Bieli, S.; Zürcher, C.; Boller, K.; Kempf, C.; Ros, C. Characterization of the early steps of human parvovirus B19 infection. *J. Virol.* **2012**, *86*, 9274–9284. [CrossRef]
21. McMahon, H.T.; Boucrot, E. Molecular mechanism and physiological functions of clathrin-mediated endocytosis. *Nat. Rev. Mol. Cell Biol.* **2011**, *12*, 517. [CrossRef] [PubMed]
22. Garcin, P.O.; Panté, N. The minute virus of mice exploits different endocytic pathways for cellular uptake. *Virology* **2015**, *482*, 157–166. [CrossRef] [PubMed]
23. Suikkanen, S.; Aaltonen, T.; Nevalainen, M.; Välilehto, O.; Lindholm, L.; Vuento, M.; Vihinen-Ranta, M. Exploitation of microtubule cytoskeleton and dynein during parvoviral traffic toward the nucleus. *J. Virol.* **2003**, *77*, 10270–10279. [CrossRef] [PubMed]
24. Di Piazza, M.; Mader, C.; Geletneky, K.; y Calle, M.H.; Weber, E.; Schlehofer, J.; Deleu, L.; Rommelaere, J. Cytosolic activation of cathepsins mediates parvovirus H-1-induced killing of cisplatin and TRAIL-resistant glioma cells. *J. Virol.* **2007**, *81*, 4186–4198. [CrossRef]

25. El-Andaloussi, N.; Endele, M.; Leuchs, B.; Bonifati, S.; Kleinschmidt, J.; Rommelaere, J.; Marchini, A. Novel adenovirus-based helper system to support production of recombinant parvovirus. *Cancer Gene Ther.* **2011**, *18*, 240–249. [CrossRef]
26. El-Andaloussi, N.; Leuchs, B.; Bonifati, S.; Rommelaere, J.; Marchini, A. Efficient recombinant parvovirus production with the help of adenovirus-derived systems. *J. Vis. Exp.* **2012**, *62*, e3518. [CrossRef]
27. Leuchs, B.; Roscher, M.; Muller, M.; Kurschner, K.; Rommelaere, J. Standardized large-scale H-1PV production process with efficient quality and quantity monitoring. *J. Virol. Methods* **2016**, *229*, 48–59. [CrossRef]
28. Rodal, S.K.; Skretting, G.; Garred, Ø.; Vilhardt, F.; Van Deurs, B.; Sandvig, K. Extraction of cholesterol with methyl-β-cyclodextrin perturbs formation of clathrin-coated endocytic vesicles. *Mol. Biol. Cell* **1999**, *10*, 961–974. [CrossRef]
29. Anderson, H.; Chen, Y.; Norkin, L. Bound simian virus 40 translocates to caveolin-enriched membrane domains, and its entry is inhibited by drugs that selectively disrupt caveolae. *Mol. Biol. Cell* **1996**, *7*, 1825–1834. [CrossRef]
30. Locker, J.K.; Schmid, S.L. Integrated electron microscopy: Super-duper resolution. *PLoS Biol.* **2013**, *11*, e1001639.
31. Short, B. A cell-free screen of caveolae interactions. *J. Cell Biol.* **2018**, *217*, 1883. [CrossRef]
32. Heuser, J.E.; Anderson, R. Hypertonic media inhibit receptor-mediated endocytosis by blocking clathrin-coated pit formation. *J. Cell Biol.* **1989**, *108*, 389–400. [CrossRef]
33. Wang, L.-H.; Rothberg, K.G.; Anderson, R. Mis-assembly of clathrin lattices on endosomes reveals a regulatory switch for coated pit formation. *J. Cell Biol.* **1993**, *123*, 1107–1117. [CrossRef]
34. Von Kleist, L.; Stahlschmidt, W.; Bulut, H.; Gromova, K.; Puchkov, D.; Robertson, M.J.; MacGregor, K.A.; Tomilin, N.; Pechstein, A.; Chau, N. Role of the clathrin terminal domain in regulating coated pit dynamics revealed by small molecule inhibition. *Cell* **2011**, *146*, 471–484. [CrossRef] [PubMed]
35. Mayle, K.M.; Le, A.M.; Kamei, D.T. The intracellular trafficking pathway of transferrin. *Biochim. Biophys. Acta* **2012**, *1820*, 264–281. [CrossRef] [PubMed]
36. Kadlecova, Z.; Spielman, S.J.; Loerke, D.; Mohanakrishnan, A.; Reed, D.K.; Schmid, S.L. Regulation of clathrin-mediated endocytosis by hierarchical allosteric activation of AP2. *J. Cell Biol.* **2017**, *216*, 167–179. [CrossRef] [PubMed]
37. Kilsdonk, E.P.; Yancey, P.G.; Stoudt, G.W.; Bangerter, F.W.; Johnson, W.J.; Phillips, M.C.; Rothblat, G.H. Cellular cholesterol efflux mediated by cyclodextrins. *J. Biol. Chem.* **1995**, *270*, 17250–17256. [CrossRef] [PubMed]
38. Singh, M.; Jadhav, H.R.; Bhatt, T. Dynamin functions and ligands: Classical mechanisms behind. *Mol. Pharmacol.* **2017**, *91*, 123–134. [CrossRef]
39. Hill, T.A.; Gordon, C.P.; McGeachie, A.B.; Venn-Brown, B.; Odell, L.R.; Chau, N.; Quan, A.; Mariana, A.; Sakoff, J.A.; Chircop, M. Inhibition of dynamin mediated endocytosis by the *Dynoles*—Synthesis and functional activity of a family of indoles. *J. Med. Chem.* **2009**, *52*, 3762–3773. [CrossRef]
40. Robertson, M.J.; Deane, F.M.; Robinson, P.J.; McCluskey, A. Synthesis of Dynole 34-2, Dynole 2-24 and Dyngo 4a for investigating dynamin GTPase. *Nat. Protoc.* **2014**, *9*, 851–870. [CrossRef]
41. Jordens, I.; Marsman, M.; Kuijl, C.; Neefjes, J. Rab proteins, connecting transport and vesicle fusion. *Traffic* **2005**, *6*, 1070–1077. [CrossRef] [PubMed]
42. Wilson, J.M.; De Hoop, M.; Zorzi, N.; Toh, B.-H.; Dotti, C.G.; Parton, R.G. EEA1, a tethering protein of the early sorting endosome, shows a polarized distribution in hippocampal neurons, epithelial cells, and fibroblasts. *Mol. Biol. Cell* **2000**, *11*, 2657–2671. [CrossRef] [PubMed]
43. Eskelinen, E.-L.; Tanaka, Y.; Saftig, P. At the acidic edge: Emerging functions for lysosomal membrane proteins. *Trends Cell Biol.* **2003**, *13*, 137–145. [CrossRef]
44. Eskelinen, E.-L. Roles of LAMP-1 and LAMP-2 in lysosome biogenesis and autophagy. *Mol. Asp. Med.* **2006**, *27*, 495–502. [CrossRef]
45. Misinzo, G.; Delputte, P.L.; Nauwynck, H.J. Inhibition of endosome-lysosome system acidification enhances porcine circovirus 2 infection of porcine epithelial cells. *J. Virol.* **2008**, *82*, 1128–1135. [CrossRef]
46. Yoshimori, T.; Yamamoto, A.; Moriyama, Y.; Futai, M.; Tashiro, Y. Bafilomycin A1, a specific inhibitor of vacuolar-type $H^{(+)}$-ATPase, inhibits acidification and protein degradation in lysosomes of cultured cells. *J. Biol. Chem.* **1991**, *266*, 17707–17712.

47. Li, J.; Bonifati, S.; Hristov, G.; Marttila, T.; Valmary-Degano, S.; Stanzel, S.; Schnolzer, M.; Mougin, C.; Aprahamian, M.; Grekova, S.P.; et al. Synergistic combination of valproic acid and oncolytic parvovirus H-1PV as a potential therapy against cervical and pancreatic carcinomas. *EMBO Mol. Med.* **2013**, *5*, 1537–1555. [CrossRef]
48. Hueffer, K.; Palermo, L.M.; Parrish, C.R. Parvovirus infection of cells by using variants of the feline transferrin receptor altering clathrin-mediated endocytosis, membrane domain localization, and capsid-binding domains. *J. Virol.* **2004**, *78*, 5601–5611. [CrossRef]
49. Simmons, G.E., Jr.; Taylor, H.E.; Hildreth, J.E. Caveolin-1 suppresses Human Immunodeficiency virus-1 replication by inhibiting acetylation of NF-κB. *Virology* **2012**, *432*, 110–119. [CrossRef]
50. Lin, S.; Nadeau, P.E.; Wang, X.; Mergia, A. Caveolin-1 reduces HIV-1 infectivity by restoration of HIV Nef mediated impairment of cholesterol efflux by apoA-I. *Retrovirology* **2012**, *9*, 1–16. [CrossRef]
51. Bohm, K.; Sun, L.; Thakor, D.; Wirth, M. Caveolin-1 limits human influenza A virus (H1N1) propagation in mouse embryo-derived fibroblasts. *Virology* **2014**, *462*, 241–253. [CrossRef] [PubMed]
52. Swanson, J.A.; Watts, C. Macropinocytosis. *Trends Cell Biol.* **1995**, *5*, 424–428. [CrossRef]
53. Preta, G.; Cronin, J.G.; Sheldon, I.M. Dynasore-not just a dynamin inhibitor. *Cell Commun. Signal.* **2015**, *13*, 24. [CrossRef]
54. Zhang, Y.-N.; Liu, Y.-Y.; Xiao, F.-C.; Liu, C.-C.; Liang, X.-D.; Chen, J.; Zhou, J.; Baloch, A.S.; Kan, L.; Zhou, B. Rab5, Rab7, and Rab11 are required for Caveola-dependent endocytosis of classical swine fever virus in porcine alveolar macrophages. *J. Virol.* **2018**, *92*, e00797-18. [CrossRef] [PubMed]
55. Mani, B.; Baltzer, C.; Valle, N.; Almendral, J.M.; Kempf, C.; Ros, C. Low pH-dependent endosomal processing of the incoming parvovirus minute virus of mice virion leads to externalization of the VP1 N-terminal sequence (N-VP1), N-VP2 cleavage, and uncoating of the full-length genome. *J. Virol.* **2006**, *80*, 1015–1024. [CrossRef]
56. Suikkanen, S.; Sääjärvi, K.; Hirsimäki, J.; Välilehto, O.; Reunanen, H.; Vihinen-Ranta, M.; Vuento, M. Role of recycling endosomes and lysosomes in dynein-dependent entry of canine parvovirus. *J. Virol.* **2002**, *76*, 4401–4411. [CrossRef]
57. Zádori, Z.; Szelei, J.; Lacoste, M.-C.; Li, Y.; Gariépy, S.; Raymond, P.; Allaire, M.; Nabi, I.R.; Tijssen, P. A viral phospholipase A2 is required for parvovirus infectivity. *Dev. Cell* **2001**, *1*, 291–302. [CrossRef]
58. Farr, G.A.; Zhang, L.-g.; Tattersall, P. Parvoviral virions deploy a capsid-tethered lipolytic enzyme to breach the endosomal membrane during cell entry. *Proc. Natl. Acad. Sci. USA* **2005**, *102*, 17148–17153. [CrossRef]
59. Canaan, S.; Zádori, Z.; Ghomashchi, F.; Bollinger, J.; Sadilek, M.; Moreau, M.E.; Tijssen, P.; Gelb, M.H. Interfacial enzymology of parvovirus phospholipases A2. *J. Biol. Chem.* **2004**, *279*, 14502–14508. [CrossRef] [PubMed]
60. Vihinen-Ranta, M.; Wang, D.; Weichert, W.S.; Parrish, C.R. The VP1 N-terminal sequence of canine parvovirus affects nuclear transport of capsids and efficient cell infection. *J. Virol.* **2002**, *76*, 1884–1891. [CrossRef] [PubMed]
61. Vihinen-Ranta, M.; Kalela, A.; Mäkinen, P.; Kakkola, L.; Marjomäki, V.; Vuento, M. Intracellular route of canine parvovirus entry. *J. Virol.* **1998**, *72*, 802–806. [CrossRef] [PubMed]
62. Basak, S.; Turner, H. Infectious entry pathway for canine parvovirus. *Virology* **1992**, *186*, 368–376. [CrossRef]
63. Ros, C.; Burckhardt, C.J.; Kempf, C. Cytoplasmic trafficking of minute virus of mice: Low-pH requirement, routing to late endosomes, and proteasome interaction. *J. Virol.* **2002**, *76*, 12634–12645. [CrossRef] [PubMed]

Article

Binding of CCCTC-Binding Factor (CTCF) to the Minute Virus of Mice Genome Is Important for Proper Processing of Viral P4-Generated Pre-mRNAs

Maria Boftsi [1], Kinjal Majumder [2], Lisa R. Burger [2] and David J. Pintel [2,*]

1. Pathobiology Area Graduate Program, Christopher S. Bond Life Sciences Center, School of Medicine, University of Missouri-Columbia, Columbia, MO 65211, USA; boftsim@umsystem.edu
2. Christopher S. Bond Life Sciences Center, Department of Molecular Microbiology and Immunology, School of Medicine, University of Missouri-Columbia, Columbia, MO 65211, USA; kmajumder@wisc.edu (K.M.); burgerlr@missouri.edu (L.R.B.)

* Correspondence: pinteld@missouri.edu

Academic Editor: Giorgio Gallinella
Received: 23 October 2020; Accepted: 25 November 2020; Published: 30 November 2020

Abstract: Specific chromatin immunoprecipitation of salt-fractionated infected cell extracts has demonstrated that the CCCTC-binding factor (CTCF), a highly conserved, 11-zinc-finger DNA-binding protein with known roles in cellular and viral genome organization and gene expression, specifically binds the genome of Minute Virus of Mice (MVM). Mutations that diminish binding of CTCF to MVM affect processing of the P4-generated pre-mRNAs. These RNAs are spliced less efficiently to generate the R1 mRNA, and definition of the NS2-specific exon upstream of the small intron is reduced, leading to relatively less R2 and the generation of a novel exon-skipped product. These results suggest a model in which CTCF is required for proper engagement of the spliceosome at the MVM small intron and for the first steps of processing of the P4-generated pre-mRNA.

Keywords: parvovirus; minute virus of mice; RNA processing; gene expression

---

1. Introduction

Parvoviruses are small (20 nm) non-enveloped icosahedral viruses that infect and cause disease in many vertebrate hosts. They are unique among all known animal viruses in that they contain ~5 kb single-stranded linear DNA genomes, with inverted terminal repeats at both ends, which form hairpin structures and serve as origins of replication [1].

The viral genome is organized into two overlapping transcription units, producing three major transcript classes, R1, R2, and R3 [2]. Transcripts R1 (4.8 kb) and R2 (3.3 kb), generated from a promoter (P4) near the left-hand end of the genome, encode the two non-structural proteins NS1 (83 kDa) and NS2 (24 kDa), respectively [3]. Transcripts R3 (3.0 kb), generated from P38 promoter, encode the two viral capsid proteins, VP1 (83 kDa) and VP2 (64 kDa), utilizing the open reading frame (ORF) in the right half of the genome [4,5]. A small, overlapping intron, common to both P4- and P38-generated transcripts, utilizes two donors, D1 (nt 2280) and D2 (nt 2317) and two acceptors, A1 (nt 2377) and A2 (nt 2399), which are alternatively used to produce nine different spliced mRNA species [2,6–8]. A large upstream intron, present in the pre-mRNAs generated from the P4 promoter, is either retained (R1 transcript class) or excised (R2 transcript class) in mature mRNAs [9].

Upon initiation of parvovirus replication, MVM forms distinct foci in the nucleus termed autonomous parvovirus-associated replication (APAR) bodies where active transcription of viral genes and viral replication takes place [10]. The viral replicator protein NS1 co-localizes with the replicating viral genome in APAR bodies, where DDR sensor and response proteins, host replication factors, and cell cycle regulators also reside [10–12].

Using a novel adaptation of high-throughput chromosome conformation capture assay, V3C (Viral Chromosome Conformation), sites on the cellular genome where MVM localizes for replication have been mapped. These cellular regions preferentially accrue DNA damage in uninfected as well as MVM infected cells, and are also constituent parts of chromosomal substructures called Topologically Associating Domains (TADs) [13,14]. These are large, megabase-sized genomic regions, which are defined by preferential interactions within them and thus are relatively insulated from neighboring regions [15,16]. The boundaries of TADs are enriched for binding sites of CCCTC-binding factor (CTCF), a highly conserved, 11-zinc-finger DNA-binding protein, which along with cohesin, play a key role in the formation and maintenance of topological domains [15,17,18]. In addition to its role in genome organization, CTCF regulates key aspects of gene expression, including transcriptional activation/repression, and enhancer/promoter insulation, by facilitating long-range chromatin interactions via looping [19,20]. Apart from its DNA-binding activity, it was reported that CTCF can bind RNA and that CTCF-RNA interactions can participate in CTCF-mediated chromatin loop formation and subsequent regulation of gene expression [21,22]. Emerging evidence suggests that CTCF also regulates gene expression at the level of mRNA splicing. More specifically, CTCF has been shown to promote inclusion of weak upstream exons in the mRNA of CD45 gene by mediating local RNA polymerase II pausing [23]. Moreover, a more recent study showed that CTCF-mediated intragenic chromatin looping facilitates inclusion of exons in spliced mRNA by bringing exons in physical proximity, providing a functional link between chromatin organization and regulation of splicing [24].

It has become clear that a number of viruses, including Kaposi's sarcoma-associated herpesvirus (KHSV), Epstein-Barr virus (EBV), and human cytomegalovirus (HCMV), utilize CTCF to control viral gene expression [25–27]. It was demonstrated that CTCF associates with several regions within the KHSV genome, and that the CTCF-cohesin protein complex regulates the cell cycle control of viral gene expression during latency [25]. In a later study, it was also shown that CTCF and cohesin play important roles in regulating KHSV reactivation from latency by modulating viral gene transcription [28]. CTCF binding on EBV genome was shown to negatively affect transcription [26] and in the case of HCMV, binding of CTCF to the first intron of the Major Immediate Early (MIE) gene repressed MIE gene expression [27]. In addition, CTCF recruitment to the small DNA genome of human papillomavirus (HPV) was shown to regulate viral gene expression and transcript processing [29].

In this report, we show that CTCF can play an important role in parvovirus gene expression. Mutations that diminish binding of CTCF to the MVM genome affect processing of P4-generated pre-mRNAs; R1 is spliced less efficiently, and definition of the NS2-specific exon upstream of the small intron is reduced, leading to relatively less R2 and the generation of a novel exon-skipped product. These results implicate a requirement for CTCF in engagement of the spliceosome at the MVM small intron and the first steps of processing of the P4-generated pre-mRNA.

2. Materials and Methods

2.1. Cell Lines and Viruses

Murine A9 and human NB324K cells were maintained in Dulbecco's modified Eagle's medium (DMEM) supplemented with 5% fetal bovine serum (FBS) and incubated at 37 °C with 5% CO_2. Wild-type MVMp for infections was produced in A9 cells as previously described [11].

2.2. Transfections and Viral Infections

For transfections, cells were grown on 60-mm tissue culture dishes until they reached ~80% confluency. Cells were transfected with plasmids using LipoD293 transfection reagent (SignaGen Laboratories, Baltimore, MD, USA) according to the manufacturer's instructions. For RNA isolation, cells were co-transfected with wild-type or mutant plasmids (2 μg) and 3xFLAG-eGFP (p3xFeGFP) expression vector (0.5 μg), and harvested at 48 h post transfection (hpt). For Chromatin

Immunoprecipitation (ChIP) assays, cells were transfected with 2 μg of wilt-type or mutant plasmids and harvested at 20 hpt. Viral infections were carried out at a Multiplicity of Infection (MOI) of 10 unless otherwise stated and infected cells were harvested at the indicated timepoints.

2.3. Cell Synchronization

For infection experiments, A9 cells were parasynchronized in G0 phase by isoleucine deprivation for 36–42 h prior to infection as previously described [11]. Following synchronization, cells were released into complete media containing 5% FBS and infected with MVMp.

2.4. Plasmids

The infectious plasmid clone of MVM (pWT), which expresses the full length viral genome, was previously described [30]. gBlocks Gene Fragments of MVM, gbNSc and gbVPc, containing the mutated CTCF binding sites at the NS and VP region, respectively, were synthesized by Integrated DNA Technologies (IDT, Coralville, IA, USA). pNSc plasmid was constructed by replacing the XcmI-BsrGI fragment of pWT (nt 644–1253) with the gbNSc gene block, so that MVMp with a mutated CTCF binding site at the NS region was expressed. Similarly, the pVPc plasmid was constructed by replacing the PmlI-XbaI fragment of pWT (3636–4347) with the gbVPc gene block, so that MVMp with a mutated CTCF binding site at the VP region was expressed. In order to make the double CTCF-binding site mutant plasmid, pDc, in which both CTCF binding sites were mutated, the PmlI-XbaI fragment in pNSc was replaced with the gene block gbVPc. To generate the marker rescue of the pNSc plasmid, pNScMR, the XcmI-BsrGI fragment in pNSc was replaced with the XcmI-BsrGI fragment from pWt. The marker rescue of the pDc plasmid, pDcMR, was constructed by replacing both the XcmI-BsrGI and PmlI-XbaI fragments in pDc with the corresponding fragments from pWT. The p4Tppt plasmid, with improved polypyrimidine tract at the large intron 3′ splice site, was previously described [31]. To generate the double CTCF-binding site mutant construct with improved polypyrimidine tract (pDc4Tppt), the BsrGI-XhoI fragment (nt 1248–2075) from p4Tppt was cloned into the pDc plasmid between the BsrGI and XhoI sites. The pD3 plasmid was constructed by replacing both the XcmI-BsrGI and PmlI-XbaI fragments in pWT with the gene blocks gbNS3 and gbVP3, respectively.

2.5. Extraction of MVMp Nucleoprotein Complexes

MVMp nucleoprotein complexes were isolated from infected cells as previously described with modifications [32]. At the indicated timepoints, cells were washed with phosphate-buffered saline (PBS), harvested into HBE buffer (10 mM HEPES, 5 mM KCl, 1 mM EDTA), and collected by centrifugation at 1000× *g* for 3 min. Cell pellets were resuspended in 500 μL HBE buffer and lysed on ice for 10 min by addition of 1% NP-40 (to a final concentration of 0.1%). To pellet the nuclei, the lysate was centrifuged for 5 min at 1000× *g*. The supernatant (cytoplasmic extract) was transferred to a clean tube and the nuclei was resuspended in 500 μL buffer HBE. Sodium chloride (NaCl) was added to the suspension to a final concentration of 100, 200, or 400 mM and incubated on ice for 2 h. The remaining chromatin (chromatin pellet) was pelleted at 10,000× *g* for 10 min while the supernatant contained the MVMp nucleoprotein complexes (salt-wash extract).

2.6. Total RNA Isolation

Total RNA was extracted from transfected or infected cells as previously described with minor modifications [33]. Briefly, for total RNA isolation, cells were lysed in TRIzol reagent (Invitrogen, Carlsbad CA, USA) and RNA was prepared according to the manufacturer's protocol.

2.7. RNase Protection (RPA) Assay

Total RNA was extracted from transfected or infected cells using TRIzol reagent (Invitrogen) according to the manufacturer's protocol and RNase protection assays were performed on 25 μg RNA as previously described [34]. The probes used for the RPAs were α-^{32}P-UTP-labeled Sp6-generated antisense RNAs. The MVM HaeIII probe, extended from before the acceptor site of the large intron (nt 1852) to within the small intron (nt 2378), was used to analyze all MVM pre-mRNAs generated during wild-type MVMp infection. The HaeIII fragment (nt 1852–2378), cloned into a pGEM-3Z cloning vector between the XbaI and SphI restriction sites, was used as a template for the preparation of the HaeIII probe. Appropriate homologous probe (HaeIII 4Tppt) was used to analyze the RNA species generated from the Dc4Tppt mutant. The MVM P4 probe (spanning nt 201 to 652) was produced to analyze the P4-generated RNA products. The MVM 201–652 fragment was cloned into the pGEM-3Z vector between the BamHI and HindIII restriction sites and it was used as a template for the synthesis of the P4 probe.

To make the 3xFeGFP antisense RNA probe, the 3xFeGFP fragment with a SP6 promoter sequence at the 3′ end was amplified from the 3xFeGFP expression vector (p3xFeGFP) by PCR with primers 5′ ATC ATG CGG CCG CCG TCA GAA TTA ACC ATG GAC TAC AAA GAC 3′ and 5′ CTA TAT TTA GGT GAC ACT ATA GTT AAT TTT ATT AGG ACA AGG CTG GTG 3′.

2.8. Northern Blotting

For Northern blot analysis, 10 µg of total RNA, prepared as described above, was resolved on a formaldehyde—1.4% agarose gel at 35 mA for 24 h. After staining with ethidium bromide for 30 min, the gel was washed in DEPC-treated water for 4 h and transferred to a nitrocellulose membrane overnight. Blots were baked for 2 h at 80 °C and hybridized with randomly primed radiolabeled MVM probes. A HaeIII probe (nt 1852–2378) was used to detect all full-length viral mRNAs and a whole genome probe (Bam) was used to specifically detect the exon-skipped product generated from the double CTCF-binding site mutant construct.

2.9. Chromatin Immunoprecipitation (ChIP) Assay in Whole Cell Lysates

ChIP assays were conducted on parasynchronized murine A9 cells infected with MVMp at an MOI of 10 or human NB324K cells transfected with the wild-type or the CTCF-binding site mutant constructs as described previously [13]. Briefly, cells were cross-linked by addition of 1% formaldehyde directly to the culture media and incubated with shaking at room temperature for 10 min. The reaction was quenched with 0.125 M glycine for 5 min and cells were collected and lysed for 20 min on ice in ChIP lysis buffer (1% SDS, 10 mM EDTA, 50 mM Tris-HCl pH 8.0, protease inhibitors). Cell lysates were sonicated with a Diagenode Bioruptor for 75 cycles (30 s on and 30 s off) and debris was pelleted by centrifugation (8000× *g*, 15 min, 4 °C). The supernatant was then added to the indicated antibody-bound Protein A Dynabeads (Invitrogen) and samples were incubated overnight with rotation at 4 °C in ChIP dilution buffer (0.01% SDS, 1.1% Triton X-100, 1.2 mM EDTA, 16.7 mM Tris-HCl pH 8.0, 167 mM NaCl). The next day, the following washes were performed (3 min each at 4 °C with rotation): once in low salt wash (0.01% SDS, 1% Triton X-100, 2 mM EDTA, 20 mM Tris-HCl pH 8.0, 150 mM NaCl), once in high salt wash (0.01% SDS, 1% Triton X-100, 2 mM EDTA, 20 mM Tris-HCl pH 8.0, 500 mM NaCl), once in lithium chloride (LiCl) wash (0.25 M LiCl, 1% NP40, 1% DOC, 1 mM EDTA, 10 mM Tris-HCl pH 8.0) and twice in TE buffer, followed by elution in SDS elution buffer (1% SDS, 0.1 M Sodium bicarbonate). The DNA-antibody complexes and input DNA were reverse cross-linked overnight at 65 °C in the presence of NaCl and proteinase K. The DNA was purified using a PCR purification kit (Qiagen, Baltimore, MD, USA) and analyzed by quantitative PCR (qPCR) with iTaq universal SYBR green master mix (Bio-Rad) and primers 5′ CGC CTT CGG ACG TCA CAC GTC 3′ (MVM nt 60–80) and 5′ CCA GCC ATG GTT AGT TGG TTA C 3′ (MVM nt 268–247). Data are presented as percent input, calculated as described previously, ref. [35] or relative to IgG.

2.10. Chromatin Immunoprecipitation (ChIP) Assay on Viral Nucleoprotein Complexes

Following the extraction of MVMp nucleoprotein complexes as described above, the salt-wash extract was cross-linked with 0.1% formaldehyde for 5 min at room temperature and the reaction was quenched with 0.125 M glycine. The sample was then loaded onto an Amicon Ultra-0.5 Centrifugal Filter Device placed in filtrate collection tube and centrifuged for 30 s at 10,000× *g* to remove salt. PBS was added to the remaining sample and centrifuged for 30 s at 10,000× *g* to exchange buffer and concentrate. The purified sample was recovered from the Amicon filter by reverse spin (1000× *g*, 2 min). The viral genome-protein complexes were incubated with the indicated antibodies bound to Dynabeads Protein A (Invitrogen) and the ChIP assay performed as described above.

2.11. Immunoblot Analysis

Infected cells were harvested at the indicated timepoints, lysed in 1× dye (25 mM Tris pH 7.5, 2% SDS, 2 mM EDTA, 6% glycerol, 20 mM DTT, bromophenol blue) and sheared using a 25 G × 5/8-inch, 1-mL needle-syringe (BD Biosciences San Jose, CA, USA). The whole-cell lysates were boiled for 10 min at a 100 °C-heat block and equal volumes of samples were loaded per well for Western blot analysis. For Western blot analysis of the salt-wash extracts, 1× dye was added directly to the samples and processed as described above. Chromatin pellet, prepared during the salt-wash extraction procedure, was resuspended in 1× dye, sheared, and processed as described above.

2.12. Southern Blot Analysis

Infected cells were harvested at the indicated timepoints, pelleted and resuspended in Southern lysis buffer (2% SDS, 150 mM NaCl, 10 mM Tris pH 8.0, 1 mM EDTA). Cells were proteinase K treated for 2 h at 37 °C, and sheared using 25 G × 5/8-inch, 1-mL needle-syringe (BD Biosciences). Total DNA content in the samples was quantified using Nanodrop, equal amount of DNA loaded per well and electrophoresed on a 1% agarose gel for 16 h at 35 V. Samples were transferred to a nitrocellulose membrane and hybridized with randomly primed radiolabeled MVM probe (Bam) or genomic DNA probe (SINE). For Southern blot analysis of the chromatin pellet and salt-wash extracts, samples were resuspended in Southern lysis buffer and processed as described above.

2.13. Reverse Transcription-Polymerase Chain Reaction (RT-PCR) and TA Cloning

Total RNA was extracted using TRIzol reagent (Invitrogen) from cells transfected with the wild-type (pWT) and the double CTCF mutant (pDc) vectors, respectively, and subjected to DNase I (Thermo Fisher Scientific, Waltham MA, USA) treatment for 1 h at 37 °C to remove genomic DNA contamination. First-strand cDNA synthesis was performed on 1 μg DNase I-treated RNA using SMART®MMLV Reverse Transcriptase (Clontech, Mountain View, CA, USA) according to the manufacturer's instructions with primer 5′ GTT TTT TTT TAG CTC TGG CTT GG 3′ (MVM 2758–2736). The cDNA product was used for downstream PCR amplification using Platinum™ Taq DNA Polymerase High Fidelity (Invitrogen) with primers 5′ GTA TTG ATC ATA GGC CTC GTC G 3′ (MVM 2514–2493) and 5′ GTA ACC AGG AAG TGT TCT CAT TTG 3′ (MVM 322–345). The PCR products were analyzed by agarose gel electrophoresis and individual bands were extracted from the gel for downstream analysis using the QIAquick Gel Extraction kit (Qiagen). The small product, generated from the double CTCF-binding site mutant construct, was cloned into a PCR®2.1 vector using the TA Cloning® Kit (Invitrogen) according to the manufacturer's protocol. The construct was transformed into competent *Escherichia coli* DH5α cells and a number of individual clones were analyzed by Sanger Sequencing. The large product, generated from both the wild-type and the mutant construct, was submitted directly for sequencing analysis.

2.14. Immunofluorescence Assay

Immunofluorescence assays were performed in human NB324K cells infected with MVMp at an MOI of 10. At 24 h post infection (hpi), cells were harvested and processed as previously described [13]. Samples were incubated with the indicated antibodies for 1 h followed by the Alexa Fluor® conjugated secondary antibodies 488 and 568 for 1 h. Samples were mounted on slides with ProLong Diamond

Artifade Mountant with DAPI (Invitrogen) and images were acquired using a Leica TCP SP8 confocal microscope and a 10 × 1.4 NA objective lens.

3. Results

3.1. CTCF Specifically Binds the Viral Genome and Localizes to MVM Replication Compartments

We previously showed that MVM replicates in close association with sites on the cellular genome, taking advantage of the fact that these sites are replete with factors involved in gene expression and DNA damage signaling [13]. Therefore, in order to identify and characterize factors that specifically bound the MVM genome during replication, we developed a nuclear salt-wash extraction protocol, which could effectively separate the replicating viral genome from cellular DNA prior to cross-linking. This protocol, based on previous strategies designed to purify soluble nuclear protein complexes of MVM [32], was here further optimized across both time and salt gradients. Two-hour incubations proved best, and as can be seen in Figure 1A, MVM replicative forms were efficiently extracted beginning at approximately 200 mM NaCl. Both histone H3 and γ-H2AX, typically associated with cellular DNA [36], were used to monitor the purity of fractionation following extraction. As can be seen in Figure 1B, γ-H2AX appeared in the 400 mM salt-wash (the γ-H2AX band in the 100 mM salt-wash was not reproducible, and likely was an overflow from the adjacent lane), and so subsequent experiments were performed using extraction conditions of 200 mM NaCl for 2 h.

In silico inspection of the MVM genome suggested a potential interaction with the multifunctional cellular DNA-binding zinc finger protein CTCF, and thus this potential interaction was investigated by ChIP assays of salt-wash extracts of MVM infected cells [37]. First, the purity of the salt-wash extraction as assayed by ChIP was confirmed. While γ-H2Ax was found to associate strongly with MVM by ChIP following cross-linking within total cell extracts (Figure 1C, left panel), γ-H2Ax did not bind the MVM genome significantly over background when cross-linking and ChIP assays were performed in optimized salt-wash extracts (Figure 1C, center panel). These results were consistent with the results of the Western blot analysis shown above in Figure 1B, and further highlighted the importance of separating viral genomes from cellular DNA prior to attempts to identify specific viral binding factors. Following separation, ChIP assays demonstrated strong and specific CTCF binding to the MVM viral genome (Figure 1C; Rad 21, another cellular chromosome binding factor [38] was used as a negative control). Consistent with the ChIP results, we found that, while CTCF displayed a punctate pattern and was found throughout the nucleus of cells at both time points shown, NS1 co-localized with CTCF in both early- and late-stage APAR bodies (Figure 1D, middle and bottom panel respectively). MVM has two potential CTCF binding sites in its genome, one within the NS1 gene and one in the capsid gene (Figure 1E). Inspection of the autonomous parvoviruses H1 and Minute Virus of Canine (MVC), revealed potential CTCF binding sites in the same relative position as MVM, and the dependovirus AAV has a CTCF binding site within its Rep gene (Figure 1E).

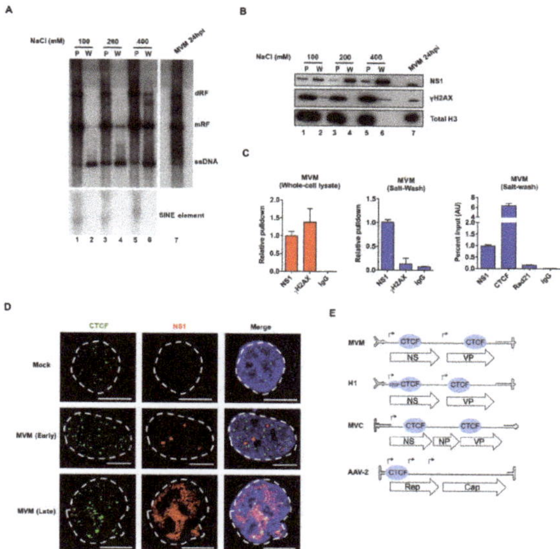

Figure 1. CTCF is associated with the viral genome and localizes to MVM replication compartments. (**A**) Murine A9 cells were parasynchronized by isoleucine deprivation and infected with MVMp at an MOI of 10 at the time of release into complete medium. At 24 hpi, viral nucleoprotein complexes were extracted from infected cells with various NaCl concentrations, and both the chromatin pellet (P) and the salt-wash extracts (W) were subsequently analyzed by Southern blotting as described in Materials and Methods. DNA extracted directly from infected cells served as a positive control (lane 7). The blot was hybridized with a radiolabeled MVM whole genome probe (top panel) and replicative intermediates of single-stranded DNA, ssDNA; monomer, mRF; and dimer, dRF; are indicated to the right. The blot was also hybridized with a genomic DNA probe against the SINE element (bottom panel). (**B**) Salt wash extracts and chromatin pellet described in (**A**) were assayed by Western blotting using antibodies directed against the indicated proteins. Whole-cell lysates of MVM-infected parasynchronized murine A9 cells were also analyzed by Western blotting with the indicated antibodies and served as a positive control (lane 7). (**C**) Murine A9 cells were parasynchronized by isoleucine deprivation and infected with MVMp at an MOI of 10 at the time of release into complete medium. At 16 hpi, cells were processed, as described in Materials and Methods for whole-cell lysate ChIP (left panel) or ChIP on salt wash extracts (middle and right panel) with the indicated antibodies. Samples were analyzed by qPCR as described in Materials and Methods. Data are presented as mean ± standard error of the means (SEM) of two individual experiments. Background binding levels were determined using mouse IgG pulldowns. (**D**) Representative confocal images of Mock versus MVM infected, non-synchronized human NB324K cells at 24 hpi, probing MVM-NS1 (red) and the host cellular factor CTCF (green). CTCF co-localized with NS1 in both early and late stage APAR bodies designated as previously described [12] (middle and bottom panel respectively). Blue corresponds to DAPI staining. Nuclear border is indicated by dashed white line. (**E**) Schematic representation of the protoparvoviruses MVM, MVC, H1, and the dependovirus AAV2 genome showing the positions of transcriptional promoters (solid black arrows), the major open reading frames that encode the viral non-structural and capsid proteins (arrowed boxes), and the relative positions of CTCF binding sites (blue oval shapes).

The sequences of the consensus CTCF binding sites in the MVM genome (RefSeq: NC_001510.1) are shown (Figure 2A). Interestingly, the consensus signals lie on opposite strands of the double stranded transcription template: the consensus NS motif lies in 5′-3′ polarity on the virus minus strand, while the VP motif lies 5′-3′ on the plus strand. To confirm that CTCF bound to the genome at these sites on the double strand replicative form, a series of mutations were made, and these were used as targets for CTCF ChIP experiments. As we could not reproducibly shear the replicating MVM genome during the ChIP procedure, this step was omitted, and so ChIP pull-downs revealed binding to the complete MVM genome. Originally, we attempted to mutate the CTCF binding sites by third nucleotide substitutions, which left the amino acid sequences unchanged; however, these mutations only partially prevented CTCF binding, and so more complete mutations were introduced. These severe mutations of both sites together led to significant loss of CTCF binding over background (Dc; Figure 2A,B). Mutation of the NS site alone reduced CTCF to nearly Dc levels (NSc; Figure 2A,B), while mutation of the VP site retained intermediate binding (VPc; Figure 2A,B). Binding in the single mutants was likely due to binding at the remaining unaltered site, which suggested that CTCF could bind independently to either site, and that binding to the NS site appeared stronger. Unfortunately, mutations needed to prevent CTCF binding destroyed the NS1 open reading frame precluding assessment of their replication. Prior to the further analyses described below, all mutants were marker rescued with wild-type MVM sequences as described in the Materials and Methods to ensure no additional mutations were present.

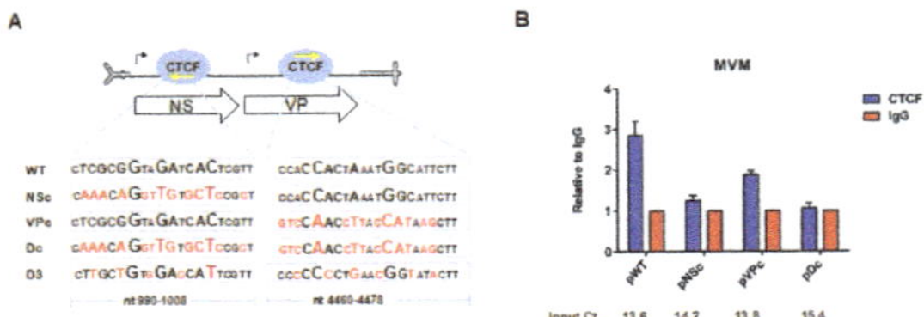

Figure 2. CTCF binding to the single DNA-binding site mutants is reduced, while it is almost completely abolished on the double mutant (**A**) schematic representation of MVM genome showing the positions of transcriptional promoters (solid black arrows), the major open reading frames (arrowed boxes), the relative positions of CTCF binding sites (blue oval shapes) and the nucleotide sequences of the WT and the mutant CTCF binding sites in the NS and VP region. The nucleotides that were mutated within the CTCF binding motifs are shown in red. The yellow arrows show the orientation of the CTCF motifs on the viral genome. (**B**) Human NB324K cells were transfected with the WT or the indicated CTCF-binding site mutants, and harvested at 20 hpt as described in Materials and Methods for whole-cell lysate ChIP using antibodies directed against the cellular factor CTCF or the mouse IgG protein. Samples were analyzed by qPCR as described in Materials and Methods and presented relative to IgG isotype control. Data are presented as mean ± SEM of two individual experiments. Ct values (average of the two experiments) for each plasmid transfection are shown and indicate similar levels of target DNA in the samples.

3.2. *CTCF-Binding Site Mutants Exhibited a Decrease in Levels of Spliced to Unspliced R1, as Well as Reduced Levels of R2 Relative to R1*

Following transfection of human NB324K cells, both the Dc double mutant, and the NSc single mutant, were found to generate significantly reduced levels of spliced R1 relative to unspliced R1 RNAs, and reduced levels of R2 relative to R1, as assayed by RNase protection assays [Figure 3B, lanes 4 and 6, respectively (ratios represent an average of two independent experiments)], using the HaeIII probe, which spans the small intron (Figure 3A). As the mutations in the NS region changed the amino acid sequence of NS1, R3 was not generated by either of these mutants. Control transfections of an eGFP expressing plasmid confirmed similar levels of transfection efficiency in these experiments (Figure 3B, bottom panel). Mutants reducing binding within the VP region site alone (VPc), in which CTCF binding to the NS region remained, showed a decrease in splicing to R1, but essentially wild-type patterns of R2 expression (Figure 3B compare lanes 5 with lane 3 and 2). A similar phenotype for the three mutants was also observed following transfection of murine A9 cells. Together, these results indicated that the phenotype of the Dc mutant was primarily due to the mutation in the NS region, and even though CTCF bound at both the NS and VP sites (Figure 2), the individual mutations exhibited different effects. While both mutants exhibited decreased splicing of R1 from the

P4-generated pre-mRNA, only the NSc mutation affected subsequent appearance or R2. It is important to note that the NSc mutations fell outside of the affected R2 RNA itself, and did not lie close to any known RNA regulatory element. Additionally, mutation of multiple nucleotides within the NS and VP motifs that did not efficiently disrupt CTCF binding (D3, diagrammed in Figure 2A) had no deleterious effect on RNA processing (Figure 3B, lanes 7, 8). RNase protection assays with a P4 probe, which specifically detects the P4 promoter-generated R1 and R2 transcripts individually, also showed an increased ratio of R2 relative to R1, while the total P4 products were similar for the two. Splicing of R1 pre-mRNA depends upon engagement of the spliceosome at the small intron [30], which was also necessary for exon definition of the upstream NS2-specific exon required for splicing of the large intron and generation of R2 [39]. Thus, our results suggested that CTCF engagement of its MVM binding sites may play a role in processes functioning at the small intron.

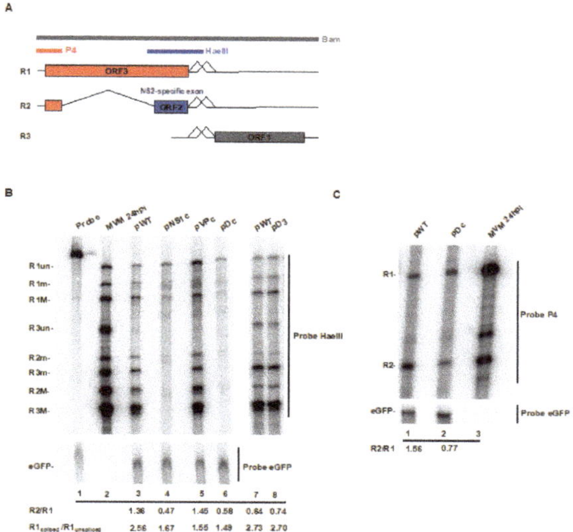

Figure 3. CTCF-binding site mutants exhibit a decrease in levels of spliced to un-spliced R1, and levels of R2 relative to R1. (**A**) Genetic map of MVM showing the three major transcript classes (R1, R2, and R3), the open reading frames that encode the two non-structural (ORF2 and ORF3) and the capsid (ORF1) proteins, and the relative positions of the small and large introns. Approximate locations of the RNase protection viral probes (Bam, P4 and HaeIII) are also indicated. (**B**) Human NB324K cells were infected with MVMp at an MOI of 10 or co-transfected with the indicated plasmids and eGFP. Infected cells were harvested at 24 hpi while transfected cells were harvested at 48 hpt and total RNA was isolated using TRIzol reagent. Samples were processed for RNase protection assay (RPA) using a HaeIII (that detects all viral transcripts) or an eGFP probe, as described in Materials and Methods. The protected bands representing all nine viral mRNA species are indicated to the left. The ratios of total R2 to total R1, and spliced to unspliced R1 are indicated at the bottom of the panel. Ratios represent an average of two independent experiments. (**C**) RPA of total RNA extracted from NB324K cells 24 h post infection or 48 h post transfection with the indicated plasmids using the P4 viral probe, detecting specifically the P4-generated transcripts, or the eGFP probe. The identities of the protected bands are shown on the left. The ratio of R2 to R1 is indicated at the bottom of the panel. Ratios represent an average of two independent experiments.

3.3. CTCF-Binding Site Mutants Resulted in Skipping of the NS2-Specific Exon and Joining of the Large Intron Donor to the Small Intron Acceptors

Northern blot analysis of RNA generated in NB324K cells by the double CTCF-binding site mutant Dc revealed a transcript, approximately the size of R3, that hybridized with a whole-genome probe

(Figure 4A, lane 2). This was surprising since this mutant, which does not produce wild-type NS1, did not generate the R3 mRNA, as was demonstrated in Figure 3B. A mutant containing a translation termination signal immediately downstream of the NS1 AUG is shown for comparison (Figure 4A, lane 3). Interestingly, the R3-size RNA generated by Dc was not detected in Northern blots using the HaeIII probe, which covers the NS2-specific exon (Figure 4A, lane 5). These results confirmed that this band was not R3, and suggested that the approximate 3 kb size RNA generated by Dc might have been an RNA product spliced at the large intron donor (nt 514) that was joined to a small intron acceptor (Although the large amount of transfected plasmid DNA in these samples makes R1 poorly visible on these gels, it was clearly apparent on the RNase protection gels of these RNAs shown in Figure 3).

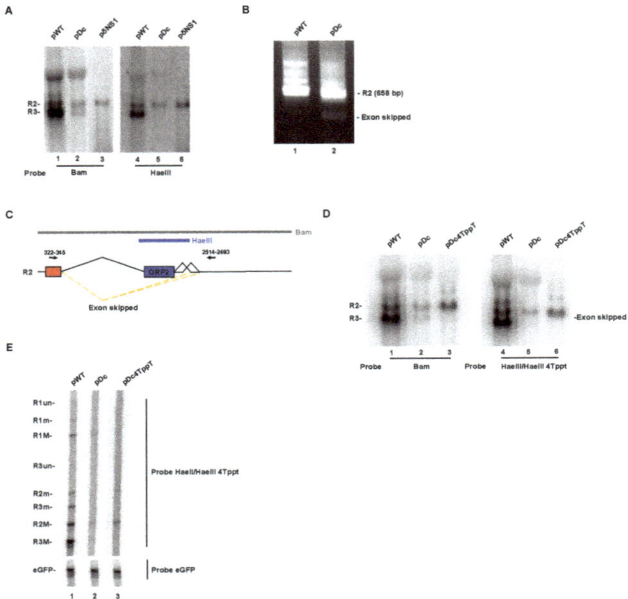

Figure 4. CTCF-binding site mutants resulted in skipping of the NS2-specific exon and joining of the large intron donor to the small intron acceptors, which can be overcome by mutations that improve the polypyrimidine tract of the upstream large intron. (**A**) Human NB324K cells, transfected with the indicated plasmids, were harvested at 48 hpt and total RNA was isolated using TRIzol reagent. Samples were processed for Northern blot analysis using the probes indicated at the bottom of the panel. The identity of the viral RNA species is shown on the left. (**B**) Total RNA extracted from NB324K cells, transfected with the WT or the double CTCF-binding site mutant, was subjected to DNase I treatment for 1 h at 37 °C. First-strand cDNA synthesis was performed on the DNase I-treated RNA samples and the cDNA product was used as a template for downstream PCR analysis as described in Materials and Methods. The identities of the amplified bands were determined by sequencing analysis and are shown to the right. (**C**) Schematic representation of R2 (top) and the exon-kipped product (bottom), generated from the double CTCF-binding site mutant construct. The exon-skipped product skips the NS2-specific exon (ORF2) and joins the large intron donor at nt 514 to the small intron acceptor A1 (nt 2377) or A2 (nt 2399). The PCR primers used to detect the exon-skipped product are indicated at the top of the panel (black arrows). The numbers on top of the arrows represent the location on the viral genome where the primers anneal. Approximate locations of the RNase protection viral probes (Bam and HaeIII) are also indicated. Total RNA extracted from NB324K cells, transfected with the indicated plasmids as well as eGFP, was subjected to Northern blot analysis (**D**) or RNase protection assay (**E**) as described in Materials and Methods using the probes shown at the bottom (**D**) or to the right (**E**) of the panel. The identity of the RNA species are also depicted.

To reveal whether such an RNA was in fact made by Dc, we performed non-quantitative RT-PCR analysis of Dc-generated RNA using primers shown in Figure 4C. As shown in Figure 4B, the Dc mutant did generate such a novel spliced product, which is diagrammed in Figure 4C. These cDNAs were cloned and sequence analysis revealed that these spliced products joined the large intron donor at nt 514 to either the small intron acceptor, A1, at nt 2377, or the small intron acceptor, A2, at nt 2399. Inspection of the more quantitative Northern results in Figure 4A suggests that this NS2-specific exon-skipped product was present at approximately half the concentration of R2.

3.4. Improvement of the Large Intron Splice Acceptor in the Dc Mutant Led to Increased NS2-Specific Exon Definition and Increased Levels of R2 RNA

If lack of CTCF binding to the MVM genome led to weakening of the large intron acceptor due to loss of definition, for splicing purposes, of the NS2 specific exon, we would expect that improving the large intron acceptor would overcome this deficiency. As can be seen in a Northern blot analysis using the whole genomic probe, strengthening the large intron acceptor polypyrimidine tract with the addition of 4 additional thymidine residues, previously shown to overcome mutations that reduced NS2-specific exon definition [39], led to both a decrease in the exon skipped product and an increase in authentic R2 generated by Dc (Figure 4D, compare lanes 2 and 3). Northern analysis of this RNA using the HaeIII probe confirmed the authenticity of the exon skipped product lost in the left panel of Figure 4 (Figure 4D, compare lane 2 to 5), and revealed enhanced levels of R2. An increase in R2 RNA generated by pDc4Tppt was confirmed by quantitative RNase protection analysis in which expression of an eGFP gene was included as a transfection control (Figure 4E, compare lane 2 to 3).

4. Discussion

In surveying the MVM genome for the binding sites of known cellular factors, we noticed consensus CTCF binding sites in the NS and the VP regions of MVM that were conserved in a number of other parvoviruses. Because we previously showed that the replicating MVM genome associates with particular sites of DNA damage on the cellular genome [13], determining whether CTCF specifically bound to MVM required that we separate the viral genome from the cellular genome prior to the cross-linking step during chromatin immunoprecipitation assays. Upon doing so, we could demonstrate specific binding of CTCF to these sites on MVM.

Full disruption of CTCF binding to MVM required destruction of both sites together; destruction of the NS site individually (which retained the VP binding site) reduced binding similarly to the double mutant, while destruction of the VP site (leaving the NS site) retained an intermediate binding phenotype. Because the mutations required to disrupt CTCF binding could not be made without disrupting the NS1 ORF, the mutants could not be assessed directly for replication. However, both the double CTCF binding site mutant Dc, and the NS-alone mutation NSc, showed a dramatic defect in gene expression. These mutants were both deficient in the splicing of the R1 RNA, and they generated relatively less R2 at the expense of a new product, an RNA that joined the large intron donor at nt 514 to one or the other of the small intron acceptors. As previously mentioned, the NSc mutations falls outside of the affected R2 RNA itself, and does not lie close to any known RNA regulatory element. Additionally, mutation of multiple nucleotides within the NS motif that did not efficiently disrupt CTCF binding had no deleterious effect on RNA processing (Figure 3), further implying that this region did not contain a previously unrecognized *cis*-acting RNA processing element. It is interesting that the VPc single mutant, although apparently not deficient in NS2-specific exon definition, still generated less relative spliced R1 RNA. This perhaps suggests that the role of CTCF binding at the individual sites, and their potential interaction, is complex, and may be related to their different orientations on the viral chromosome. Our preliminary results have shown that the exon-skipped RNA product does transit to the cytoplasm, but we could detect no protein product that it generates.

Interestingly, we observed the newly spliced exon-skipped RNA product before [39,40]. In previous studies that characterized splicing of the P4-generated pre-mRNA, we found that when the NS2-specific exon was poorly defined—either by virtue of its weak large intron acceptor [30], by certain mutations within the NS2-specific exon itself [40], or importantly, by mutation of the downstream small intron [30], an RNA was generated in which the NS2-specific exon was skipped. Because definition of the NS2-specific exon functions to strengthen the adjacent upstream large intron acceptor at nt 1989 [39], improvement, in those mutants, of the large intron acceptor by the addition of four thymidine residues in its polypyrimidine tract overcame the defect in NS2-specific exon definition [39]. These observations, as well as the absence in infected cells, of P4-generated RNAs lacking only the large intron but not the small intron [31], led us to propose a model (diagrammed in Figure 5) in which the spliceosome first engages the R1 pre-mRNA at the small intron, allowing its splicing as well as facilitating its interaction with the upstream large intron acceptor to define the NS2-specific exon allowing splicing of the large intron [39].

In the light of these previous results, the results presented here—that the Dc and NSc mutants exhibited reduced splicing of R1 and generated an exon-skipped product at the expense of R2, which could be suppressed by improvement of the large intron polypyrimidine track—suggested a model in which CTCF binding likely plays a role in proper engagement of the spliceosome at the small intron. In its absence, R1 would be poorly spliced, and the NS2-specific exon poorly defined, leading to the generation of the new exon-skipped product we observe. How interruption of interaction of the spliceosome at the small intron may affect our general model of P4-generated pre-mRNA processing is shown in Figure 5. Binding of CTCF to the site in the NS1 gene appears to play a more significant role in this effect than binding to the site in the capsid gene.

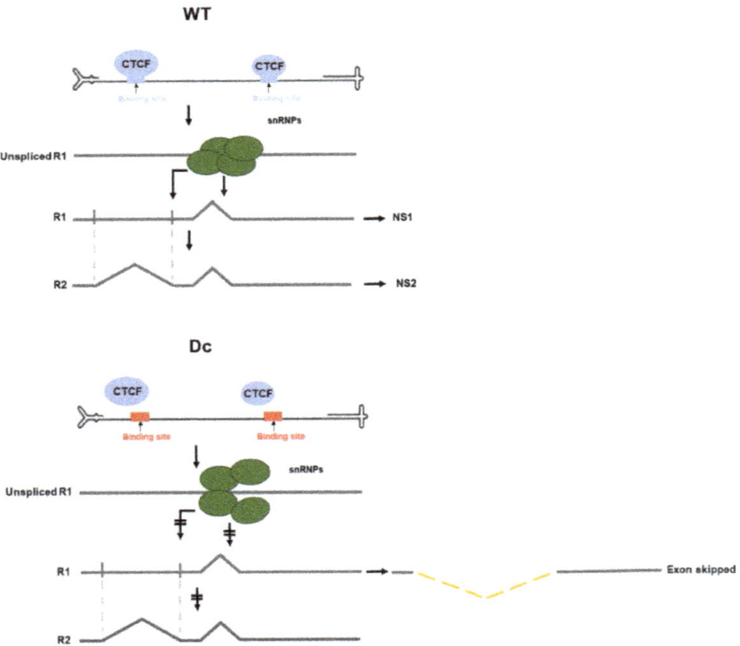

Figure 5. Model depicting a potential mechanism by which CTCF regulates splicing of MVM P4-generated transcripts. Top: A general model of processing of P4-generated pre-mRNAs; CTCF binding to the viral genome likely plays a role in proper engagement of the spliceosome at the small intron, allowing its splicing as well as facilitating its interaction with the upstream large intron acceptor, to define the NS2-specific exon allowing splicing of the large intron. Bottom: In the absence of CTCF binding, R1 would be poorly spliced, and the NS2-specific exon poorly defined, leading to the generation of a novel exon-skipped product.

How CTCF binding to MVM functions to play its role in MVM RNA processing is not yet known. CTCF has been shown to have a role in chromosomal architecture, specifically looping of DNA, as well as transcriptional activation, and has been shown to have RNA binding activity [19–22]. RNA immunoprecipitation experiments, done as we have previously described [41] did not demonstrate CTCF binding to MVM RNA. It was first reported that DNA-bound CTCF regulates alternative pre-mRNA splicing by mediating RNA Polymerase II pausing, allowing the inclusion of upstream weak exons [23]. A more recent report, however, suggested that CTCF regulation of alternative splicing of human papillomavirus early genes was more complicated. Specifically, it was found, similar to the results reported here, that loss of CTCF binding to the viral genome resulted in both increased levels of unspliced transcripts and an alteration of splice site usage upstream of the CTCF binding site, with a significant reduction of a specific alternatively spliced product [29]. Thus, it is possible that CTCF binding affects RNA processing through modulation of the elongating transcription complex. ChIP assays of RNA pol II on the MVM P4 promoter showed no reproducible difference between the Dc mutant and wildtype MVM. However, it is well known that RNA processing factors bind to the extending RNA polymerase at its CTD [42], and our assays would not have distinguished if the composition of the complexes engaging the Dc P4 promoter and wild-type P4 differed. Perhaps relevantly, we have previously shown that when the AAV2 P40 promoter and the AAV5 P5 promoter were replaced with either the HIV LTR or the CMV promoter, the RNA generated by these constructs was processed differently [43,44].

Importantly, it has been shown that the ratio of R2:R1 is exquisitely critical for successful MVM infection. Even small changes in this ratio can have large effects on replication [45]. Thus, the role of CTCF in controlling the ratio of R2 to R1 would be predicted to have significant effects on replication.

Author Contributions: Conceptualization, M.B., K.M. and D.J.P.; methodology, M.B., K.M. and L.R.B.; validation, M.B., K.M., L.R.B. and D.J.P.; resources, K.M. and D.J.P.; writing—original draft preparation, M.B.; writing—review and editing, M.B., K.M. and D.J.P.; funding acquisition, K.M. and D.J.P. All authors have read and agreed to the published version of the manuscript.

Funding: This work was supported by NIH grants R01AI 046458 and R01AI116595 to DJP. K.M. was supported by Ruth L. Kirschstein Postdoctoral Individual National Research Service Award AI 131468, and NIH K99/00 AI 148511.

Conflicts of Interest: The authors declare no conflict of interest. The funders had no role in the design of the study; in the collection, analyses, or interpretation of data; in the writing of the manuscript, or in the decision to publish the results.

References

1. Cotmore, S.F.; Tattersall, P. Parvoviruses: Small Does Not Mean Simple. *Annu. Rev. Virol.* **2014**, *1*, 517–537. [CrossRef] [PubMed]
2. Pintel, D.; Dadachanji, D.; Astell, C.R.; Ward, D.C. The genome of minute virus of mice, an autonomous parvovirus, encodes two overlapping transcription units. *Nucleic Acids Res.* **1983**, *11*, 1019–1038. [CrossRef] [PubMed]
3. Cotmore, S.F.; Tattersall, P. Organization of nonstructural genes of the autonomous parvovirus minute virus of mice. *J. Virol.* **1986**, *58*, 724–732. [CrossRef] [PubMed]
4. Labieniec-Pintel, L.; Pintel, D. The minute virus of mice P39 transcription unit can encode both capsid proteins. *J. Virol.* **1986**, *57*, 1163–1167. [CrossRef] [PubMed]
5. Clemens, K.E.; Cerutis, D.R.; Burger, L.R.; Yang, C.Q.; Pintel, D.J. Cloning of minute virus of mice cDNAs and preliminary analysis of individual viral proteins expressed in murine cells. *J. Virol.* **1990**, *64*, 3967–3973. [CrossRef] [PubMed]
6. Jongeneel, C.V.; Sahli, R.; McMaster, G.K.; Hirt, B. A precise map of splice junctions in the mRNAs of minute virus of mice, an autonomous parvovirus. *J. Virol.* **1986**, *59*, 564–573. [CrossRef]
7. Morgan, W.R.; Ward, D.C. Three splicing patterns are used to excise the small intron common to all minute virus of mice RNAs. *J. Virol.* **1986**, *60*, 1170–1174. [CrossRef]

8. Schoborg, R.V.; Pintel, D.J. Accumulation of MVM gene products is differentially regulated by transcription initiation, RNA processing and protein stability. *Virology* **1991**, *181*, 22–34. [CrossRef]
9. Pintel, D.J.; Gersappe, A.; Haut, D.; Pearson, J. Determinants that govern alternative splicing of parvovirus pre-mRNAs. *Semin. Virol.* **1995**, *6*, 283–290. [CrossRef]
10. Bashir, T.; Rommelaere, J.; Cziepluch, C. In Vivo Accumulation of Cyclin A and Cellular Replication Factors in Autonomous Parvovirus Minute Virus of Mice-Associated Replication Bodies. *J. Virol.* **2001**, *75*, 4394–4398. [CrossRef]
11. Adeyemi, R.O.; Landry, S.; Davis, M.E.; Weitzman, M.D.; Pintel, D.J. Parvovirus minute virus of mice induces a DNA damage response that facilitates viral replication. *PLoS Pathog.* **2010**, *6*, e1001141. [CrossRef] [PubMed]
12. Ruiz, Z.; Mihaylov, I.S.; Cotmore, S.F.; Tattersall, P. Recruitment of DNA replication and damage response proteins to viral replication centers during infection with NS2 mutants of Minute Virus of Mice (MVM). *Virology* **2011**, *410*, 375–384. [CrossRef] [PubMed]
13. Majumder, K.; Wang, J.; Boftsi, M.; Fuller, M.S.; Rede, J.E.; Joshi, T.; Pintel, D.J. Parvovirus minute virus of mice interacts with sites of cellular DNA damage to establish and amplify its lytic infection. *Elife* **2018**, *7*. [CrossRef]
14. Rao, S.S.; Huntley, M.H.; Durand, N.C.; Stamenova, E.K.; Bochkov, I.D.; Robinson, J.T.; Sanborn, A.L.; Machol, I.; Omer, A.D.; Lander, E.S.; et al. A 3D map of the human genome at kilobase resolution reveals principles of chromatin looping. *Cell* **2014**, *159*, 1665–1680. [CrossRef] [PubMed]
15. Dixon, J.R.; Selvaraj, S.; Yue, F.; Kim, A.; Li, Y.; Shen, Y.; Hu, M.; Liu, J.S.; Ren, B. Topological domains in mammalian genomes identified by analysis of chromatin interactions. *Nature* **2012**, *485*, 376–380. [CrossRef]
16. Nora, E.P.; Lajoie, B.R.; Schulz, E.G.; Giorgetti, L.; Okamoto, I.; Servant, N.; Piolot, T.; van Berkum, N.L.; Meisig, J.; Sedat, J.; et al. Spatial partitioning of the regulatory landscape of the X-inactivation centre. *Nature* **2012**, *485*, 381–385. [CrossRef]
17. Phillips-Cremins, J.E.; Sauria, M.E.; Sanyal, A.; Gerasimova, T.I.; Lajoie, B.R.; Bell, J.S.; Ong, C.T.; Hookway, T.A.; Guo, C.; Sun, Y.; et al. Architectural protein subclasses shape 3D organization of genomes during lineage commitment. *Cell* **2013**, *153*, 1281–1295. [CrossRef]
18. Zuin, J.; Dixon, J.R.; van der Reijden, M.I.; Ye, Z.; Kolovos, P.; Brouwer, R.W.; van de Corput, M.P.; van de Werken, H.J.; Knoch, T.A.; van, I.W.F.; et al. Cohesin and CTCF differentially affect chromatin architecture and gene expression in human cells. *Proc. Natl. Acad. Sci. USA* **2014**, *111*, 996–1001. [CrossRef]
19. Phillips, J.E.; Corces, V.G. CTCF: Master weaver of the genome. *Cell* **2009**, *137*, 1194–1211. [CrossRef]
20. Braccioli, L.; de Wit, E. CTCF: A Swiss-army knife for genome organization and transcription regulation. *Essays Biochem.* **2019**, *63*, 157–165. [CrossRef]
21. Saldana-Meyer, R.; Gonzalez-Buendia, E.; Guerrero, G.; Narendra, V.; Bonasio, R.; Recillas-Targa, F.; Reinberg, D. CTCF regulates the human p53 gene through direct interaction with its natural antisense transcript, Wrap53. *Genes Dev.* **2014**, *28*, 723–734. [CrossRef] [PubMed]
22. Saldana-Meyer, R.; Rodriguez-Hernaez, J.; Escobar, T.; Nishana, M.; Jacome-Lopez, K.; Nora, E.P.; Bruneau, B.G.; Tsirigos, A.; Furlan-Magaril, M.; Skok, J.; et al. RNA Interactions Are Essential for CTCF-Mediated Genome Organization. *Mol. Cell* **2019**, *76*, 412–422.e5. [CrossRef]
23. Shukla, S.; Kavak, E.; Gregory, M.; Imashimizu, M.; Shutinoski, B.; Kashlev, M.; Oberdoerffer, P.; Sandberg, R.; Oberdoerffer, S. CTCF-promoted RNA polymerase II pausing links DNA methylation to splicing. *Nature* **2011**, *479*, 74–79. [CrossRef] [PubMed]
24. Ruiz-Velasco, M.; Kumar, M.; Lai, M.C.; Bhat, P.; Solis-Pinson, A.B.; Reyes, A.; Kleinsorg, S.; Noh, K.M.; Gibson, T.J.; Zaugg, J.B. CTCF-Mediated Chromatin Loops between Promoter and Gene Body Regulate Alternative Splicing across Individuals. *Cell Syst.* **2017**, *5*, 628–637.e6. [CrossRef] [PubMed]
25. Kang, H.; Lieberman, P.M. Cell Cycle Control of Kaposi's Sarcoma-Associated Herpesvirus Latency Transcription by CTCF-Cohesin Interactions. *J. Virol.* **2009**, *83*, 6199–6210. [CrossRef] [PubMed]
26. Chau, C.M.; Zhang, X.-Y.; McMahon, S.B.; Lieberman, P.M. Regulation of Epstein-Barr virus latency type by the chromatin boundary factor CTCF. *J. Virol.* **2006**, *80*, 5723–5732. [CrossRef] [PubMed]
27. Martínez, F.P.; Cruz, R.; Lu, F.; Plasschaert, R.; Deng, Z.; Rivera-Molina, Y.A.; Bartolomei, M.S.; Lieberman, P.M.; Tang, Q. CTCF binding to the first intron of the major immediate early (MIE) gene of human cytomegalovirus (HCMV) negatively regulates MIE gene expression and HCMV replication. *J. Virol.* **2014**, *88*, 7389–7401. [CrossRef] [PubMed]

28. Li, D.-J.; Verma, D.; Mosbruger, T.; Swaminathan, S. CTCF and Rad21 act as host cell restriction factors for Kaposi's sarcoma-associated herpesvirus (KSHV) lytic replication by modulating viral gene transcription. *PLoS Pathog.* **2014**, *10*, e1003880. [CrossRef]
29. Paris, C.; Pentland, I.; Groves, I.; Roberts, D.C.; Powis, S.J.; Coleman, N.; Roberts, S.; Parish, J.L. CCCTC-binding factor recruitment to the early region of the human papillomavirus 18 genome regulates viral oncogene expression. *J. Virol.* **2015**, *89*, 4770–4785. [CrossRef]
30. Zhao, Q.; Schoborg, R.V.; Pintel, D.J. Alternative splicing of pre-mRNAs encoding the nonstructural proteins of minute virus of mice is facilitated by sequences within the downstream intron. *J. Virol.* **1994**, *68*, 2849–2859. [CrossRef]
31. Naeger, L.K.; Cater, J.; Pintel, D.J. The small nonstructural protein (NS2) of the parvovirus minute virus of mice is required for efficient DNA replication and infectious virus production in a cell-type-specific manner. *J. Virol.* **1990**, *64*, 6166–6175. [CrossRef] [PubMed]
32. Doerig, C.; McMaster, G.; Sogo, J.; Bruggmann, H.; Beard, P. Nucleoprotein complexes of minute virus of mice have a distinct structure different from that of chromatin. *J. Virol.* **1986**, *58*, 817. [CrossRef] [PubMed]
33. Blissenbach, M.; Grewe, B.; Hoffmann, B.; Brandt, S.; Überla, K. Nuclear RNA Export and Packaging Functions of HIV-1 Rev Revisited. *J. Virol.* **2010**, *84*, 6598. [CrossRef] [PubMed]
34. Venkatesh, L.K.; Fasina, O.; Pintel, D.J. RNAse Mapping and Quantitation of RNA Isoforms. In *RNA Abundance Analysis: Methods and Protocols*; Jin, H., Gassmann, W., Eds.; Humana Press: Totowa, NJ, USA, 2012; pp. 121–129.
35. Fuller, M.S.; Majumder, K.; Pintel, D.J. Minute Virus of Mice Inhibits Transcription of the Cyclin B1 Gene during Infection. *J. Virol.* **2017**, *91*. [CrossRef]
36. Kinner, A.; Wu, W.; Staudt, C.; Iliakis, G. Gamma-H2AX in recognition and signaling of DNA double-strand breaks in the context of chromatin. *Nucleic Acids Res.* **2008**, *36*, 5678–5694. [CrossRef]
37. Fornes, O.; Castro-Mondragon, J.A.; Khan, A.; van der Lee, R.; Zhang, X.; Richmond, P.A.; Modi, B.P.; Correard, S.; Gheorghe, M.; Baranašić, D.; et al. JASPAR 2020: Update of the open-access database of transcription factor binding profiles. *Nucleic Acids Res.* **2020**, *48*, D87–D92. [CrossRef]
38. Nasmyth, K.; Haering, C.H. Cohesin: Its Roles and Mechanisms. *Annu. Rev. Genet.* **2009**, *43*, 525–558. [CrossRef]
39. Zhao, Q.; Gersappe, A.; Pintel, D.J. Efficient excision of the upstream large intron from P4-generated pre-mRNA of the parvovirus minute virus of mice requires at least one donor and the 3′ splice site of the small downstream intron. *J. Virol.* **1995**, *69*, 6170–6179. [CrossRef]
40. Zhao, Q.; Mathur, S.; Burger, L.R.; Pintel, D.J. Sequences within the parvovirus minute virus of mice NS2-specific exon are required for inclusion of this exon into spliced steady-state RNA. *J. Virol.* **1995**, *69*, 5864–5868. [CrossRef]
41. Dong, Y.; Fasina, O.O.; Pintel, D.J. Minute Virus of Canines NP1 Protein Interacts with the Cellular Factor CPSF6 To Regulate Viral Alternative RNA Processing. *J. Virol.* **2019**, *93*. [CrossRef]
42. Hsin, J.P.; Manley, J.L. The RNA polymerase II CTD coordinates transcription and RNA processing. *Genes Dev.* **2012**, *26*, 2119–2137. [CrossRef] [PubMed]
43. Qiu, J.; Pintel, D.J. The adeno-associated virus type 2 Rep protein regulates RNA processing via interaction with the transcription template. *Mol. Cell. Biol.* **2002**, *22*, 3639–3652. [CrossRef] [PubMed]
44. Farris, K.D.; Pintel, D.J. Improved splicing of adeno-associated viral (AAV) capsid protein-supplying pre-mRNAs leads to increased recombinant AAV vector production. *Hum. Gene Ther.* **2008**, *19*, 1421–1427. [CrossRef] [PubMed]
45. Choi, E.-Y.; Newman, A.E.; Burger, L.; Pintel, D. Replication of minute virus of mice DNA is critically dependent on accumulated levels of NS2. *J. Virol.* **2005**, *79*, 12375–12381. [CrossRef]

Article

No G-Quadruplex Structures in the DNA of Parvovirus B19: Experimental Evidence versus Bioinformatic Predictions

Gloria Bua †, Daniele Tedesco †,‡, Ilaria Conti §, Alessandro Reggiani, Manuela Bartolini and Giorgio Gallinella *,||

Department of Pharmacy and Biotechnology, University of Bologna, 40126 Bologna, Italy; gloria.bua2@unibo.it (G.B.); daniele.tedesco@isof.cnr.it (D.T.); ilaria.conti@unife.it (I.C.); alessandro.reggiani5@unibo.it (A.R.); manuela.bartolini3@unibo.it (M.B.)

* Correspondence: giorgio.gallinella@unibo.it

† These authors contributed equally to this work.

‡ Current address: Institute for Organic Synthesis and Photoreactivity (ISOF), National Research Council (CNR), 40129 Bologna, Italy.

§ Current address: Department of Morphology, Surgery and Experimental Medicine, University of Ferrara, 44121 Ferrara, Italy.

|| Current address: Unit of Microbiology, Alma Mater Studiorum-University of Bologna, S. Orsola-Malpighi Hospital, Via Massarenti 9, 40138 Bologna, Italy.

Received: 27 July 2020; Accepted: 21 August 2020; Published: 25 August 2020

Abstract: Parvovirus B19 (B19V), an ssDNA virus in the family Parvoviridae, is a human pathogenic virus, responsible for a wide range of clinical manifestations, still in need of effective and specific antivirals. DNA structures, including G-quadruplex (G4), have been recognised as relevant functional features in viral genomes, and small-molecule ligands binding to these structures are promising antiviral compounds. Bioinformatic tools predict the presence of potential G4 forming sequences (PQSs) in the genome of B19V, raising interest as targets for antiviral strategies. Predictions locate PQSs in the genomic terminal regions, in proximity to replicative origins. The actual propensity of these PQSs to form G4 structures was investigated by circular dichroism spectroscopic analysis on synthetic oligonucleotides of corresponding sequences. No signature of G4 structures was detected, and the interaction with the G4 ligand BRACO-19 (*N,N′*-(9-{[4-(dimethylamino) phenyl]amino}acridine-3,6-diyl)bis(3-pyrrolidin-1-ylpropanamide) did not appear consistent with the stabilisation of G4 structures. Any potential role of PQSs in the viral lifecycle was then assessed in an in vitro infection model system, by evaluating any variation in replication or expression of B19V in the presence of the G4 ligands BRACO-19 and pyridostatin. Neither showed a significant inhibitory activity on B19V replication or expression. Experimental challenge did not support bioinformatic predictions. The terminal regions of B19V are characterised by relevant sequence and symmetry constraints, which are functional to viral replication. Our experiments suggest that these impose a stringent requirement prevailing over the propensity of forming actual G4 structures.

Keywords: parvovirus B19; G-quadruplex; bioinformatics; antivirals; BRACO-19; pyridostatin

---

1. Introduction

Parvovirus B19 (B19V), an ssDNA virus in the family Parvoviridae [1], is a human pathogenic virus, widely circulating in the population, responsible for an ample spectrum of clinical manifestations [2]. The genome is a 5.6 kb ssDNA molecule of either polarity, with a coding repertoire comprising a non-structural (NS) protein, functional to virus replication, and two structural proteins, VP1 and VP2, constituting a T = 1, 22 nm icosahedral capsid [3,4]. The virus is characterised by a selective but not exclusive tropism for erythroid progenitor cells (EPCs) in the bone marrow and by a strict dependence on the cellular machinery and environment for its replication [5,6].

The selective tropism of B19V for EPCs in the bone marrow and the ability to induce cell cycle arrest and apoptosis in productively infected cells can cause a partial block in erythropoiesis. This may manifest as a transient or persistent erythroid aplasia, clinically acute and severe in patients with underlying haematological disorders, or chronic in patients with immune system deficits [7]. The virus is capable of infecting and maintaining long-term persistence in disparate tissues, mostly within endothelial or stromal cells, and can establish a complex relationship with the immune system, whose efficacy in innate and adaptive responses is crucial to the course of infection and the development of pathological processes [8,9]. In addition to haematological consequences, B19V infection can commonly manifest as erythema infectiosum and cause post-infection arthropathies. Further, a wide range of other different pathologies have been reported, among them mainly myocarditis [10] and autoimmune processes [11]. Infection in pregnancy may be transmitted to the foetus, posing a risk of foetal death and/or foetal hydrops [12–14].

B19V infection requires diagnostic awareness to lead and support clinical care in severe cases [15]. The development of antiviral strategies directed against B19V as compared to other viruses is still lagging, although recent work has identified a few compounds that show a selective inhibition of B19V replication in vitro [16]. Such compounds include hydroxyurea (HU), a ribonucleotide reductase inhibitor, also used for the treatment of sickle-cell disease and known to have "virostatic" properties [17]; the nucleotide analogues cidofovir (CDV) and its lipid derivative brincidofovir (BCV), broad-spectrum anti-viral agents mostly active against dsDNA viruses [18–20]; and a few coumarin derivatives [21]. Some flavonoid compounds can inhibit the endonuclease activity of viral NS protein, a function critical to the replicative process of B19V [22].

The genome of B19V has a limited coding potential, and its replication depends largely on the cellular environment. Consequently, a deeper understanding of the viral lifecycle and virus–cell interactions are required to identify further targets and agents for an effective antiviral strategy. Unconventional DNA structures have been recognised as relevant features for the regulation of several biological processes, including replication, recombination, and transcription [23]. Particular emphasis has been given to the potential of G-rich sequences to adopt G-quadruplex (G4) planar structures disrupting the regular double-helix structure of DNA [24]. These structures are characterised by stacks of guanine tetrads, which are bound via Hoogsteen-type hydrogen bonds, and can typically form when runs of 2–4 guanine bases are regularly spaced on the DNA sequence. Small-molecule ligands recognising and binding to these structures, either with interfering or with stabilising effects, may act as modulators in the biological process involved, raising interest as compounds of pharmacological interest [25].

Methodological developments have allowed the in silico prediction of specific G4 structures directly from primary sequences, and the number of studies reporting genome-wide G4 exploration across species has rapidly increased [26], including viruses [27]. A recent survey of viral genomes by a regular expression patterns search has led to assembly of a comprehensive database (G4-virus) reporting the presence, distribution, and statistical significance of potential quadruplex sequences (PQSs) in reference genomes and genome sets for all viral families [28]. In some cases, the presence and a biological role of PQS structures in viral genomes have been validated in experimental models, and the role as antiviral agents of specific G4-ligands such as pyridostatin (PDS) and BRACO-19 (*N,N*′-(9-{[4-(dimethylamino)phenyl]amino}acridine-3,6-diyl)bis(3-pyrrolidin-1-ylpropanamide) demonstrated in relevant instances [29]. Within the G4-virus database, indication for the presence of PQSs in the B19V genome was reported, raising the need for an experimental challenge of the bioinformatic prediction and, as a consequence, for the investigation of any possible relevance of these structures as targets for antivirals against B19V.

On these grounds, we carried out a closer bioinformatic inspection of the B19V genome for the presence of PQSs, comparing the results reported in the G4-virus database to targeted predictions obtained by a different computational method for G-quadruplex prediction, the QGRS (Quadruplex forming G-Rich Sequences) mapper [30], a method based on a scoring algorithm. By this analysis, we identified two sequence stretches located in the genomic terminal regions, close to the origins of replication of viral DNA, as a potentially relevant PQSs. Experiments were carried out to test the prediction. Synthetic oligonucleotides corresponding to the PQSs were investigated by circular dichroism (CD) spectroscopy, which can provide information on the propensity to form G4 structures. Then, any potential role of PQSs in the viral lifecycle was assessed by using G4 ligands in a model virus–cell system and evaluating the occurrence of a dose-dependent variation in replication or expression levels of B19V.

2. Materials and Methods

2.1. Bioinformatic Analysis

The B19V sequence used in bioinformatic analysis is a derived consensus sequence, referred to as B19V EC [GenBank KY940273] [31]. The G4-virus PQS database [28] was accessed at http://www.medcomp.medicina.unipd.it/main_site/doku.php?id=g4virus. The QGRS Mapper web server [30] was accessed at http://bioinformatics.ramapo.edu/QGRS/index.php.

2.2. Chemicals

Oligonucleotides used in CD analysis (Table 1) were obtained from Eurofins Genomics (Ebersberg, Germany) (https://www.eurofinsgenomics.eu/). BRACO-19 and pyridostatin were obtained from Merck-Sigma (Milan, Italy). Stock solutions were prepared in H_2O at 1 mM and further diluted for subsequent experiments.

Table 1. Oligonucleotides used for the CD analysis on PQSs in the DNA of B19V.

Oligo	Nts	Sequence	Molecular Weight (Da)	G-Score
HIV LTR-II	33	GGGGACTTTCCAGGGAGGCGTGGCCTGGGCGGG	10,349	68
PQS 140	44	GGGCCAGCTTGCTTGGGGTTGCCTTGACACTAAGACAAGCGGCG	13,630	14
PQS 113	45	GGGACTTCCGGAATTAGGGTTGGCTCTGGGCCAGCTTGCTTGGGG	14,012	67
PQS 068	45	TCATTTCCTGTGACGTCATTTCCTGTGACGTCACTTCCGGTGGGC	13,752	-

2.3. CD Analysis

Circular dichroism (CD) studies on oligonucleotides were carried out on a Jasco (Tokyo, Japan) J-810 spectropolarimeter equipped with a PTC-423S Peltier-type temperature control system. Measurements were performed using a micro-volume QS quartz cell with black walls (1 cm path length, 500 µL volume; Hellma Italia, Milan, Italy). Oligonucleotides and BRACO-19 were diluted from stock solution into an analysis buffer (KCl 70 mM, potassium acetate 20 mM, pH 6.8) at 2 and 10 µM, respectively. PDS was not used in CD studies because its addition to oligonucleotides caused precipitation in the samples, making it unsuitable for spectroscopic analysis. CD spectra (330–230 nm) were recorded at 17 different temperatures (every 5 °C between 15 and 95 °C) applying a 0.25 °C/min gradient for both heating and cooling ramps. A 4 nm spectral bandwidth, a 0.2 nm data interval, a 100 nm/min scanning speed and a 2 s data integration time were employed for measurements; solvent-corrected spectra were then converted to molar units per residue ($\Delta\varepsilon_{res}$, in M^{-1} cm^{-1}). CD melting curves were determined by plotting the $\Delta\varepsilon_{res}$ values as a function of temperature (T) for each oligonucleotide, using the wavelength at which the difference between their CD signals at 15 and 95 °C was maximum ($\lambda_{\Delta_{max}}$). Mid-transition temperatures (T_m) for both heating and cooling ramps were then derived by non-linear regression on the CD melting curves using a 6-parameter logistic function [32,33].

2.4. Cells

Erythroid progenitor cells (EPCs) were generated in vitro from peripheral blood mononuclear cells (PBMC), as described [5]. Blood donations were made available for institutional research purposes from the Immunohaematology and Transfusion Service, S. Orsola-Malpighi University Hospital, Bologna (authorisation 0070755/1980/2014). Availability was granted under conditions complying with Italian privacy law. Neither specific ethics committee approval nor written consent from donors was required for this research project.

2.5. Cytotoxicity

The effects of tested compounds on cell viability were monitored by the Cell Counting Kit 8 (WST-8/CCK8) assay (Dojindo Molecular Technologies, Microtech, Italy), as described [20]. DMSO at 10% was used as a cytotoxicity positive control. The assay is based on a production of a formazan dye in response to cellular metabolic activity, measured as absorbance (OD) values. Replicate net OD values were normalised with respect to the control samples and expressed as mean percentage values for cell viability.

2.6. Infection

B19V was obtained from a cloned synthetic genome, first transfected into UT7/EpoS1 cells, then propagated by serial passage in EPCs, as described [31]. For infection, EPCs were incubated at a density of 10^7 cell/mL, in the presence of B19V to a multiplicity of infection (moi, expressed as geq/cell) of 10^3 geq/cell, for 2 h at 37 °C. After removal of inoculum virus, EPCs were incubated at 37 °C in 5% CO_2 in complete growth medium, at the different concentrations of tested compounds, at an initial density of 10^6 cells/mL.

2.7. Molecular Analysis

Equal amounts of cell cultures, corresponding to 1.5×10^5 cells, were collected as appropriate at 2 or 48 h post-infection (hpi) and processed by using the Maxwell Viral Total Nucleic Acid kit on a Maxwell MDx platform (Promega), to obtain a total nucleic acid fraction in elution volumes of 150 μL. The quantitative evaluation of target nucleic acids was carried out by qPCR assays in a Rotor-Q system (Qiagen, Hilden, Germany). For the analysis of B19V DNA, aliquots of the eluted nucleic acids (corresponding to ~500 cells) were directly amplified in a qPCR assay (Maxima SYBR Green qPCR Master Mix, Thermo Scientific, Life Technologies, Monza, Italy). For the analysis of B19V RNA, parallel aliquots were first treated with the Turbo DNAfree reagent (Ambion, Life Technologies) before amplification in a qRT-PCR assay (Express One-step SYBR GreenER Kit, Invitrogen, Life Technologies). Standard cycling programs were used, followed by a melting curve analysis to define the T_m of amplified products. The primer pair R2210–R2355, located in the central exon of B19V genome, was used to amplify both viral DNA and total RNA, and a target sequence in the region of genomic DNA coding for 5.8S rRNA (rDNA) was amplified in parallel reactions for normalisation [5,31].

3. Results

3.1. Sequence, Symmetry, and Higher-Order Structures in B19V Genome

The B19V reference sequence used for bioinformatic analysis is a derived consensus sequence, resulting from the alignment of a selected, non-redundant set of complete genomic sequences, referred to as B19V EC [GenBank KY940273]. Such a sequence provides the basis to a synthetic genetic system for B19V, able to yield virus with full replicative competence used for subsequent experiments [31]. The whole genome is 5596 nts long, and its arrangement presents two levels of symmetry. On a genomic scale, a unique internal region, 4830 nts, containing all the coding sequences, is flanked by inverted terminal regions, each 383 nts, serving as replicative origins. Within the terminal regions, the distal 365 nts are disposed as a palindromic sequence around a central site of dyad symmetry. The palindrome is imperfect, presenting a few base mismatches leading to two different sequences, one the inverse complement of the other and usually referred to as "flip-flop", which can combine independently at each end, thus producing four different sequence isomers.

Superimposed on these symmetries, the B19V genome presents signatures of higher-order structures such as PQSs (Figure 1). The prediction on the presence and distribution of PQSs in B19V genome reported in the G4-virus database was compared to predictions obtained by the QGRS mapper program. Predictions were only partially concordant. The G4-virus database reports a list of all PQSs identified in the genome, their position on positive and negative strands, their degree of conservation among isolates included in the dataset expressed as frequency, and the statistical significance of their abundance [28]. For B19V, on a dataset of 13 sequences, only the presence of dinucleotide PQS, and not of tri- or tetra-nucleotide PQS, was considered statistically significant over a random distribution. A disperse dinucleotide (GG) PQS distribution was reported, including 22 GG-PQSs in the plus strand and 18 GG-PQSs in the in minus strand, at frequencies in the range 0.08–1.00. On the other hand, the QGRS mapper [30] uniquely identified, at a relevant score (G-score > 60), two sequence stretches with features of a PQS ($G_3N_{13}G_3N_8G_3N_{11}G_3$) located within the terminal regions, which are characteristically GC-rich. In particular, these PQSs are located on either plus or minus strand, in close 5′ proximity to the axis of dyad symmetry, partially overlapping with the sequence asymmetries and, thus, in different relative positions with respect to "flip" and "flop" isomers (Figure 2).

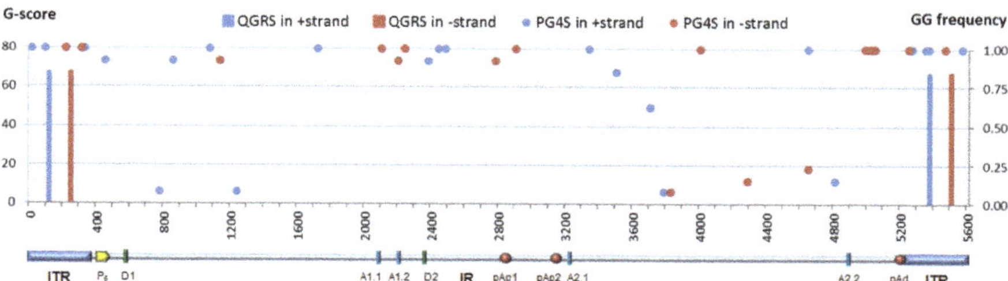

Figure 1. Parvovirus B19 (B19V) genome symmetry and potential quadruplex sequences (PQSs). On a genomic scale, the overall symmetry arrangement includes the two terminal regions flanking the internal unique region. The map of B19V genome shows the two inverted terminal regions (ITR) and the internal region (IR) with the distribution of cis-acting functional sites (P6, promoter; pAp1, pAp2, proximal cleavage-polyadenylation sites; pAd, distal cleavage-polyadenylation site; D1, D2, splice donor sites; A1.1, A1.2, A2.1, A2.2, splice acceptor sites). Superimposed on this arrangement, the B19V genome presents signatures of higher-order structures such as PQSs. Dots (PG4S): GG-PQS in + and − strands, genomic distribution and frequency as reported in the G4-virus database (GG-frequency plotted on right *y*-axis) [28]. Bars (QGRS (Quadruplex forming G-Rich Sequences)): PQSs in + and − strands, genomic distribution and relevance score for PQSs uniquely identified by the QGRS mapper (G-score score plotted on left *y*-axis) [30].

The palindromic sequences in the terminal regions allow intra-strand base pairing, leading to a hairpin configuration, as well as inter-strand base pairing leading to an extended configuration. Hairpins can provide priming for second-strand synthesis, whereas strands in the extended configuration need to separate and fold back into hairpin structures for reinitiating replication. Predictions locate PQSs within a functional replicative origin, so that the strand unwinding and folding mechanisms occurring during genome replication can offer the opportunity for DNA strands to assume a G4 structure. This, in turn, may play a role in the regulation of viral genome replication or expression. The following experiments analysed the actual propensity of predicted PQSs to assume a G4 configuration, and any potential relevant role of these in the viral lifecycle.

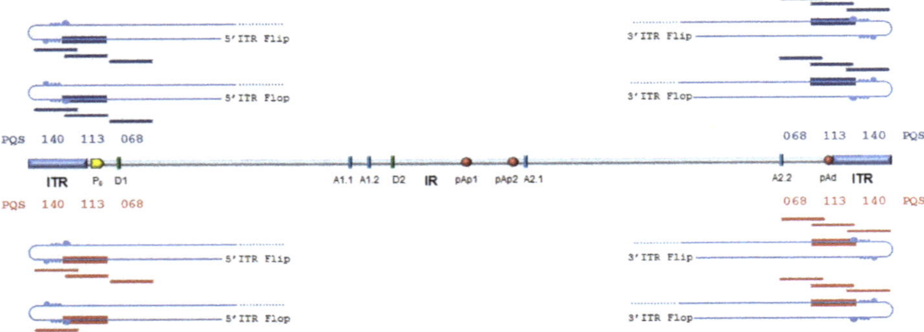

Figure 2. B19V genome symmetry and PQSs. ITRs are shown in the hairpin configuration for the positive and negative strands in the different "flip-flop" isomers. The potential G4 structures predicted by QGRS Mapper are located within the terminal sequences (shown as boxes), 5′ to the dyad symmetry, either on the plus strand (blue) or minus strand (red). PQSs partially overlap with the asymmetries leading to the flip/flop isomers (bubbles). Oligonucleotides used in circular dichroism (CD) experiments (PQSs, Table 1) are shown in context (blue/red stripes). See also Supplementary Figure S1.

3.2. *PQSs in B19V DNA: CD Analysis*

The propensity of the PQSs identified by the bioinformatic analysis to form G4 structures, which can occur in one of 26 folding arrangements [34], was first investigated by CD spectroscopic analysis on synthetic oligonucleotides of the corresponding sequence (Table 1). CD spectroscopy is routinely employed to investigate the secondary structure of nucleic acids, thanks to its sensitivity to chirality [35], and in this framework it can be used to evaluate the presence and geometry of G4 structures. Each geometry is characterised by different angles for the glycosidic bonds of guanosines and a different topology for the loops linking the stacked tetrads of the G4 stem, defining the coupling among the guanine chromophores of the bases and giving rise to peculiar CD signatures that can be used as an indicator for the presence of G4 structures [36]. Further, CD melting curve analysis [32,33] can yield information on thermal stability and binding of small molecules, in this case G4 ligands such as BRACO-19, a 3,6,9-trisubstitued acridine derivative designed to bind and stabilise quadruplex DNA structures.

The oligonucleotide HIV LTR-II (Table 1) was chosen as a positive control for the formation of G4 structures [37]. The CD spectrum of this oligonucleotide at low temperature (Figure 3A) can be interpreted as the overlap between the contribution of a parallel G4 structure, which gives a strong positive band centred at around 265 nm [36], and the profile of a GC-rich (76%) ssDNA in B-form, which gives a positive band at around 280 nm and a negative band at around 245 nm [35]. The CD melting curves of the oligonucleotide at 265 nm (Figure 4A), both in the absence and in the presence of BRACO-19, show a clear decrease in intensity for the positive band at 265 nm during the heating ramp, indicative of the disruption of the G4 structure upon thermal denaturation, and a fully reversible profile upon renaturation due to the reorganisation of the G4 structure during the cooling ramp. As expected for a G4 ligand, BRACO-19 stabilises the G4 structure of HIV LTR-II, as the T_m of the melting curves is shifted towards higher values (~+7 °C; Table 2), although the degree of stabilisation was found to be smaller than previously reported in the literature [37]. All these observations confirm the presence of a G4 structure in HIV LTR-II.

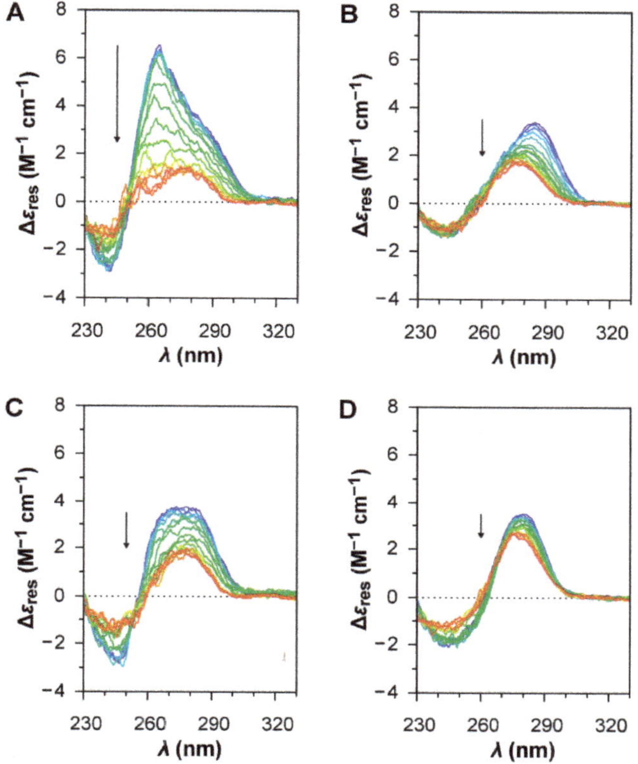

Figure 3. CD spectra of the oligonucleotides under investigation (2 μM) during the heating ramps of the melting assays. (**A**) HIV LTR-II; (**B**) PQS 113; (**C**) PQS 140; (**D**) PQS 068. The arrows indicate the evolution along the heating ramp from 15 to 95 °C.

Table 2. Mid-transition temperatures (T_m, in °C) for the oligonucleotides under investigation, both in the absence or in the presence of BRACO-19 (10 μM), as determined by CD melting assays.

	$\lambda_{\Delta max}$ (nm)	**Isolated Oligonucleotide (2 μM)**		**Oligonucleotide (2 μM) + BRACO-19 (10 μM)**	
		15 °C → 95 °C	**95 °C → 15 °C**	**15 °C → 95 °C**	**95 °C → 15 °C**
HIV LTR-II	265	55.0 ± 1.4	55.9 ± 0.4	61.8 ± 1.5	62.4 ± 1.6
PQS 113	290	37.4 ± 2.4	0.3 ± 81.5	37.3 ± 2.5	45.3 ± 16.7
PQS 140	285	54.6 ± 2.3	56.2 ± 1.8	64.7 ± 1.7	57.8 ± 2.8
PQS 068	285	67.3 ± 2.8	65.0 ± 6.6	73.3 ± 3.3	58.1 ± 1.4

BRACO-19—(*N,N*′-(9-{[4-(dimethylamino)phenyl]amino}acridine-3,6-diyl)bis(3-pyrrolidin-1-ylpropanamide).

For B19V, three different oligonucleotides were investigated. Oligo PQS 113 has a sequence matching the most probable PQS in the B19 genome, showing the highest G-score. Oligo PQS 140 has a sequence of corresponding length located in 5′ proximity to the dyad symmetry, upstream and partially overlapping with PQS 113, showing a low G-score. Oligo PQS 068 is also a sequence of corresponding length, located downstream to PQS 113 and showing a null G-score. The oligonucleotides PQS 113, PQS 140, and PQS 068 all display the CD profiles of ssDNA in B-form with no clear contribution from G4 structures (Figure 3B–D); CD signatures peculiar to G4 structures were not observed, while the differences in the CD profiles are most probably due to different primary structures [35].

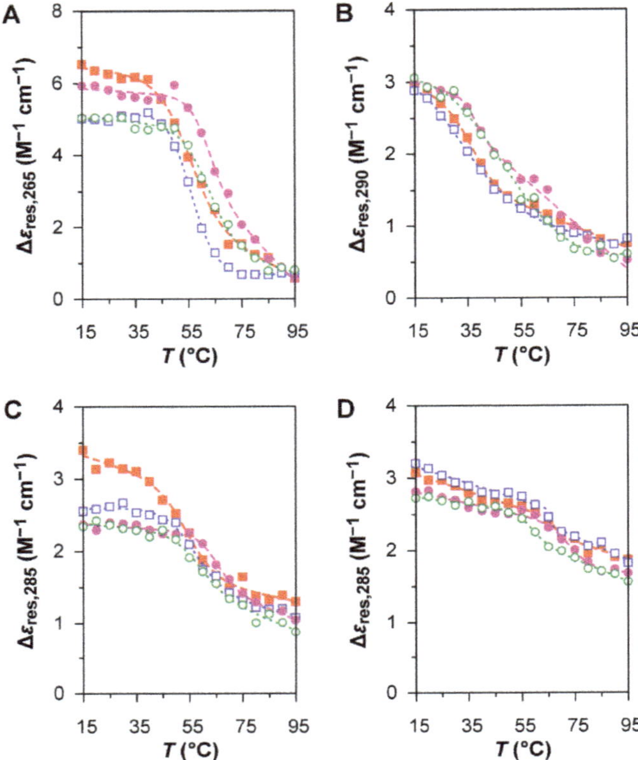

Figure 4. CD melting curves for the oligonucleotides under investigation (2 μM), both in the absence or in the presence of BRACO-19 (10 μM). (**A**) HIV LTR-II; (**B**) PQS 113; (**C**) PQS 140; (**D**) PQS 068. Filled squares: heating ramps (15 to 95 °C) in the absence of BRACO-19. Empty squares: cooling ramps (95 to 15 °C) in the absence of BRACO-19. Filled circles: heating ramps (15 to 95 °C) in the presence of BRACO-19. Empty circles: cooling ramps (95 to 15 °C) in the presence of BRACO-19. BRACO-19—(*N*, *N*′-(9-{[4-(dimethylamino)phenyl]amino}acridine-3,6-diyl)bis(3-pyrrolidin-1-ylpropanamide).

The CD melting profiles of PQS 113 at 290 nm in the absence of BRACO-19 (Figure 4B) show a broad conformational transition at low temperature after both denaturation and renaturation, revealing a high degree of instability in solution; the large uncertainty of the T_m value determined on the cooling ramp (Table 2) is a result of such instability. The CD melting profiles of PQS 140 and PQS 068 in the absence of BRACO-19 (Figure 4C,D) both display narrower, reversible thermal transitions; in both cases, the temperature-dependent variation in CD response at 285 nm has a smaller magnitude than that of HIV LTR-II. The CD melting curves of the oligonucleotides in the presence of BRACO-19 provide an indication of binding, although the underlying mechanisms of these binding interactions appear to be quite different from those observed with HIV LTR-II. Once again, the behaviour of PQS 113 (Figure 4B) is more complex: the trend of the melting curves suggests the possibility of a two-state conformational transition, which is not accurately described by the non-linear regression model used to analyse the melting profiles of G4 structures. On the other hand, the melting curves of PQS 140 and PSQ 068 (Figure 4C,D) are not reversible, since the T_m value of the heating ramp is higher than that of the cooling ramp (Table 2). This phenomenon of hysteresis may be explained by a slower kinetics of denaturation and renaturation due to the presence of BRACO-19. Overall, BRACO-19 appears to interact with all the oligonucleotides under investigation, although the mechanism of binding is not consistent with the stabilisation of eventual G4 structures.

3.3. PQSs in B19V DNA: Biological Analysis

To extend the results of CD studies and investigate a possible role of putative G4 structures in B19V DNA, we tested the biological effects on the virus–cell system of two reported G4 ligands, BRACO-19 and pyridostatin (PDS) [29]. BRACO-19 has been shown to inhibit telomerase activity, to possess antitumour activity and antiviral activity on different viruses in vitro, including HIV-1. PDS is a very selective G4 DNA-binding small molecule designed to form a complex with and stabilise G4 structures. It has been shown to strongly stabilise telomeric G4, triggering a DNA-damage response at telomeres. As an antiviral agent, PDS has been used to study the role of G4 in Epstein Barr Virus (EBV). As a model cell system, we used primary EPCs, which constitute a heterogeneous cellular population mimicking the natural target cells in in vivo infection and that present full permissiveness to viral replication at the appropriate differentiation stage [5,31]. Effects on cell viability and any possible activity on B19V were assessed in a time course of infection, by evaluating any dose-dependent effects of BRACO-19 and PDS.

Effects on cell viability. EPCs were cultured for 48 h at 37 °C in medium containing different concentrations of each compound (0.1–100 μM range), then cell viability was assessed by a WST-8-based colorimetric assay. Results are reported in Figure 5 and are expressed as percentage viability with respect to control cells incubated without compounds. A reduction in cell viability below 50% of control was observed starting from 10 and 5 μM for BRACO-19 and PDS, respectively. At higher concentrations, 50 and 100 μM, the metabolic activity of cells was totally inhibited.

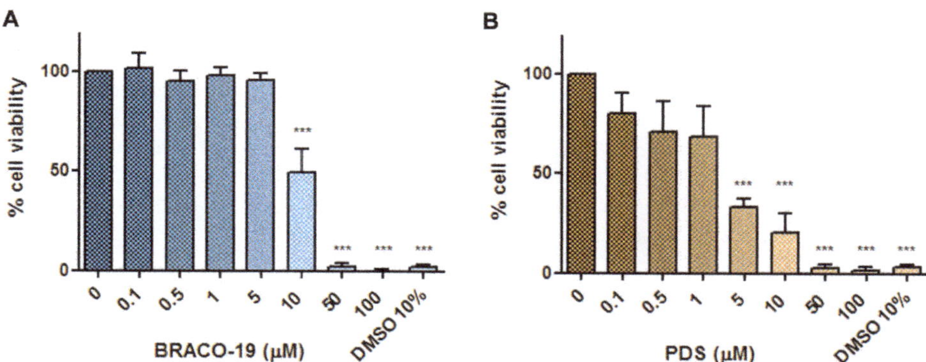

Figure 5. Percentage of viability of erythroid progenitor cells (EPCs) cultured for 48 h in presence of different concentrations of BRACO-19 (**A**) and pyridostatin (PDS) (**B**). DMSO at 10% was used as a cytotoxicity positive control. Values are expressed as mean percentage compared to the control with medium only. Data were collected from triplicate wells in two different experiments. Statistical analysis was performed by one-way ANOVA (analysis of variance) followed by Dunnett's multiple comparison test. *** p value < 0.001.

By expressing a dose-dependent relationship between compound concentration and percentage cell viability, non-linear regression curves allowed determining 50% cytotoxic concentration (CC50) values: 9.99 μM for BRACO-19 (95% confidence interval: 8.9–11.53 μM; $R^2 = 0.95$); 4.01 μM for PDS (95% confidence interval: 0.47–15.52 μM; $R^2 = 0.84$).

Antiviral activity against B19V: The effects of BRACO-19 and PDS on B19V were evaluated by quantitative determination of viral replication and expression in a time course of infection. EPCs were infected with B19V at the multiplicity of infection of 10^3 geq/cell and cultured either in the absence or in the presence of each compound, at 0.5, 5, and 50 μM. Cells were collected at 2 and 48 h post-infection (hpi). The extent of viral replication and expression was assessed by qPCR and RT-qPCR evaluation of the number of viral genomes and total transcripts respectively, at 48 hpi compared to that at the baseline at 2 hpi (Figure 6).

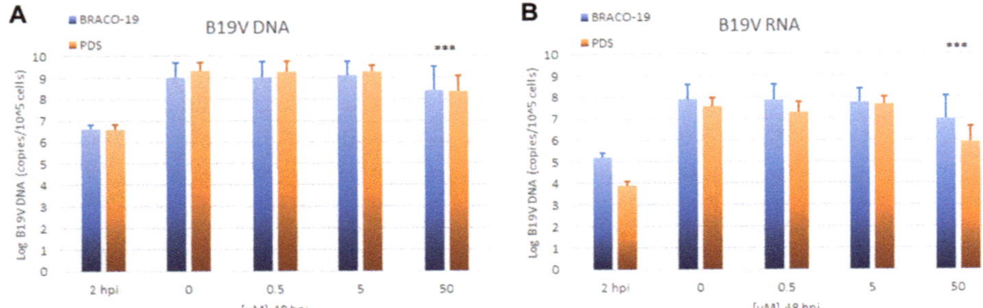

Figure 6. Amount of B19V DNA (**A**) and RNA (**B**) at 2 hpi, before addition of tested compounds, and at 48 hpi in infected cells cultured in the presence of BRACO-19 and PDS at the indicated concentrations (Log copies/10^5 cells). Data were collected from triplicate qPCR and RT-qPCR reactions in duplicate experiments. Statistical analysis was performed by one-way ANOVA (analysis of variance) followed by Dunnett's multiple comparison test among 48 hpi samples. *** p value < 0.001.

Viral DNA increased from 2 to 48 hpi on average 2.5 Log in the control samples, indicating productive viral replication. Inhibition of viral replication, displayed as a significant net reduction in viral DNA from 2 to 48 hpi, was not observed in cells treated with BRACO-19 and PDS. Compared to control samples, a partial inhibition of viral replication was evident only at the highest concentrations, with percentage values of 65% and 85% for BRACO-19 and PDS, respectively. Transcription of the viral genome, as determined by the increase of total viral mRNAs from 2 to 48 hpi, was also unaffected unless at the highest concentrations of BRACO-19 and PDS, with percentage values of 88% and 98%, respectively. The absence of dose-dependent effects on virus and the concomitant and prevalent effects on cell viability suggest that any marginal antiviral activity of BRACO-19 and PDS is likely due to the inhibition of the cellular metabolism rather than to a specific inhibitory activity on the virus, whose replication and expression appears unaffected by these G4 ligands.

4. Discussion

B19V is a virus with distinctive features that induces interest, not least in the characterisation of its lifecycle and of virus–host interaction [6]. B19V is a widely circulating human pathogenic virus, although its clinical impact is often underestimated, and the development of specific antiviral tools still suffers from a striking gap. In addition to the propensity to enhanced diagnostic awareness [15], the development of effective antiviral strategies against B19V should be considered a relevant goal in the field [16]. A better understanding of the viral lifecycle and virus–cell interactions are required to identify relevant targets for more efficient and specific antiviral strategies. Unconventional DNA structures, in particular G-quadruplex planar structures disrupting the regular double helix structure of DNA, are increasingly recognised as relevant features for the regulation of critical biological processes. Viruses can include G4-forming sequences in their genomes as part of their interaction network within the cellular environment, and in many instances, these structures can provide targets for small-molecule ligands that can provide an antiviral effect by interfering with the normal viral regulation pathways [27–29].

The comprehensive survey in the G4-virus database provides a framework overview of PQS elements in viral genomes, aiming, in its statement, at expediting research on G-quadruplex in viruses, and at finding novel therapeutic opportunities. Out these PQSs, the presence and relevance of G4s as functional elements have been validated in some cases [29], or only predicted otherwise, requiring experimental evidence as in the present case for B19V. For B19V genome, the G4-virus database reports a disperse presence of dinucleotide "GG" PQSs that can be considered statistically significant over a random distribution. An independent prediction on the presence of G4 structures can also be obtained by the QGRS mapper, which identifies at a high score two PQSs within the genomic terminal regions. In our work, we sought to validate these predictions on the presence of PQSs, but neither chemical nor biological evidence could lend experimental support to bioinformatics.

In B19V, predicted PQSs are mainly located within the terminal regions (ITRs), which are critically involved in the viral lifecycle under several aspects [6]. First, ITRs serve as origins of replication of the viral genome. A palindromic sequence is required to allow strand fold-back to form hairpin structures, in turn necessary for priming second-strand synthesis. The sequence asymmetries in the palindrome (flip/flop heterogeneity) are also strictly required [31], as they can possibly induce distortions in the hairpin secondary structure or determine the exact placement of sequence motifs recognised by binding moieties. Moreover, the ITRs are populated by binding motifs for the viral NS and several cellular proteins, relevant for both replication and transcription of the viral genome [38–40]. Finally, ITRs have the characteristics of CpG islands and are a possible target for epigenetic modifications such as CpG methylation, in turn able to regulate expression of the viral genome [41]. It should be also mentioned that for viruses in the family, the sole indirect evidence of the presence of PQSs forming G4 structures has been presented for Adeno-associated viruses (AAV) ITRs [42].

Our experimental challenge of bioinformatic predictions analysed the actual propensity of PQSs in B19V DNA to assume G4 structures and any possible inhibitory activity of G4 ligands on the viral lifecycle. Results did not lend support to the bioinformatic predictions on the occurrence of G4 structures in B19V genome and did not show any antiviral role for G4 ligands such as BRACO-19 and PDS. The reason for this discrepancy is possibly due to the sequence and symmetry constraints imposed on the sequence of B19V prevailing over the propensity of forming G4 structures, so that preservation of the secondary hairpin structures within ITRs is likely a more stringent functional requirement than the possibility of forming actual G4 structures. The QGRS mapper is reported to predict G4 structures with high accuracy (>0.95) [26], but for B19V DNA the evidence classifies the predicted PQSs as false positives. Within the ITRs, a high overall GC content may introduce a sequence bias and just increase the probability of detecting PQS-like signatures by bioinformatic tools. Actually, predictions reported in the G4-virus database indicate only a moderate statistical significance of the presence of dinucleotide PQSs. The patterns identified by the QGRS mapper program are not stringent, although the G-score obtained for the PQS regions in B19V ITRs matches that of a validated G4 structure such as HIV LTR-II. Overall, the formation and/or any relevant biological role of unconventional DNA structures such as G4 are unlikely in B19V ITRs, as determined by both CD and in vitro biological studies. On the contrary, it can be hypothesised that the formation of unconventional structures would possibly interfere with many processes crucial to viral replication without conferring any discernible selective advantage. Based on the present data, the development of antiviral strategies directed at perturbing the replicative origins in B19V DNA ITRs cannot include G4 structures as specific targets or G4 ligands as antiviral agents. In this respect, the characteristic combination of hairpin structures and sequence asymmetries appears to be a more relevant feature.

5. Conclusions

As a concluding remark, our work highlights how the enormous potential for structural or functional predictions provided by bioinformatic tools must be used with caution and results subjected to critical scrutiny. Computational methods tend to be assertive and additive, especially when aiming at the construction of comprehensive databases based on pattern search algorithms, as in the present case. Experimental validation/falsification of such predictions is required for a correct understanding of the biological systems as well as for the assessment of the computational algorithms' reliability. As in the present case, a negative experimental evidence is constructive both to avoid misconceptions and to provide a benchmark to evaluate the performance of computational methods.

Supplementary Materials: The following are available online at http://www.mdpi.com/1999-4915/12/9/935/s1. Figure S1: Inverted Terminal Regions (ITR) in B19V genome.

Author Contributions: Conceptualisation, M.B. and G.G.; data curation, G.B. and D.T.; investigation, G.B., D.T., I.C. and A.R.; methodology, G.B. and D.T.; resources, M.B. and G.G.; supervision, M.B. and G.G.; writing—original draft, G.B. and D.T.; writing—review and editing, M.B. and G.G. All authors have read and agreed to the published version of the manuscript.

Funding: This research received no specific grant from any funding agency in the public, commercial, or not-for-profit sectors. G.B. is a recipient of a post-doctoral fellowship from the Fondazione Luisa Fanti Melloni, University of Bologna.

Conflicts of Interest: The authors declare no conflict of interest. The funders had no role in the design of the study; in the collection, analyses, or interpretation of data; in the writing of the manuscript; or in the decision to publish the results.

References

1. Cotmore, S.F.; Agbandje-McKenna, M.; Canuti, M.; Chiorini, J.A.; Eis-Hubinger, A.M.; Hughes, J.; Mietzsch, M.; Modha, S.; Ogliastro, M.; Penzes, J.J.; et al. ICTV Virus Taxonomy Profile: Parvoviridae. *J. Gen. Virol.* **2019**, *100*, 367–368. [CrossRef] [PubMed]
2. Gallinella, G. Parvovirus B19 Achievements and Challenges. *ISRN Virol.* **2013**. [CrossRef]
3. Qiu, J.; Soderlund-Venermo, M.; Young, N.S. Human Parvoviruses. *Clin. Microbiol. Rev.* **2017**, *30*, 43–113. [CrossRef] [PubMed]
4. Mietzsch, M.; Penzes, J.J.; Agbandje-McKenna, M. Twenty-Five Years of Structural Parvovirology. *Viruses* **2019**, *11*, 362. [CrossRef]
5. Bua, G.; Manaresi, E.; Bonvicini, F.; Gallinella, G. Parvovirus B19 Replication and Expression in Differentiating Erythroid Progenitor Cells. *PLoS ONE* **2016**, *11*, e0148547. [CrossRef]
6. Ganaie, S.S.; Qiu, J. Recent Advances in Replication and Infection of Human Parvovirus B19. *Front. Cell. Infect. Microbiol.* **2018**, *8*, 166. [CrossRef]
7. Kerr, J.R. A review of blood diseases and cytopenias associated with human parvovirus B19 infection. *Rev. Med. Virol.* **2015**, *25*, 224–240. [CrossRef]
8. Adamson-Small, L.A.; Ignatovich, I.V.; Laemmerhirt, M.G.; Hobbs, J.A. Persistent parvovirus B19 infection in non-erythroid tissues: Possible role in the inflammatory and disease process. *Virus Res.* **2014**, *190*, 8–16. [CrossRef]
9. Bua, G.; Gallinella, G. How does parvovirus B19 DNA achieve lifelong persistence in human cells? *Future Virol.* **2017**, *12*, 549–553. [CrossRef]
10. Verdonschot, J.; Hazebroek, M.; Merken, J.; Debing, Y.; Dennert, R.; Brunner-La Rocca, H.P.; Heymans, S. Relevance of cardiac parvovirus B19 in myocarditis and dilated cardiomyopathy: Review of the literature. *Eur. J. Heart Fail.* **2016**, *18*, 1430–1441. [CrossRef]
11. Kerr, J.R. The role of parvovirus B19 in the pathogenesis of autoimmunity and autoimmune disease. *J. Clin. Pathol.* **2016**, *69*, 279–291. [CrossRef]
12. Bonvicini, F.; Puccetti, C.; Salfi, N.C.; Guerra, B.; Gallinella, G.; Rizzo, N.; Zerbini, M. Gestational and fetal outcomes in B19 maternal infection: A problem of diagnosis. *J. Clin. Microbiol.* **2011**, *49*, 3514–3518. [CrossRef] [PubMed]
13. Puccetti, C.; Contoli, M.; Bonvicini, F.; Cervi, F.; Simonazzi, G.; Gallinella, G.; Murano, P.; Farina, A.; Guerra, B.; Zerbini, M.; et al. Parvovirus B19 in pregnancy: Possible consequences of vertical transmission. *Prenat. Diagn.* **2012**, *32*, 897–902. [CrossRef] [PubMed]
14. Bonvicini, F.; Bua, G.; Gallinella, G. Parvovirus B19 infection in pregnancy-awareness and opportunities. *Curr. Opin. Virol.* **2017**, *27*, 8–14. [CrossRef] [PubMed]
15. Gallinella, G. The clinical use of parvovirus B19 assays: Recent advances. *Expert Rev. Mol. Diagn.* **2018**, *18*, 821–832. [CrossRef] [PubMed]
16. Manaresi, E.; Gallinella, G. Advances in the Development of Antiviral Strategies against Parvovirus B19. *Viruses* **2019**, *11*, 659. [CrossRef]
17. Bonvicini, F.; Bua, G.; Conti, I.; Manaresi, E.; Gallinella, G. Hydroxyurea inhibits parvovirus B19 replication in erythroid progenitor cells. *Biochem. Pharmacol.* **2017**, *136*, 32–39. [CrossRef]

18. Bonvicini, F.; Bua, G.; Manaresi, E.; Gallinella, G. Antiviral effect of cidofovir on parvovirus B19 replication. *Antivir. Res.* **2015**, *113*, 11–18. [CrossRef]
19. Bonvicini, F.; Bua, G.; Manaresi, E.; Gallinella, G. Enhanced inhibition of parvovirus B19 replication by cidofovir in extendedly exposed erythroid progenitor cells. *Virus Res.* **2016**, *220*, 47–51. [CrossRef]
20. Bua, G.; Conti, I.; Manaresi, E.; Sethna, P.; Foster, S.; Bonvicini, F.; Gallinella, G. Antiviral activity of brincidofovir on parvovirus B19. *Antivir. Res.* **2019**, *162*, 22–29. [CrossRef]
21. Conti, I.; Morigi, R.; Locatelli, A.; Rambaldi, M.; Bua, G.; Gallinella, G.; Leoni, A. Synthesis of 3-(Imidazo[2,1-b]thiazol-6-yl)-2H-chromen-2-one Derivatives and Study of Their Antiviral Activity against Parvovirus B19. *Molecules* **2019**, *24*, 1037. [CrossRef] [PubMed]
22. Xu, P.; Ganaie, S.S.; Wang, X.; Wang, Z.; Kleiboeker, S.; Horton, N.C.; Heier, R.F.; Meyers, M.J.; Tavis, J.E.; Qiu, J. Endonuclease Activity Inhibition of the NS1 Protein of Parvovirus B19 as a Novel Target for Antiviral Drug Development. *Antimicrob. Agents Chemother.* **2019**, *63*. [CrossRef] [PubMed]
23. Kaushik, M.; Kaushik, S.; Roy, K.; Singh, A.; Mahendru, S.; Kumar, M.; Chaudhary, S.; Ahmed, S.; Kukreti, S. A bouquet of DNA structures: Emerging diversity. *Biochem. Biophys. Rep.* **2016**, *5*, 388–395. [CrossRef] [PubMed]
24. Rhodes, D.; Lipps, H.J. G-quadruplexes and their regulatory roles in biology. *Nucleic Acids Res.* **2015**, *43*, 8627–8637. [CrossRef] [PubMed]
25. Monsen, R.C.; Trent, J.O. G-quadruplex virtual drug screening: A review. *Biochimie* **2018**, *152*, 134–148. [CrossRef] [PubMed]
26. Puig Lombardi, E.; Londono-Vallejo, A. A guide to computational methods for G-quadruplex prediction. *Nucleic Acids Res.* **2020**, *48*, 1–15. [CrossRef]
27. Puig Lombardi, E.; Londono-Vallejo, A.; Nicolas, A. Relationship Between G-Quadruplex Sequence Composition in Viruses and Their Hosts. *Molecules* **2019**, *24*, 1942. [CrossRef]
28. Lavezzo, E.; Berselli, M.; Frasson, I.; Perrone, R.; Palu, G.; Brazzale, A.R.; Richter, S.N.; Toppo, S. G-quadruplex forming sequences in the genome of all known human viruses: A comprehensive guide. *PLoS Comput. Biol.* **2018**, *14*, e1006675. [CrossRef]
29. Ruggiero, E.; Richter, S.N. G-quadruplexes and G-quadruplex ligands: Targets and tools in antiviral therapy. *Nucleic Acids Res.* **2018**, *46*, 3270–3283. [CrossRef]
30. Kikin, O.; D'Antonio, L.; Bagga, P.S. QGRS Mapper: A web-based server for predicting G-quadruplexes in nucleotide sequences. *Nucleic Acids Res.* **2006**, *34*, W676–W682. [CrossRef]
31. Manaresi, E.; Conti, I.; Bua, G.; Bonvicini, F.; Gallinella, G. A Parvovirus B19 synthetic genome: Sequence features and functional competence. *Virology* **2017**, *508*, 54–62. [CrossRef] [PubMed]
32. Mergny, J.L.; Lacroix, L. Analysis of thermal melting curves. *Oligonucleotides* **2003**, *13*, 515–537. [CrossRef] [PubMed]
33. Mergny, J.L.; Lacroix, L. UV Melting of G-Quadruplexes. *Curr. Protoc. Nucleic Acid Chem.* **2009**, *37*. [CrossRef]
34. Webba da Silva, M. Geometric formalism for DNA quadruplex folding. *Chemistry* **2007**, *13*, 9738–9745. [CrossRef] [PubMed]
35. Kypr, J.; Kejnovska, I.; Renciuk, D.; Vorlickova, M. Circular dichroism and conformational polymorphism of DNA. *Nucleic Acids Res.* **2009**, *37*, 1713–1725. [CrossRef] [PubMed]
36. Karsisiotis, A.I.; Hessari, N.M.A.; Novellino, E.; Spada, G.P.; Randazzo, A.; Webba da Silva, M. Topological Characterization of Nucleic Acid G-Quadruplexes by UV Absorption and Circular Dichroism. *Angew. Chem. Int. Ed.* **2011**, *50*, 10645–10648. [CrossRef]
37. Perrone, R.; Nadai, M.; Frasson, I.; Poe, J.A.; Butovskaya, E.; Smithgall, T.E.; Palumbo, M.; Palu, G.; Richter, S.N. A dynamic G-quadruplex region regulates the HIV-1 long terminal repeat promoter. *J. Med. Chem.* **2013**, *56*, 6521–6530. [CrossRef]
38. Tewary, S.K.; Zhao, H.; Deng, X.; Qiu, J.; Tang, L. The human parvovirus B19 non-structural protein 1 N-terminal domain specifically binds to the origin of replication in the viral DNA. *Virology* **2014**, *449*, 297–303. [CrossRef]
39. Luo, Y.; Qiu, J. Human parvovirus B19: A mechanistic overview of infection and DNA replication. *Future Virol.* **2015**, *10*, 155–167. [CrossRef]
40. Zou, W.; Wang, Z.; Xiong, M.; Chen, A.Y.; Xu, P.; Ganaie, S.S.; Badawi, Y.; Kleiboeker, S.; Nishimune, H.; Ye, S.Q.; et al. Human Parvovirus B19 Utilizes Cellular DNA Replication Machinery for Viral DNA Replication. *J. Virol.* **2018**, *92*. [CrossRef]

41. Bonvicini, F.; Manaresi, E.; Di Furio, F.; De Falco, L.; Gallinella, G. Parvovirus B19 DNA CpG dinucleotide methylation and epigenetic regulation of viral expression. *PLoS ONE* **2012**, *7*, e33316. [CrossRef] [PubMed]
42. Satkunanathan, S.; Thorpe, R.; Zhao, Y. The function of DNA binding protein nucleophosmin in AAV replication. *Virology* **2017**, *510*, 46–54. [CrossRef] [PubMed]

Article

Enhanced Cell-Based Detection of Parvovirus B19V Infectious Units According to Cell Cycle Status

Céline Ducloux [1,†], Bruno You [1,†], Amandine Langelé [1,2], Olivier Goupille [2], Emmanuel Payen [2], Stany Chrétien [2] and Zahra Kadri [2,*]

1 Laboratoire Français du Fractionnement et des Biotechnologies (LFB), 3 Avenue des Tropiques, BP 305, Courtabœuf CEDEX, 91958 Les Ulis, France; celine.ducloux@gmail.com (C.D.); you@lfb.fr (B.Y.); angeleamandine@lfb.fr (A.L.)

2 Division of Innovative Therapies, UMR-1184, IMVA-HB and IDMIT Center, CEA, INSERM and Paris-Saclay University, F-92265 Fontenay-aux-Roses, France; olivier.goupille@cea.fr (O.G.); emmanuel.payen@cea.fr (E.P.); stany.chretien@cea.fr (S.C.)

* Correspondence: zahra.kadri@cea.fr

† These authors contributed equally to this work.

Academic Editor: Giorgio Gallinella
Received: 27 July 2020; Accepted: 29 September 2020; Published: 18 December 2020

Abstract: Human parvovirus B19 (B19V) causes various human diseases, ranging from childhood benign infection to arthropathies, severe anemia and fetal hydrops, depending on the health state and hematological status of the patient. To counteract B19V blood-borne contamination, evaluation of B19 DNA in plasma pools and viral inactivation/removal steps are performed, but nucleic acid testing does not correctly reflect B19V infectivity. There is currently no appropriate cellular model for detection of infectious units of B19V. We describe here an improved cell-based method for detecting B19V infectious units by evaluating its host transcription. We evaluated the ability of various cell lines to support B19V infection. Of all tested, UT7/Epo cell line, UT7/Epo-STI, showed the greatest sensitivity to B19 infection combined with ease of performance. We generated stable clones by limiting dilution on the UT7/Epo-STI cell line with graduated permissiveness for B19V and demonstrated a direct correlation between infectivity and S/G2/M cell cycle stage. Two of the clones tested, B12 and E2, reached sensitivity levels higher than those of UT7/Epo-S1 and CD36$^+$ erythroid progenitor cells. These findings highlight the importance of cell cycle status for sensitivity to B19V, and we propose a promising new straightforward cell-based method for quantifying B19V infectious units.

Keywords: B19 parvovirus; detection; erythroid cells; cell cycle; permissivity

1. Introduction

Human Parvovirus B19 (B19V), a member of the genus *Erythroparvovirus* of the Parvoviridae family, is a widespread virus that is pathogenic to humans [1]. The genome of B19V is a linear 5.6-kb single-stranded DNA, packaged into a 23–28 nm non-enveloped icosahedral capsid [2]. Replication occurs in the nucleus of infected cells, via a double-stranded replicative intermediate and a rolling hairpin mechanism. B19V infection has been associated with a wide spectrum of diseases, ranging from erythema infectiosum during childhood (known as the "fifth disease" and characterized by a common "slapped-cheek" rash) [3,4], to arthropathies [5], severe anemia [6] and systemic manifestations involving the central nervous system, heart and liver, depending on the immune competence of the host [7]. Productive B19V is restricted to human erythroid progenitor cells [8], and its clinical manifestations are linked to the destruction of infected cells [9]. Indeed acute B19V infection can cause pure red-cell aplasia in patients with pre-existing hematologic disorders leading to high levels of erythrocyte turnover (e.g., in sickle cell disease or thalassemia patients) [10–12], and in immunocompromised or transplanted patients [13]. The virus is transmitted via respiratory secretions and feto-maternal blood transfers. During pregnancy, infection with B19V can cause non-immune fetal hydrops, congenital anemia, myocarditis and terminal heart failure, leading to spontaneous abortion or stillbirth of the fetus [14,15]. The high prevalence of B19V infection in the general population and the large number of blood donations used in the manufacture of plasma-derived factor concentrates favors high levels of contamination. Few reports [16–18] of clinical B19V infection resulting from the transfusion of contaminated blood components or infusions of plasma-derived medicinal products suggests that measures to reduce the transmission risk (e.g., nucleic acid testing (NAT) and/or virus removal/inactivation steps) are effective. This efficacy has led to NAT becoming the gold standard for testing products of biological origin [19]. Indeed, to counteract B19V blood-borne patient's contamination, evaluation of B19 DNA in plasma pools and viral inactivation/removal steps are performed, but nucleic acid testing did not correctly reflect B19V infectivity [17,20].

Reducing the risk of B19V infection is mandatory for suppliers of blood-derived products worldwide [21]. The elimination of viruses must be assessed in processes for the production of plasma-derived medical products, but B19V DNA quantification may be inadequate: viral DNA can persist in the serum for months after acute infection, and its levels are therefore not necessarily correlated with infectivity [20,22]. The use of titration-based B19V infectivity assays is therefore essential. Moreover, the last few decades have seen the development of regenerative therapies based on Hematopoietic or Mesenchymal Stem Cells (HSC or MSC) from bone marrow and synovium donors, respectively. According to the guidelines ensuring clinical grade of human stem cells, one of the major safety concerns is detecting latent viruses in cell sources [13,23]. Stem cells seem to act as a latent reservoir for B19 infection [24]. If viral contamination is overlooked at initial screening, then the virus may be amplified during culture before transplantation, through the reactivation of latent B19V [25]. For all these reasons, a practical and sensitive in vitro method for assessing B19V infectivity is required. However, efforts to develop such methods have been hampered by the lack of suitable B19-sensitive cell lines.

B19V displays a marked tropism for erythroid progenitor cells (EPC), but there is still no well-established cell line for B19V infection. The UT7/Epo-S1 cell line [26], an erythropoietin (Epo)-dependent subclone derived from the mega-karyoblastoid cell line UT-7 [27], is the most widely used cell model, because of its high sensitivity to B19V replication and transcription [28]. However, B19V infection is limited to a small number of cells (1%–9%, versus 30%–40%% for primary or immortalized erythroid progenitor cells) [29–31].

Here, we compared the sensitivities of a number of different erythroid cell lines to B19 infection with that of UT7/Epo-S1. We generated stable clones with graduated permissivity to B19V from a single parental cell line. Using the FUCCI (fluorescent ubiquitination cell cycle indicator) system to analyze cell cycle, we demonstrated a direct correlation between infectivity and S/G2/M cell cycle stage, and characterized two clones, B12 and E2, with sensitivity to B19V up to 35 time higher than that of UT7/Epo-S1.

2. Materials and Methods

2.1. Cell Lines

Three distinct UT-7/Epo cell lines were used: (1) UT7/Epo-S1, a clone of UT7/Epo [32], was obtained from Dr Kazuo Sugamura (Tohoku University Graduate School of Medicine, Japan). (2) UT7/Epo-APHP and UT7/Epo-Cl3, a subclone isolated from UT7/Epo-APHP, were a gift from Dr Morinet (APHP: Assistance Publique - Hôpitaux de Paris, Saint Louis Hospital). (3) UT7/Epo-STI cells were derived from UT-7/GM cell line and were maintained at low passage, with stringency for erythroid features [33]. UT7 cell lines were maintained at 37 °C, under an atmosphere containing 5% CO_2, in alpha minimum essential medium (αMEM) supplemented with 10% fetal calf serum (FCS), 2 mM L-glutamine (Hyclone), 100 U/mL penicillin, 100 μg/mL streptomycin and 2 U/mL recombinant human (rh) Erythropoietin (rh-Epo, Euromedex, RC213-15). Where specified, 0.5 μM JQ1 (Sigma-Aldrich, France) or 2 ng/mL TGF-β (Peprotech, France) was added to the culture medium for two days before B19V infection. TF1 and TF1-ER erythroleukemia cells [34] were maintained in Roswell Park Memorial Institute (RPMI) 1640 medium supplemented with 10% FCS, 2 mM L-glutamine (Hyclone), 100 U/mL penicillin, 100 μg/mL streptomycin and 2 U/mL rh-Epo or 25 ng/mL human granulocyte macrophage colony-stimulating factor (GM-CSF, Peprotech). KU812Ep6 cells [35], a gift from Dr Sakai (Health Science Research Resources Bank, Tokyo, Japan), were maintained in RPMI-1640, 2 U/mL rh-Epo, 10% FCS, 100 U/mL penicillin, 100 μg/mL streptomycin and Insulin Transferrin Selenium-X supplement (ITS-X, Gibco), at 37 °C, 5% CO_2. Human embryonic kidney (HEK) 293T and NIH-3T3 cells were maintained in Dulbecco's modified Eagle's medium (DMEM) supplemented with 10% FCS, 2 mM L-glutamine, 100 U/mL penicillin and 100 μg/mL streptomycin.

2.2. CD36$^+$ Erythroid Progenitor Cell (EPC) Line Generation

Umbilical cord blood (CB) units from normal full-term deliveries were obtained, with the informed consent of the mothers, from the Obstetrics Unit of Saint Louis Hospital, Paris, and collected in placental blood collection bags (Maco Pharma, Tourcoing, France). Blood mononuclear cells were purified by Ficoll density gradient separation (Leucosep, Greiner Bio-one) and Hanks medium (Thermo-Fisher). Low-density cells were recovered and enriched for CD34$^+$ cells by automated cell sorting (CD34 isolation kit and autoMACS System, Miltenyi Biotec). CD34$^+$ cells were cultured in serum-free expansion medium: Iscove's Modified Dulbecco's Medium (IMDM), 15% BIT 9500 (BSA-Insulin-Transferrin, Stem Cell Technologies), 60 ng/mL rh-Stem Cell Factor (SCF), 10 ng/mL rh- interleukin-3 (IL-3), 10 ng/mL rh-IL-6, 2 U/mL rh-Epo, 100 U/mL penicillin and 100 μg/mL streptomycin. After seven days of culture, CD36$^+$ cells were isolated with biotin-coupled anti-CD36 antibody and anti-biotin microbeads on an autoMACS System. CD36$^+$ EPCs were obtained by lentivirus-mediated transduction with the *hTERT* and *E6/E7* genes from human papillomavirus type 16, as previously described [36], and were grown in expansion medium to generate a continuous CD36$^+$ EPC line.

2.3. B19 Virus Stock and Cell Inoculation

Plasma samples containing high titers of infectious B19V from asymptomatic donors at blood donation were provided by the Etablissement Français du Sang (EFS). Determination of anti-B19V IgG and IgM in plasma samples was performed using respectively LIAISON® Biotrin Parvovirus B19 IgG (cat. N° 317000) and IgM (cat. N° 317010) (Diasorin S.A., Antony, France). According to manufacturer's instructions, samples were analyzed by Chemiluminescence ImmunoAssay (CLIA) and obtained results were compared to provided negative and positive controls. Plasma samples were determined to be qualitatively negative for both B19V IgG and IgM, with a viral titer of 10^{11} B19V DNA genome equivalent (ge)/mL. The infection assay was performed using a protocol similar to that previously described [37]. Briefly, cells were maintained in exponential growth condition by dilution to 0.3×10^6 cells/mL the day before infection. On the day of infection, cells were washed and diluted in FCS-free medium without Epo, at a density of 10^7 cells/mL. B19V inoculation was carried out in a 96-well plate, with 10 μL of cell suspension (10^5 cells) and 50 μL of a 100-fold dilution of B19V plasma (10^9 ge/mL), corresponding to a mean of 500 ge/cell. The cells were then incubated at 4 °C for 2 h, and then at 37 °C, for 1 h, under an atmosphere containing 5% CO_2. We added 140 μL of complete medium and maintained the cells in culture until 72 h. Where specified, we added chloroquine (CQ) to the complete medium, at a final concentration of 25 μM. Cell viability was assessed by Trypan blue exclusion test (0.4% in PBS, Thermo Fisher Scientific), by counting blue and total cells under a microscope, with a hemocytometer. After correction for the dilution factor, viability was calculated as follows: percent of viable cells = (1 − (number of blue cells/number of total cells)) × 100. At 24, 48 or 72 h post infection (hpi), cells were centrifuged (8 min at 300× *g*), supernatants were discarded and cell pellets were frozen at −80 °C until analysis.

2.4. Fucci2a Lentivirus Production and Cell Transduction

The Fucci2a DNA sequence [38] (RDB13080, RIKEN BioSource Center) was synthetized into the LTGCPU7 lentiviral vector backbone [39] without the puromycin resistance-gene cassette, and under the control of the EF1α promoter and enhancer (GenScript). Lentiviral particles were produced by the transient transfection of HEK293T cells with the five-plasmid packaging system, by PEIpro (Polyplus transfection), as previously described [40]. These particles were then concentrated by ultracentrifugation. Infectious titers were determined in NIH-3T3 cells. We transduced 0.5×10^6 UT7/Epo-STI cells with FUCCI particles at a mean of infection of 10 in 200 µL of complete medium, and the cells were kept at 37 °C for 4 h. Cells were subsequently diluted at 0.1×10^6 cells/mL. On days 6 and 9 post-transduction, cells were analyzed by cytometry for the expression of FUCCI proteins.

2.5. UT7/Epo-FUCCI Clones Generation

UT7/Epo-FUCCI refers further to a UT7/Epo-STI pool expressing FUCCI. UT7/Epo-FUCCI clones were isolated in a U-bottom 96-well plate, by limiting dilution, with one seeded cell per well in 100 µL of complete medium. Cells were visualized by microscopy, and wells containing more than one cell or non-fluorescent cells were excluded. Clones were then separately expanded with an assigned name corresponding to their location on the plate. After expansion, each clone was considered further as a new cell line. A cell bank of 156 isolated clones was constituted (stored at 80 °C in 90% FCS, 10% DiMethyl SulfOxide (DMSO)). To reach exponential growth, cells were diluted at 0.3×10^6 cells/mL. The day after dilution, isolated clones were subjected to FUCCI expression profiling. The stability of the cell cycle profiles of the isolated clones was controlled both sequentially, for at least five independent cultures, and for 10 passages of the same culture.

2.6. Flow Cytometry Analysis of Cell Cycle Status

FUCCI-2a stable transduction in UT7-FUCCI cells allows the expression of two fluorescent proteins (m-Venus and m-Cherry) fused with specific cell cycle proteins (respectively Geminin and Cdt1). Fusion proteins are continuously expressed, but as Cdt-1 and Geminin are ubiquitinated and degraded by the proteasome at specific cell cycle stages, resting fluorescence reflects cell cycle status: during G_1 phase, the geminin-mVenus fusion protein is degraded, while the Cdt1–mCherry fusion is expressed resulting in red-fluorescence. During G_1/S transition, both proteins are present in the cells and the cell nuclei appear yellow as the green and red fluorescence overlay. In the S, G_2, and M phases, the Cdt1–m-Cherry fusion is degraded, leaving only the geminin–m-Venus fusion and resulting in a green-fluorescence signal. This dynamic color change from red to yellow to green serves as an indicator of the progression through cell cycle and division. Fucci2a bicistronic expression was monitored with an LSRFortessa cytometer (BD Biosciences, Le Pont de Claix, France). Fluorescent fusion proteins were detected with the 488 nm blue laser and a 530/30 nm bandpass filter (B530/30) for mVenus-hGeminin, and the 590 nm yellow laser and a 610/20 nm bandpass filter (Y610/20) for mCherry-hCdt1. For alternative monitoring of the cell cycle according to DNA content, cells were stained with the permeable DNA dye Hoechst 3342 (10 µg/mL) for 1 h at 37 °C, and immediately analyzed for DNA content with the 355 nm violet laser and a 450/40 nm bandpass filter (V450/40). FACSDiVa and FlowJo X software (BD Biosciences, Le Pont de Claix, France) were used to operate the instrument and for data analysis, respectively.

2.7. RNA Extraction and Duplex RT-qPCR

Total RNA was extracted from cell pellets with the RNeasy 96 QIAcubeHT kit and a QIAcubeHT machine, according to the manufacturer's instructions. The extraction step included DNase treatment for 15 min, to decrease the risk of genomic DNA amplification during PCR. Real-time reverse transcription-quantitative PCR (RT-qPCR) was performed with the *Taq*man Fast Virus one-step PCR kit (Applied Biosystems). B19 VP2 transcripts were amplified with the sense primer B19-21 5′-TGGCAGACCAGTTTCGTGAA-3′ (nts 2342-2361), the antisense primer B19-22 5′-CCGGCAAACTTCCTTGAAAA-3′ (nts 3247-3266) and the probe B19-V23 5′-VIC-CAGCTGC CCCTGTGGCCC-3′ (nts 3228-3245). For control and normalization with respect to the number of cells, we used a duplex strategy. A target sequence of the spliced beta actin transcript was selected and amplified with the sense primer actin-S 5′-GGCACCCAGCACAATGAAG-3′, the antisense primer actin-AS 5′GCCGATCCACACGGAGTACT-3′ and the probe actin-FAM 5′-FAM-TCAAG ATCATTGCTCCTCCTGAGCGC-3′. Reactions were performed on 5 µL of extracted RNA with the Quant Studio 3 PCR system. The reaction began with reverse transcription at 48 °C for 15 min, followed by inactivation of the reverse transcriptase and activation of the polymerase by heating at 95 °C for 10 min, followed by 40 cycles of 15 s at 95 °C and 30 s at 60 °C. The PCR program was optimized for amplification of the VP2 spliced transcripts rather than the VP2 genomic sequence (Figure 1A).

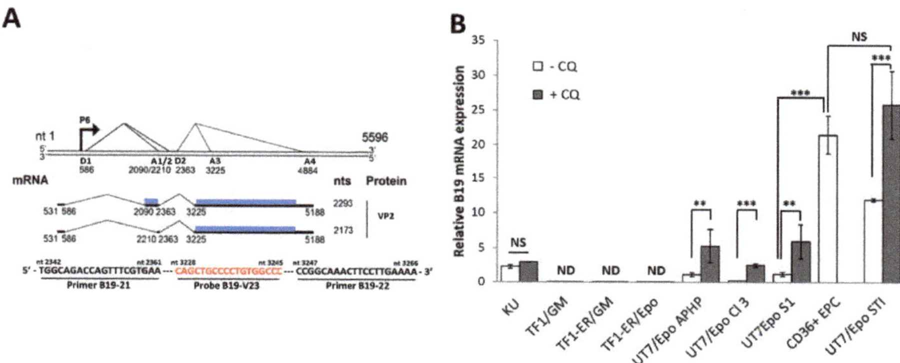

Figure 1. Comparison of the B19V sensitivity and permissiveness of hematopoietic cell lines. (**A**) B19V transcription profile (adapted from Ganaie et al., J Virol. 2018 [7]). The major transcription unit of the B19V duplex genome (GenBank accession no. AY386330) is shown to scale at the top, with the P6 promoter, 2 splice donors (D1, D2) and 4 acceptors (A1 to A4) sites. In blue, mRNA encoding the VP2 viral proteins, with nucleotides (nts). At the bottom, the primers and probe used for the RT-PCR amplification of VP2. (**B**) Bone marrow-derived primary Erythroid Progenitor Cells (CD36$^+$ EPCs), human leukemic cell lines (TF1, TF1-ER, UT7/Epo-APHP, UT7/Epo-STI) and isolated clones (KU812Ep6, UT7/Epo-cl3 and UT7/Epo-S1) were seeded in triplicate and inoculated with or without B19V in culture medium supplemented with Epo (2 U/mL)(/Epo), or granulocyte macrophage colony-stimulating factor (GM-CSF) (25 ng/mL)(/GM) for TF1 and TF1-ER. When specified, cells were cultivated with (+) or without (-) Chloroquine (CQ, 25 µM). No CQ treatment was applied to CD36+EPC. 72 h post-infection, cells were pelleted and lysed. RNA was extracted and analyzed by RT-qPCR for VP2 to quantify B19 viral genome expression, and for β-actin for cell number normalization. For each cell line, the results without B19V correspond to the negative control. Relative B19V threshold cycle (Ct) values were normalized relative to b-actin Ct and expressed according to the $2^{-\Delta\Delta Ct}$ method with normalization against mean VP2 expression for UT7/Epo-S1 cells without CQ (n = 6). Results are presented as means + SEM of 3 independent experiments. ** $p < 0.01$; *** $p < 0.001$; NS = No Significance. ND = Not detected.

2.8. Statistical Analysis

Values are expressed as mean ± SEM. Data were determined to be normally distributed as the max and the min values in each data set were <3.sd from the mean. Data were analyzed using Prism 6.0 (GraphPad Software) by one-tailed Student's t-test or one-way ANOVA with Bonferroni post-test ($\alpha < 0.05$). Parameters measured over multiple time points were analyzed with two-way ANOVA with Bonferroni post-test and time was within subject factor. The significance level displayed on figures are as follows: * p <0.05, ** p <0.01, *** p <0.001 and "NS" means no significance. Samples and experiment sizes were determined empirically to achieve sufficient statistical power.

3. Results

To assess and compare the degree of permissivity to B19V, hematopoietic cell lines were infected with B19V and maintained for 72 h. Where specified, chloroquine (CQ) was added to boost virus entry and prevent the degradation of incoming viruses through a blockade of lysosome transfer [41]. Active transcription of the B19V genome in host cells was evaluated by RT-qPCR for the *VP2* capsid gene (Figure 1A), with normalization to beta-actin gene expression. As a reference to calculate relative B19 mRNA expression, the value for UT7/Epo-S1 without chloroquine was set to 1 (Figure 1B). As previously reported, UT7/Epo-S1 and KU812Ep6 cells were less permissive to B19V than CD36$^+$ EPCs. Chloroquine treatment markedly enhanced UT7/Epo-S1 sensitivity (five-fold), but without reaching the level obtained for CD36$^+$ EPCs. VP2 expression was undetectable in both the parental TF1 erythroleukemia cell line and a TF1-ER cell line expressing a full Epo-receptor, under the control of GM-CSF or Epo, with or without chloroquine treatment. Among all UT7/Epo cell lines tested, UT7/Epo-STI was the UT7/Epo cell line tested with the highest sensitivity to B19V, with B19 mRNA levels 11.8 ± 0.2 times higher than those in UT7/EpoS1. Sensitivity was enhanced by chloroquine treatment and reaches an equivalent level compared to CD36$^+$ EPCs (UT7/Epo-STI + CQ: 25.8 ± 4.9 vs. CD36$^+$ EPCs 21.49 ± 2.7).

This increase in sensitivity was not due to resistance to B19V-induced cytotoxicity (Figure 2A). The expression kinetics of UT7/Epo-STI B19V were similar to those for CD36$^+$ EPC, with a maximum reached at 72 h post-infection for both cell lines (Figure 2B).

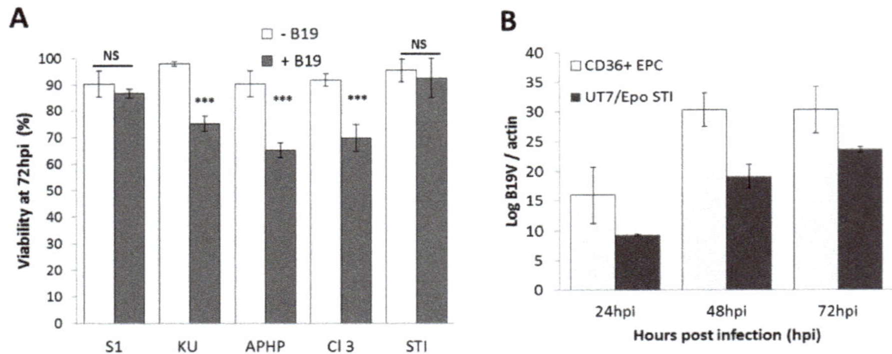

Figure 2. Comparison of B19V sensitivity of hematopoietic cell lines. (**A**) Cell viability was assessed 72 h post-infection. The results shown are the means + SD of three independent experiments. (**B**) UT7/Epo-STI cells and CD36$^+$ EPCs were cultured in triplicate, with or without B19V, for 72 h. At 24, 48 and 72 h post-inoculation, cells were collected by centrifugation. RNA was extracted from the cell pellet and VP2 mRNA levels were analyzed to quantify B19 viral DNA expression, and β-actin mRNA levels were analyzed for cell number normalization. For each cell line, results without B19V correspond to the negative control. Relative B19V threshold cycle (Ct) values were normalized relatively to the β-actin (log B19V/actin). The results shown are the means + SEM of three independent experiments. *** p <0.001; NS = No Significance.

As sensitivity to B19V is directly linked to maturation stage, we therefore subjected UT7/Epo-STI cells to the chemical (JQ1) or hormonal (TGF-β) induction of erythroid differentiation two days before B19V infection. Both treatments decreased B19V infection by a factor of about 10, to levels similar to those obtained for UT7/Epo-S1 (Figure 3).

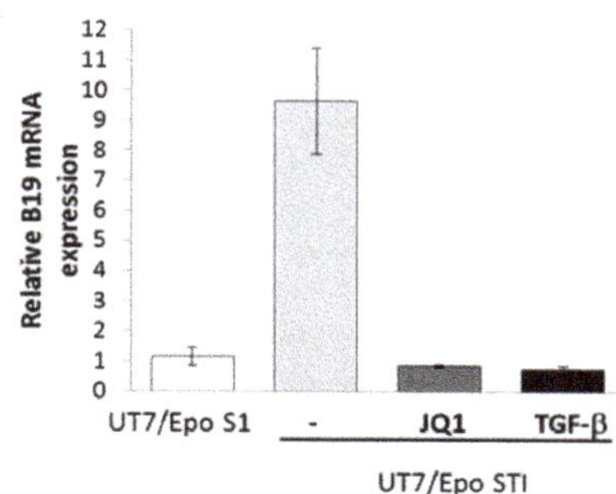

Figure 3. B19V-sensitivity of UT7/Epo-STI cells is linked to maturation stage. UT7/Epo-STI cells were cultured for 48 h before inoculation with B19V, without (-) or with JQ1 (0.5 μM) or TGF-β (2 ng/mL). 72 h post-inoculation, relative levels of B19V VP-2 mRNA were evaluated with UT7/Epo-S1 cells as the reference.

We chose to select clones according to cell cycle status, to increase sensitivity to B19V. UT7/Epo-STI cells were transduced with FUCCI (Fluorescence Ubiquitination Cell Cycle Indicator) lentiviral particles to generate the UT7/Epo-FUCCI cell line (Figure 4A). The FUCCI cell cycle sensor allows cell cycle analysis of living cells. The UT7/Epo-FUCCI cell line presents three different color profiles, from green, corresponding to the S, G2 and M phases, to red, consequent to G1 phase, with a green plus red (yellow) overlay indicating the G1-to-S transition. We checked that these dynamic color changes correctly represented progression through the cell cycle and division, by staining the DNA content of UT7/Epo-FUCCI cells with Hoechst stain (Figure 4B). An overlay of the DNA staining and FUCCI profiles resulted in a perfect match between the cell cycle status assigned by the FUCCI technique and that assigned on the basis of DNA content: G1 (red) FUCCI cells were detected at a DNA content of 2N, whereas cells at S-G2-M (green) had DNA content peaks of 2N to 4N, consistent with the expected replication of the DNA replication before mitosis. The G1/S transition phase (yellow) population was located at the 2N peak, with a slight shift from G1 cells. Overall, these results confirm that FUCCI is an appropriate cell cycle indicator for UT7/Epo cells.

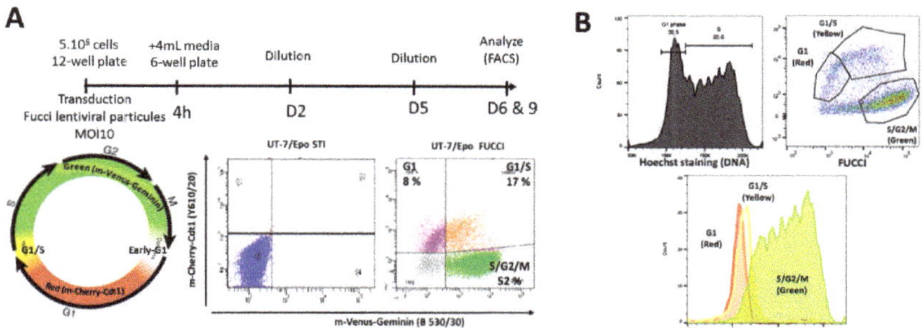

Figure 4. Generation of a UT7/Epo-STI cell line with stable expression of the Fluorescence Ubiquitination Cell Cycle Indicator (FUCCI). (**A**) Experimental design for the generation of the UT7/Epo-FUCCI cell line. Bottom: Two-color cell cycle mapping with the FUCCI2a Cell Cycle Sensor and right, flow cytometry analysis of exponentially growing UT7/Epo-STI and UT7/Epo-FUCCI cells. The profile shown corresponds to one representative experiment. (**B**) DNA content and FUCCI profiles for the same sample. Exponentially growing UT7/Epo-FUCCI cells were stained with Hoechst 33342. DNA content (Hoechst on the *x*-axis; cell count on the *y*-axis) and FUCCI proteins (m-Venus on the *x*-axis; m-Cherry on the *y*-axis) were concomitantly evaluated by flow cytometry. Bottom: Overlay of gated cell cycle populations, as determined by FUCCI analysis with DNA content profile.

We then generated different UT7/Epo-FUCCI clones, each obtained by limiting dilution and culture from a single fluorescent cell. Unlike the UT7/Epo-FUCCI pool, these clones were generated from single cells and 100% of the cells were therefore transduced: the colorless cells of the FUCCI profile correspond to the early G1 (eG1) phase and were included in the G1 phase for the purposes of this analysis. We isolated 156 independent clones and expanded each as new sub-lines. FUCCI-negative clones, accounting for one third of the cells isolated, were excluded. We studied the cell cycle status of FUCCI-positive clones. We defined three types of cell cycle profile in a total of 97 clones: (1) 54 clones presented a cell cycle with more than 60% of the cells in G1 phase (55.7% of clones); (2) 29 clones presented a balanced distribution of cells between the G1 and S/G2/M phases (29.9% of clones); (3) 14 clones had a high percentage of cells in the S/G2/M phases (14.4% of clones). With the aim of analyzing these three types of cell cycle profile, we selected 11 isolated clones in regard to the diversity of their cell cycle patterns at exponential growth (Figure S1).

We evaluated sensitivity to B19V of exponentially growing selected clones as previously described (Figure 5A). Permissivity ranged from 1-fold to 35-fold relative to UT7/Epo-S1. Three populations were assigned: group I, with a sensitivity close to that of UT7/Epo-S1 (six clones); group II, gathering clones with UT7/Epo-FUCCI-like permissivity (three clones); group III, containing clones B12 and E2, displaying remarkable sensitivity to B19V infection. Interestingly, classification based on B19V sensitivity seemed to group together clones with similar cell cycle patterns (Figure 5B). The cell cycle profiles of group I clones displayed a predominance of the G1 phase. Group II clones displayed a balance between the G1 and S/G2/M phases, as observed for the original UT7/Epo-FUCCI pool. Finally, the S/G2/M cell population predominantly represents the group III profile, with 82% and 75.8% for B12 and E2, respectively.

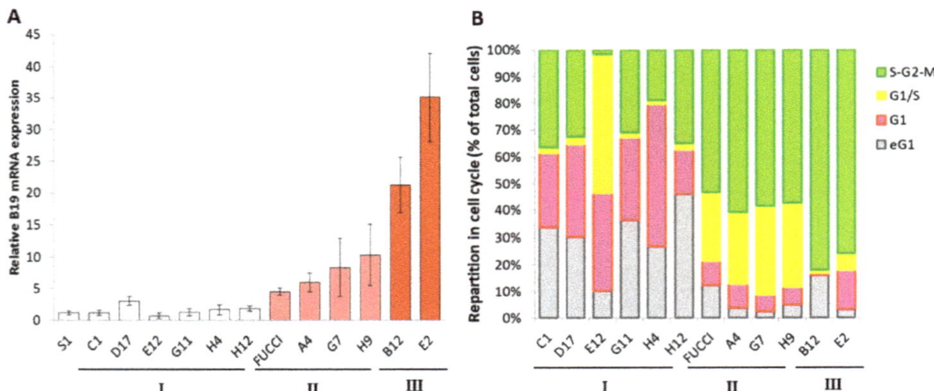

Figure 5. Improvement of B19V sensitivity and permissiveness according to cell cycle status. (**A**) UT7/Epo-S1 cells (S1), UT7/Epo-STI cells expressing the FUCCI system (FUCCI) and 11 UT7/FUCCI-derived isolated clones were inoculated with B19V. Relative levels of B19V mRNA were determined 72h post-infection, with UT7/Epo-S1 as the reference, and cell lines were classified on the basis of B19V sensitivity as group I for S1-equivalent clones, group II for FUCCI-equivalent clones, and group III for highly permissive clones. The results shown are the means + SD of 3 independent experiments for groups I and II, and $n = 9$ for group III clones. (**B**) The cell cycle status of exponentially growing FUCCI cell lines and isolated clones was assessed by flow cytometry. The results shown are the means + SEM of three independent measurements.

We evaluated the correlation between cell cycle stage and B19V sensitivity, by analyzing the correlation of the coefficient of determination (R^2) obtained for eG1, G1, G1/S and S/G2/M with B19V mRNA levels (Figure S2 and Figure 6). For G1 cell cycle parameter, R^2 was low, with values of 0.3743 for early G1 (eG1) and 0.5148 for G1.

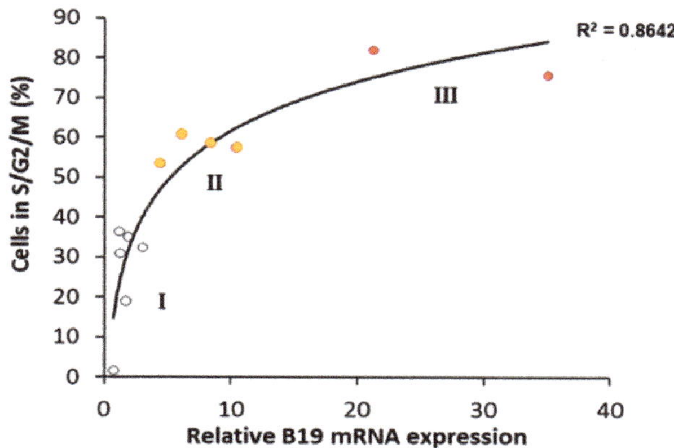

Figure 6. Comparison of relative levels of B19 mRNA (as in Figure 5A) and cell cycle status (as in Figure 5B). Each dot corresponds to the mean result for a single cell line or clone ($n = 3$), classified to groups I (white marks ○), II (orange marks •) and III (red marks •). Logarithmic regression analysis and R^2 values are presented.

The highest value was obtained for the S/G2/M phase of the cell cycle, with $R^2 = 0.8642$, demonstrating an excellent agreement between the percentage of cells in S/G2/M stage and B19V sensitivity (Figure 6).

Results obtained show that E2 and B12 are the most permissive clones. In order to assess the stability of their cell cycle profile, E2 and B12 were submitted to serial culture passage and cultivated for up to 80 days. Cell cycle profiles were analyzed by flow cytometry according to FUCCI expression (Figure 7 and Figure S3).

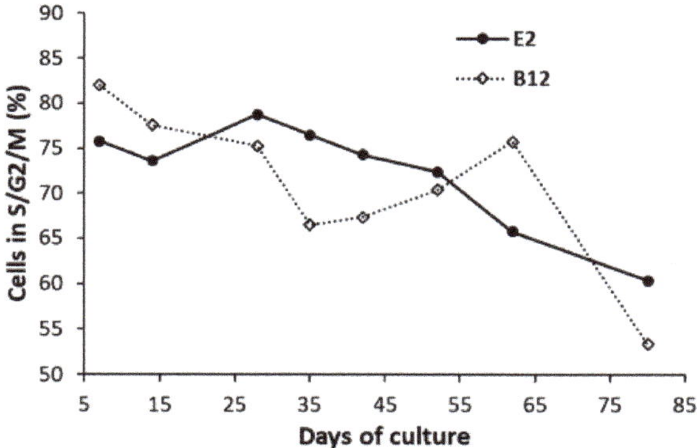

Figure 7. S/G2/M cell percentage of E2 and B12 clones during serial culture passage. UT7/Epo-FUCCI clone E2 and B12 were maintained in culture during 80 days after thawing with up to 25 culture passages. At indicated time points, cell cycle was evaluated by flow cytometry according to specific FUCCI protein expression. The graph represents the evolution of the percentage of cells in S/G2/M cell cycle status during cell culture, as calculated in Fig S3A (E2) and B (B12) Results shown are the means of three independent measurements.

E2 and B12 clone showed a decrease of percentage of S/G2/M cells from, respectively, 75.8% and 82 % at day 7, to 60.4 and 53.4% at day 80, suggesting a higher stability of E2 cell line. B19V permissivity stability was also analyzed throughout serial culture passage and compared to UT7/Epo-S1 reference (Figure 8).

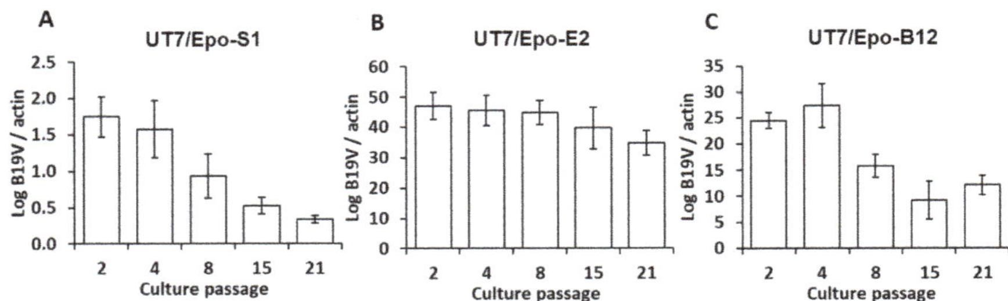

Figure 8. B19V permissivity throughout serial culture passage—UT7/Epo-S1 cell line (**A**) E2 (**B**) and B12 (**C**) clones were maintained in culture during 67 days after thawing with up to 21 culture passages. At indicated time points, B19V permissivity was evaluated by measuring levels of VP2 mRNA and normalized with β-actin mRNA by the $2^{-\Delta Ct}$ method. The results shown are the means + SD of 3 independent measurements.

While UT7/Epo-S1 and B12 clone showed a decrease of sensitivity to B19V, E2 clones' permissivity seems to be stable at high culture passage.

Overall, our results identify two highly permissive UT7 clones, B12 and E2, and show that the S/G2/M phase is essential for B19V sensitivity.

4. Discussion

Most of the currently available approaches focus on the detection of B19V DNA, but there is a need for a suitable in vitro method for the direct quantification of virion infectivity, for use in assessing neutralizing antibodies, to evaluate viral inactivation assays or in antiviral research field. However, efforts to develop such methods have been hampered by the lack of suitable B19-sensitive cell lines in vitro. We describe here a new cell model with high sensitivity to B19V infection. As expected, hematopoietic cell lines of different origins were heterogeneous but, surprisingly, our results also demonstrate considerable variability among cell lines derived from the same patient, all named UT-7/Epo. This variability of B19V sensitivity may depend on erythroid stage, B19V entry receptor expression and/or the activation of specific signaling pathways [7]. Our findings highlight the need for tracking criteria to ensure the stability of the cell line used. As we show here that B19V sensitivity is linked to S/G2/M cell cycle status, we propose the use of cell cycle status to define the optimal cells for selection and as a keeper of clone stability. This study proposes an improved cellular model for the detection of B19V infectious units, with a sensitivity up to 35 times higher than previously achieved.

B19V has an extremely strong tropism for human erythroid progenitor cells. Since the discovery that B19V inhibits erythroid colony formation in bone marrow cultures by inducing the premature apoptosis of erythroid progenitor cells [42], numerous approaches and studies attempt to find a method of virus culture in vitro. Primary [43,44] or immortalized [36] $CD36^+$ erythroid progenitor cells (EPC) derived from hematopoietic stem cells were the most permissive cell models for B19V infection. $CD36^+$ EPCs reflect the natural etiologic B19V cell host, but the main problem with the use of this model is the difficulty in obtaining a continuously homogeneous cell line, with respect to differentiation stage, proliferation rate and metabolic activity. Moreover, the reagents and cytokines required for cell culture (SCF, Il-3, Il-6, Epo) preclude the use of $CD36^+$ EPCs for routine B19V cell-based detection methods. To counteract this lack of suitability, cancer cell lines constitute a sound, practical, cost-effective alternative model, overcoming these difficulties. In recent years, many cancer cell lines have been tested, but only a few erythroid leukemic (KU812) [35] or mega-karyoblastoid cell lines (UT-7) [26] with erythroid characteristics support B19V replication. In our study, we chose also to investigate TF-1 permissivity. The TF-1 cell line is derived from the bone marrow aspirate of an erythroleukemic patient [45]. These cells display marked erythroid morphological and cytochemical features common to CD36$^+$ EPCs, and the constitutive expression of globin genes highlights the commitment of the cells to the erythroid lineage [46]. Surprisingly, Gallinella et al. showed that TF-1 cells allow only B19V entry, with impaired viral genome replication and transcription, as shown by the presence of single-stranded DNA, and the absence of double-stranded DNA and RNA in B19V-infected TF-1 cells [47]. As previously described, no B19V RNA was detectable in the TF-1 cell line. The cellular factors involved in the transcriptional activation of the B19V promoter contribute to the restriction of permissiveness. Two factors, erythropoietin (Epo) and STAT-5, are key factors involved in B19V replication and transcription [48]. TF-1 cells express a truncated and mutated form of the Epo receptor [49], leading to impaired STAT-5 activation [34]. In the TF-1-ER cell line, stable ectopic expression of a full-length Epo receptor restores Epo-induced proliferation and STAT-5 activation. Here, despite Epo receptor signaling and STAT-5 activation, we found no evidence of B19V transcription, reflecting the involvement of unknown processes in the molecular mechanisms controlling B19V permissivity.

The first cell line reported to be permissive for B19 infection was an Epo-dependent subclone of UT-7, a mega-karyoblastoid cell line [32]. In 2006, Wong et al. published a comparative study of B19V sensitivity and permissivity in various cell lines [22]. They obtained evidence for the B19V infection of UT7/Epo and KU812Ep6 cells, although the percentage of B19V-positive cells was low (<1% immunofluorescent B19V$^+$ cells). UT7/Epo-S1, a subclone of UT7/Epo obtained by limiting dilution and screening for B19V susceptibility [26], had the highest sensitivity, with approximately 15% of the cells staining positive for B19V [29]. Permissivity is restricted to a subset of cells, but the degree of viral DNA replication in these cells is similar to that in EPCs [50,51]. Since its characterization, the UT7/Epo-S1 cell line has been widely used to investigate the molecular mechanisms of B19V infection and to develop antiviral strategies against B19 [28]. We used UT7/Epo-S1 as a reference and compared the sensitivity of UT7/Epo cells from different laboratories. B19V permissivity seemed to be similar in the various UT7/Epo cells, but UT7/Epo-STI cells displayed levels of B19V gene expression almost 10 times higher than those in UT7/Epo-S1 cells. UT7/Epo-STI cells have been cultured with great care to ensure the preservation of their erythroid features, and they undergo erythroid differentiation following treatment with JQ1, a Bet-domain protein inhibitor [33] or TGF-β1 [52]. However attempts to characterize cell lines have been hampered by the heterogeneity of continually evolving multiple sub-clonal leukemic populations, as revealed at the cytogenetic level by the unstable karyotype documented at various time points for UT-7: at the admission of the patient to hospital (44 chromosomes, XY), at the first cell line characterization (92 ± 6 chromosomes, XXYY) [27], in subsequent publication (82 ± 4 chromosomes, XXYY [53] and in our own cell line in 2017 (72 ± 13 chr., XXYY; unpublished data). This karyotype heterogeneity highlights the presence of heterogeneous subclones within cell lines and might account for the variation of B19V sensitivity among UT-7 cell lines and clones.

The cell cycle is known to be crucial for erythroid differentiation, ensuring precise coordination of the critical differentiation process by Epo and erythroid-specific transcription factors [54,55]. We decided to select clones on the basis of cell cycle status. The FUCCI system represents a convenient approach to track cell cycle as its readability allows analysis of living cells at a single cell level [56]. By using clones with different cell cycle status, we demonstrated a strong correlation between S/G2/M cell cycle status and permissivity. A total RNA analysis, with a comparison of transcriptomes from highly permissive (group III) and less sensitive (group I & II) clones would shed more light on the molecular mechanisms involved in B19V infection. Along with karyotype, study of permissive versus restrictive clones could lead to the discovery of key molecular factors for B19V permissivity.

B19V has been shown to induce cell cycle arrest at G2 phase [26,57], but the importance of cell cycle status for B19V entry has not been investigated. A complex combination of multiple factors, including differentiation stage, specific cell cycle status, surface receptor and co-receptor, signaling pathways and transcription factors, may account for the difficulty of identifying the best cellular model for completion of the B19 viral cycle. We describe here two clones, E2 and B12, with a permissivity for B19 up to 35 times higher than that of the previously described references. By comparison with their less sensitive counterparts (groups I & II), these new highly permissive cell models (group III) constitute a potential advance towards understanding the crucial molecular determinants of B19V infectivity.

In addition to the use of E2 and B12 clones to investigate the molecular mechanisms of B19 infection, cell-based methods can be used for the detection/quantification of B19 infectious units, at low levels (<10^4 DNA geq), in human fluids and tissues. There is a need for a practical in vitro method for the direct quantification of virion infectivity, as applied to the screening and/or assessment of neutralizing antibodies, antiviral drugs, and viral inactivation assays.

In the context of plasma-derived medicinal products, due to the lack of a suitable in vitro culture assay for B19, animal parvoviruses are currently used as a model for B19V, to assess B19 viral reduction during manufacturing processes. However, it remains unclear whether these models accurately reflect the behavior of B19V. Yunoki et al. reported an unexpected greater resistance of B19V than of canine parvoviruses (CPVs) to heating [58]. Mani et al. pointed up a discrepancy between animal and human parvovirus in a comparative study, describing the remarkable stability of the B19 viral genome in its encapsidated state [59]. Animal model parvoviruses display a certain resistance to heat inactivation [60,61] and pH stability [62], but comparative studies have indicated that they may behave differently from human B19 [2]. As E2 and B12 were the most sensitive cells in our study, with a permissivity up to 35 times higher than that of previously established references, they could allow the use of human parvovirus for the testing of viral inactivation processes, and the results of these tests would reflect the behavior of the native human virus.

Given the severity of B19V infection in immunocompromised patients, the development of antiviral strategies and drugs directed against B19V is of the highest relevance [28]. Depending on the immune state of the infected patient, acute infections can be clinically severe, and an impaired immune response can lead to persistent infections. The administration of high-dose intravenous immunoglobulins (IVIG) is currently considered the only available option for neutralizing the infectious virus [23,63]. In addition to the use of IVIG, the discovery of antiviral drugs with significant activity against B19 would offer important opportunities in the treatment and management of severe clinical manifestations. Two factors have critically limited the search for compounds to date. Firstly, the lack of a standardized and sensitive in vitro cell culture model has hampered advances in this field. Due to its usefulness, practicability and sensitivity, our cell model could replace the use of $CD36^+$ EPCs and UT7/Epo-S1 cells in the discovery and evaluation of antiviral candidate compounds. Secondly, antiviral research requires native B19 infectious particles. However, B19V particles from viremic patients limit the feasibility of high-throughput screening against the available chemical libraries. No appropriate system for cell culture and in vitro virus production are available to date. UT-7 cells have been reported to produce infectious viral particles in vitro [43,64], but only a few UT-7 cells are infected and virions are produced in small numbers [65–67]. The strategies used here for the selection of clones permissive for B19V could also be used to select highly productive clones. Our most permissive clones, E2 and B12, could be challenged and assessed for in vitro viral production.

5. Conclusions

Altogether, we propose here an improved cell model with a high degree of permissivity to B19V, allowing the sensitive detection of infectious particles of B19. This finding opens up challenging new perspectives for basic research on the B19V life cycle. It may also offer opportunities for improving key steps in a number of critical applied approaches, including the sensitive evaluation of B19V virions in manufactured blood-derived products, and new strategies for B19V production in vitro.

Supplementary Materials: The following are available online at http://www.mdpi.com/1999-4915/12/12/1467/s1, Figure S1: Cell cycle profile of the exponentially growing UT7/Epo-FUCCI cell line and clones, Figure S2:

Comparison of relative levels of B19 mRNA and percentage of cells in respective cell cycle status. Figure S3: Cell cycle stability of E2 and B12 clones during serial culture passage.

Author Contributions: Conceptualization, C.D., B.Y. and Z.K.; Methodology, C.D., A.L. and Z.K.; Validation, C.D., B.Y. and Z.K.; Formal Analysis, Z.K.; Investigation, C.D., A.L. and Z.K..; Resources, C.D., A.L., B.Y., E.P. and Z.K.; Data Curation, C.D., A.L. and Z.K.; Writing—Original Draft Preparation, Z.K.; Writing—Review & Editing, C.D., B.Y., A.L., O.G., E.P., S.C. and Z.K.; Visualization, C.D. and Z.K.; Supervision, B.Y., S.C. and Z.K.; Project Administration, Z.K. All authors have read and agreed to the published version of the manuscript.

Funding: This research received no external funding. This work was supported by the *Commissariat à l'Energie Atomique et aux Energies Alternatives* (CEA).

Acknowledgments: The authors acknowledge the RIKEN BioResource Center (RIKEN BRC) for providing the pCAG-Fucci2a vector (provided by Richard Mort) and the National Heart, Lung and Blood Institute (NHLBI) for UT7/Epo-S1 cells (provided by Kazuo Sugamura and Susan Wong). We thank Morinet and Capron for expertise of B19 inoculation; Anne-Sophie Gallouët and Mario Gomez-Pacheco (Infectious Disease Models and Innovative Therapies-IDMIT-Fontenay-aux-Roses, France) for their help and support for cytometry studies, Corinne Bergeron (LFB) for developing the RNA extraction and qPCR protocol; Valérie Barlet and Isabelle Dupont (EFS Grand Est, Metz-Tessy site) for providing the B19-positive plasma sample.

Conflicts of Interest: The authors declare no conflict of interest.

References

1. Cotmore, S.F.; Agbandje-McKenna, M.; Canuti, M.; Chiorini, J.A.; Eis-Hubinger, A.M.; Hughes, J.; Mietzsch, M.; Modha, S.; Ogliastro, M.; Penzes, J.J.; et al. ICTV Virus Taxonomy Profile: Parvoviridae. *J. Gen. Virol.* **2019**, *100*, 367–368. [CrossRef] [PubMed]
2. Mietzsch, M.; Penzes, J.J.; Agbandje-McKenna, M. Twenty-Five Years of Structural Parvovirology. *Viruses* **2019**, *11*, 362. [CrossRef]
3. Brass, C.; Elliott, L.M.; Stevens, D.A. Academy rash. A probable epidemic of erythema infectiosum ('fifth disease'). *JAMA* **1982**, *248*, 568–572. [CrossRef] [PubMed]
4. Anderson, M.J.; Jones, S.E.; Fisher-Hoch, S.P.; Lewis, E.; Hall, S.M.; Bartlett, C.L.; Cohen, B.J.; Mortimer, P.P.; Pereira, M.S. Human parvovirus, the cause of erythema infectiosum (fifth disease)? *Lancet* **1983**, *1*, 1378. [CrossRef]
5. Hosszu, E.; Sallai, A. Human parvovirus B19 infection in a child suffering from chronic arthritis. *Orv. Hetil.* **1997**, *138*, 611–613. [PubMed]
6. Frickhofen, N.; Chen, Z.J.; Young, N.S.; Cohen, B.J.; Heimpel, H.; Abkowitz, J.L. Parvovirus B19 as a cause of acquired chronic pure red cell aplasia. *Br. J. Haematol.* **1994**, *87*, 818–824. [CrossRef] [PubMed]
7. Ganaie, S.S.; Qiu, J. Recent Advances in Replication and Infection of Human Parvovirus B19. *Front. Cell Infect. Microbiol.* **2018**, *8*, 166. [CrossRef]
8. Young, N.; Harrison, M.; Moore, J.; Mortimer, P.; Humphries, R.K. Direct demonstration of the human parvovirus in erythroid progenitor cells infected in vitro. *J. Clin. Investig.* **1984**, *74*, 2024–2032. [CrossRef]
9. Kurtzman, G.J.; Ozawa, K.; Cohen, B.; Hanson, G.; Oseas, R.; Young, N.S. Chronic bone marrow failure due to persistent B19 parvovirus infection. *N. Engl. J. Med.* **1987**, *317*, 287–294. [CrossRef]
10. Duncan, J.R.; Potter, C.B.; Cappellini, M.D.; Kurtz, J.B.; Anderson, M.J.; Weatherall, D.J. Aplastic crisis due to parvovirus infection in pyruvate kinase deficiency. *Lancet* **1983**, *2*, 14–16. [CrossRef]
11. Bertrand, J.; Dennis, M.; Cutler, T. It's Complicated: Parvovirus B19 in Thalassemia. *Am. J. Med.* **2017**, *130*, 1269–1271. [CrossRef] [PubMed]
12. Obeid Mohamed, S.O.; Osman Mohamed, E.M.; Ahmed Osman, A.A.; Abdellatif MohamedElmugadam, F.A.; Abdalla Ibrahim, G.A. A Meta-Analysis on the Seroprevalence of Parvovirus B19 among Patients with Sickle Cell Disease. *Biomed. Res. Int.* **2019**, *2019*, 2757450. [CrossRef] [PubMed]
13. Obeid, K.M. Infections with DNA Viruses, Adenovirus, Polyomaviruses, and Parvovirus B19 in Hematopoietic Stem Cell Transplant Recipients and Patients with Hematologic Malignancies. *Infect. Dis. Clin. N. Am.* **2019**, *33*, 501–521. [CrossRef]
14. Tolfvenstam, T.; Papadogiannakis, N.; Norbeck, O.; Petersson, K.; Broliden, K. Frequency of human parvovirus B19 infection in intrauterine fetal death. *Lancet* **2001**, *357*, 1494–1497. [CrossRef]
15. Brown, T.; Anand, A.; Ritchie, L.D.; Clewley, J.P.; Reid, T.M. Intrauterine parvovirus infection associated with hydrops fetalis. *Lancet* **1984**, *2*, 1033–1034. [CrossRef]

16. Satake, M.; Hoshi, Y.; Taira, R.; Momose, S.Y.; Hino, S.; Tadokoro, K. Symptomatic parvovirus B19 infection caused by blood component transfusion. *Transfusion* **2011**, *51*, 1887–1895. [CrossRef]
17. Nagaharu, K.; Sugimoto, Y.; Hoshi, Y.; Yamaguchi, T.; Ito, R.; Matsubayashi, K.; Kawakami, K.; Ohishi, K. Persistent symptomatic parvovirus B19 infection with severe thrombocytopenia transmitted by red blood cell transfusion containing low parvovirus B19 DNA levels. *Transfusion* **2017**, *57*, 1414–1418. [CrossRef]
18. Marano, G.; Vaglio, S.; Pupella, S.; Facco, G.; Calizzani, G.; Candura, F.; Liumbruno, G.M.; Grazzini, G. Human Parvovirus B19 and blood product safety: A tale of twenty years of improvements. *Blood Transfus.* **2015**, *13*, 184–196. [CrossRef]
19. Baylis, S.A.; Wallace, P.; McCulloch, E.; Niesters, H.G.M.; Nubling, C.M. Standardization of Nucleic Acid Tests: The Approach of the World Health Organization. *J. Clin. Microbiol.* **2019**, *57*. [CrossRef]
20. Molenaar-de Backer, M.W.; Russcher, A.; Kroes, A.C.; Koppelman, M.H.; Lanfermeijer, M.; Zaaijer, H.L. Detection of parvovirus B19 DNA in blood: Viruses or DNA remnants? *J. Clin. Virol.* **2016**, *84*, 19–23. [CrossRef]
21. Baylis, S.A.; Ma, L.; Padley, D.J.; Heath, A.B.; Yu, M.W.; Grp, C.S. Collaborative study to establish a World Health Organization International genotype panel for parvovirus B19 DNA nucleic acid amplification technology (NAT)-based assays. *Vox Sang.* **2012**, *102*, 204–211. [CrossRef] [PubMed]
22. Wong, S.; Brown, K.E. Development of an improved method of detection of infectious parvovirus B19. *J. Clin. Virol.* **2006**, *35*, 407–413. [CrossRef] [PubMed]
23. Eid, A.J.; Ardura, M.I.; The AST Infectious Diseases Community of Practice. Human parvovirus B19 in solid organ transplantation: Guidelines from the American society of transplantation infectious diseases community of practice. *Clin. Transplant.* **2019**, *33*, e13535. [CrossRef]
24. Sundin, M.; Lindblom, A.; Orvell, C.; Barrett, A.J.; Sundberg, B.; Watz, E.; Wikman, A.; Broliden, K.; Le Blanc, K. Persistence of human parvovirus B19 in multipotent mesenchymal stromal cells expressing the erythrocyte P antigen: Implications for transplantation. *Biol. Blood Marrow Transpl.* **2008**, *14*, 1172–1179. [CrossRef] [PubMed]
25. Watanabe, K.; Otabe, K.; Shimizu, N.; Komori, K.; Mizuno, M.; Katano, H.; Koga, H.; Sekiya, I. High-sensitivity virus and mycoplasma screening test reveals high prevalence of parvovirus B19 infection in human synovial tissues and bone marrow. *Stem Cell Res. Ther.* **2018**, *9*, 80. [CrossRef] [PubMed]
26. Morita, E.; Tada, K.; Chisaka, H.; Asao, H.; Sato, H.; Yaegashi, N.; Sugamura, K. Human parvovirus B19 induces cell cycle arrest at G(2) phase with accumulation of mitotic cyclins. *J. Virol.* **2001**, *75*, 7555–7563. [CrossRef]
27. Komatsu, N.; Nakauchi, H.; Miwa, A.; Ishihara, T.; Eguchi, M.; Moroi, M.; Okada, M.; Sato, Y.; Wada, H.; Yawata, Y.; et al. Establishment and characterization of a human leukemic cell line with megakaryocytic features: Dependency on granulocyte-macrophage colony-stimulating factor, interleukin 3, or erythropoietin for growth and survival. *Cancer Res.* **1991**, *51*, 341–348.
28. Manaresi, E.; Gallinella, G. Advances in the Development of Antiviral Strategies against Parvovirus B19. *Viruses* **2019**, *11*, 659. [CrossRef]
29. Manaresi, E.; Bua, G.; Bonvicini, F.; Gallinella, G. A flow-FISH assay for the quantitative analysis of parvovirus B19 infected cells. *J. Virol. Methods* **2015**, *223*, 50–54. [CrossRef]
30. Bonvicini, F.; Manaresi, E.; Bua, G.; Venturoli, S.; Gallinella, G. Keeping pace with parvovirus B19 genetic variability: A multiplex genotype-specific quantitative PCR assay. *J. Clin. Microbiol.* **2013**, *51*, 3753–3759. [CrossRef]
31. Bonvicini, F.; Mirasoli, M.; Manaresi, E.; Bua, G.; Calabria, D.; Roda, A.; Gallinella, G. Single-cell chemiluminescence imaging of parvovirus B19 life cycle. *Virus Res.* **2013**, *178*, 517–521. [CrossRef] [PubMed]
32. Shimomura, S.; Komatsu, N.; Frickhofen, N.; Anderson, S.; Kajigaya, S.; Young, N.S. First continuous propagation of B19 parvovirus in a cell line. *Blood* **1992**, *79*, 18–24. [CrossRef] [PubMed]
33. Goupille, O.; Penglong, T.; Lefevre, C.; Granger, M.; Kadri, Z.; Fucharoen, S.; Maouche-Chretien, L.; Leboulch, P.; Chretien, S. BET bromodomain inhibition rescues erythropoietin differentiation of human erythroleukemia cell line UT7. *Biochem. Biophys. Res. Commun.* **2012**, *429*, 1–5. [CrossRef] [PubMed]
34. Chretien, S.; Varlet, P.; Verdier, F.; Gobert, S.; Cartron, J.P.; Gisselbrecht, S.; Mayeux, P.; Lacombe, C. Erythropoietin-induced erythroid differentiation of the human erythroleukemia cell line TF-1 correlates with impaired STAT5 activation. *EMBO J.* **1996**, *15*, 4174–4181. [CrossRef]

35. Miyagawa, E.; Yoshida, T.; Takahashi, H.; Yamaguchi, K.; Nagano, T.; Kiriyama, Y.; Okochi, K.; Sato, H. Infection of the erythroid cell line, KU812Ep6 with human parvovirus B19 and its application to titration of B19 infectivity. *J. Virol. Methods* **1999**, *83*, 45–54. [CrossRef]
36. Wong, S.; Keyvanfar, K.; Wan, Z.; Kajigaya, S.; Young, N.S.; Zhi, N. Establishment of an erythroid cell line from primary CD36+ erythroid progenitor cells. *Exp. Hematol.* **2010**, *38*, 994–1005. [CrossRef]
37. Nguyen, Q.T.; Wong, S.; Heegaard, E.D.; Brown, K.E. Identification and characterization of a second novel human erythrovirus variant, A6. *Virology* **2002**, *301*, 374–380. [CrossRef]
38. Mort, R.L.; Ford, M.J.; Sakaue-Sawano, A.; Lindstrom, N.O.; Casadio, A.; Douglas, A.T.; Keighren, M.A.; Hohenstein, P.; Miyawaki, A.; Jackson, I.J. Fucci2a: A bicistronic cell cycle reporter that allows Cre mediated tissue specific expression in mice. *Cell Cycle* **2014**, *13*, 2681–2696. [CrossRef] [PubMed]
39. Bhukhai, K.; de Dreuzy, E.; Giorgi, M.; Colomb, C.; Negre, O.; Denaro, M.; Gillet-Legrand, B.; Cheuzeville, J.; Paulard, A.; Trebeden-Negre, H.; et al. Ex Vivo Selection of Transduced Hematopoietic Stem Cells for Gene Therapy of beta-Hemoglobinopathies. *Mol. Ther.* **2018**, *26*, 480–495. [CrossRef] [PubMed]
40. Westerman, K.A.; Ao, Z.; Cohen, E.A.; Leboulch, P. Design of a trans protease lentiviral packaging system that produces high titer virus. *Retrovirology* **2007**, *4*, 96. [CrossRef] [PubMed]
41. Bonsch, C.; Kempf, C.; Mueller, I.; Manning, L.; Laman, M.; Davis, T.M.; Ros, C. Chloroquine and its derivatives exacerbate B19V-associated anemia by promoting viral replication. *PLoS Negl. Trop. Dis.* **2010**, *4*, e669. [CrossRef] [PubMed]
42. Mortimer, P.P.; Humphries, R.K.; Moore, J.G.; Purcell, R.H.; Young, N.S. A human parvovirus-like virus inhibits haematopoietic colony formation in vitro. *Nature* **1983**, *302*, 426–429. [CrossRef] [PubMed]
43. Wong, S.; Zhi, N.; Filippone, C.; Keyvanfar, K.; Kajigaya, S.; Brown, K.E.; Young, N.S. Ex vivo-generated CD36+ erythroid progenitors are highly permissive to human parvovirus B19 replication. *J. Virol.* **2008**, *82*, 2470–2476. [CrossRef] [PubMed]
44. Bua, G.; Manaresi, E.; Bonvicini, F.; Gallinella, G. Parvovirus B19 Replication and Expression in Differentiating Erythroid Progenitor Cells. *PLoS ONE* **2016**, *11*, e0148547. [CrossRef]
45. Kitamura, T.; Tange, T.; Terasawa, T.; Chiba, S.; Kuwaki, T.; Miyagawa, K.; Piao, Y.F.; Miyazono, K.; Urabe, A.; Takaku, F. Establishment and characterization of a unique human cell line that proliferates dependently on GM-CSF, IL-3, or erythropoietin. *J. Cell Physiol.* **1989**, *140*, 323–334. [CrossRef]
46. Kitamura, T.; Tojo, A.; Kuwaki, T.; Chiba, S.; Miyazono, K.; Urabe, A.; Takaku, F. Identification and analysis of human erythropoietin receptors on a factor-dependent cell line, TF-1. *Blood* **1989**, *73*, 375–380. [CrossRef]
47. Gallinella, G.; Manaresi, E.; Zuffi, E.; Venturoli, S.; Bonsi, L.; Bagnara, G.P.; Musiani, M.; Zerbini, M. Different patterns of restriction to B19 parvovirus replication in human blast cell lines. *Virology* **2000**, *278*, 361–367. [CrossRef]
48. Ganaie, S.S.; Zou, W.; Xu, P.; Deng, X.; Kleiboeker, S.; Qiu, J. Phosphorylated STAT5 directly facilitates parvovirus B19 DNA replication in human erythroid progenitors through interaction with the MCM complex. *PLoS Pathog.* **2017**, *13*, e1006370. [CrossRef]
49. Winkelmann, J.C.; Ward, J.; Mayeux, P.; Lacombe, C.; Schimmenti, L.; Jenkins, R.B. A translocated erythropoietin receptor gene in a human erythroleukemia cell line (TF-1) expresses an abnormal transcript and a truncated protein. *Blood* **1995**, *85*, 179–185. [CrossRef]
50. Bonvicini, F.; Filippone, C.; Delbarba, S.; Manaresi, E.; Zerbini, M.; Musiani, M.; Gallinella, G. Parvovirus B19 genome as a single, two-state replicative and transcriptional unit. *Virology* **2006**, *347*, 447–454. [CrossRef]
51. Bonvicini, F.; Filippone, C.; Manaresi, E.; Zerbini, M.; Musiani, M.; Gallinella, G. Functional analysis and quantitative determination of the expression profile of human parvovirus B19. *Virology* **2008**, *381*, 168–177. [CrossRef] [PubMed]
52. Zermati, Y.; Varet, B.; Hermine, O. TGF-beta1 drives and accelerates erythroid differentiation in the epo-dependent UT-7 cell line even in the absence of erythropoietin. *Exp. Hematol.* **2000**, *28*, 256–266. [CrossRef]
53. Chretien, S.; Moreau-Gachelin, F.; Apiou, F.; Courtois, G.; Mayeux, P.; Dutrillaux, B.; Cartron, J.P.; Gisselbrecht, S.; Lacombe, C. Putative oncogenic role of the erythropoietin receptor in murine and human erythroleukemia cells. *Blood* **1994**, *83*, 1813–1821. [CrossRef] [PubMed]
54. Kadri, Z.; Shimizu, R.; Ohneda, O.; Maouche-Chretien, L.; Gisselbrecht, S.; Yamamoto, M.; Romeo, P.H.; Leboulch, P.; Chretien, S. Direct binding of pRb/E2F-2 to GATA-1 regulates maturation and terminal cell division during erythropoiesis. *PLoS Biol* **2009**, *7*, e1000123. [CrossRef] [PubMed]

55. Pop, R.; Shearstone, J.R.; Shen, Q.; Liu, Y.; Hallstrom, K.; Koulnis, M.; Gribnau, J.; Socolovsky, M. A key commitment step in erythropoiesis is synchronized with the cell cycle clock through mutual inhibition between PU.1 and S-phase progression. *PLoS Biol.* **2010**, *8*. [CrossRef] [PubMed]
56. Sakaue-Sawano, A.; Kurokawa, H.; Morimura, T.; Hanyu, A.; Hama, H.; Osawa, H.; Kashiwagi, S.; Fukami, K.; Miyata, T.; Miyoshi, H.; et al. Visualizing spatiotemporal dynamics of multicellular cell-cycle progression. *Cell* **2008**, *132*, 487–498. [CrossRef]
57. Xu, P.; Zhou, Z.; Xiong, M.; Zou, W.; Deng, X.; Ganaie, S.S.; Kleiboeker, S.; Peng, J.; Liu, K.; Wang, S.; et al. Parvovirus B19 NS1 protein induces cell cycle arrest at G2-phase by activating the ATR-CDC25C-CDK1 pathway. *PLoS Pathog.* **2017**, *13*, e1006266. [CrossRef]
58. Yunoki, M.; Tsujikawa, M.; Urayama, T.; Sasaki, Y.; Morita, M.; Tanaka, H.; Hattori, S.; Takechi, K.; Ikuta, K. Heat sensitivity of human parvovirus B19. *Vox Sang.* **2003**, *84*, 164–169. [CrossRef]
59. Mani, B.; Gerber, M.; Lieby, P.; Boschetti, N.; Kempf, C.; Ros, C. Molecular mechanism underlying B19 virus inactivation and comparison to other parvoviruses. *Transfusion* **2007**, *47*, 1765–1774. [CrossRef]
60. Blumel, J.; Schmidt, I.; Willkommen, H.; Lower, J. Inactivation of parvovirus B19 during pasteurization of human serum albumin. *Transfusion* **2002**, *42*, 1011–1018. [CrossRef]
61. Hattori, S.; Yunoki, M.; Tsujikawa, M.; Urayama, T.; Tachibana, Y.; Yamamoto, I.; Yamamoto, S.; Ikuta, K. Variability of parvovirus B19 to inactivation by liquid heating in plasma products. *Vox Sang.* **2007**, *92*, 121–124. [CrossRef] [PubMed]
62. Boschetti, N.; Niederhauser, I.; Kempf, C.; Stuhler, A.; Lower, J.; Blumel, J. Different susceptibility of B19 virus and mice minute virus to low pH treatment. *Transfusion* **2004**, *44*, 1079–1086. [CrossRef] [PubMed]
63. Crabol, Y.; Terrier, B.; Rozenberg, F.; Pestre, V.; Legendre, C.; Hermine, O.; Montagnier-Petrissans, C.; Guillevin, L.; Mouthon, L.; Groupe d'experts de l'Assistance Publique-Hopitaux de, P. Intravenous immunoglobulin therapy for pure red cell aplasia related to human parvovirus b19 infection: A retrospective study of 10 patients and review of the literature. *Clin. Infect. Dis.* **2013**, *56*, 968–977. [CrossRef] [PubMed]
64. Filippone, C.; Zhi, N.; Wong, S.; Lu, J.; Kajigaya, S.; Gallinella, G.; Kakkola, L.; Soderlund-Venermo, M.; Young, N.S.; Brown, K.E. VP1u phospholipase activity is critical for infectivity of full-length parvovirus B19 genomic clones. *Virology* **2008**, *374*, 444–452. [CrossRef] [PubMed]
65. Zhi, N.; Mills, I.P.; Lu, J.; Wong, S.; Filippone, C.; Brown, K.E. Molecular and functional analyses of a human parvovirus B19 infectious clone demonstrates essential roles for NS1, VP1, and the 11-kilodalton protein in virus replication and infectivity. *J. Virol.* **2006**, *80*, 5941–5950. [CrossRef]
66. Wolfisberg, R.; Ruprecht, N.; Kempf, C.; Ros, C. Impaired genome encapsidation restricts the in vitro propagation of human parvovirus B19. *J. Virol. Methods* **2013**, *193*, 215–225. [CrossRef]
67. Chen, A.Y.; Kleiboeker, S.; Qiu, J. Productive parvovirus B19 infection of primary human erythroid progenitor cells at hypoxia is regulated by STAT5A and MEK signaling but not HIFalpha. *PLoS Pathog.* **2011**, *7*, e1002088. [CrossRef]

Publisher's Note: MDPI stays neutral with regard to jurisdictional claims in published maps and institutional affiliations.

Review

AMDV Vaccine: Challenges and Perspectives

Nathan M. Markarian [1] and Levon Abrahamyan [2,*]

[1] Faculty of Veterinary Medicine, Université de Montréal, Saint-Hyacinthe, QC J2S 2M2, Canada; nathan.marko.markarian@umontreal.ca
[2] Swine and Poultry Infectious Diseases Research Center (CRIPA), Research Group on Infectious Diseases of Production Animals (GREMIP), Faculty of Veterinary Medicine, University of Montreal, Saint-Hyacinthe, QC J2S 2M2, Canada
* Correspondence: levon.abrahamyan@umontreal.ca

Abstract: Aleutian mink disease virus (AMDV) is known to cause the most significant disease in the mink industry. It is globally widespread and manifested as a deadly plasmacytosis and hyperglobulinemia. So far, measures to control the viral spread have been limited to manual serological testing for AMDV-positive mink. Further, due to the persistent nature of this virus, attempts to eradicate Aleutian disease (AD) have largely failed. Therefore, effective strategies to control the viral spread are of crucial importance for wildlife protection. One potentially key tool in the fight against this disease is by the immunization of mink against AMDV. Throughout many years, several researchers have tried to develop AMDV vaccines and demonstrated varying degrees of protection in mink by those vaccines. Despite these attempts, there are currently no vaccines available against AMDV, allowing the continuation of the spread of Aleutian disease. Herein, we summarize previous AMDV immunization attempts in mink as well as other preventative measures with the purpose to shed light on future studies designing such a potentially crucial preventative tool against Aleutian disease.

Keywords: AMDV; Aleutian disease; mink parvovirus; Aleutian mink disease virus; vaccine

Citation: Markarian, N.M.; Abrahamyan, L. AMDV Vaccine: Challenges and Perspectives. *Viruses* **2021**, *13*, 1833. https://doi.org/10.3390/v13091833

Academic Editor: Giorgio Gallinella

Received: 2 August 2021
Accepted: 9 September 2021
Published: 14 September 2021

1. Introduction

In the mid-1900s, some American mink farmers reported cases of a novel disease causing an enlargement of kidneys, the spleen and lymph nodes in the blue–gray variety of mink known as the Aleutian mink [1]. With time, the number of cases of this disease, called Aleutian disease (AD), rapidly increased throughout many ranches and the disease was eventually discovered to be due to a virus, as opposed to the initial false presumption of being a genetic disorder [2–5]. Today, AD is the most significant disease in the worldwide mink industry since it is responsible for causing infertility and the loss of animals, leading to low fur quality and, ultimately, significant financial losses for farmers [6]. The causative agent of this deadly and widespread disease is the Aleutian mink disease virus (AMDV), which is categorized as an *Amdoparvovirus* genus, one of the six genera of the *Parvovirinae* subfamily. This subfamily is part of the *Parvoviridae* family that belongs to the *Picovirales* order [7].

AMDV, being a parvovirus, is a single-stranded DNA virus possessing a small genome of around 4.8 kb in size with two large open reading frames (ORFs), among which are the left ORF at nucleotide positions 116–1975, the right ORF at nucleotide position 2241–4346, three smaller central ORFs and palindromic structures at both the 3′ and 5′ termini [3,8,9]. The left ORF encodes for three non-structural proteins (NS1, NS2 and NS3) that play a role in regulating gene expression and viral replication, whereas the right ORF codes for two viral capsid proteins (VP1 and VP2) serving as the key proteins for viral tropism and pathogenesis [10]. As for the palindromic structures at both 3′ and 5′ termini, these can fold into hairpin telomeres essential in the replication process [11]. Structurally, AMDV virions contain a predominantly negative strand DNA genome, measuring around 20 to 25 nm in diameter, are spherical, non-enveloped and are composed of 60 self-assembling proteins, including both VP1 and VP2 at a 1:9 ratio [11–13].

As for cell tropism, AMDV infects alveolar type II epithelial cells in mink kits and in macrophages for persisting infections, as well as B and T lymphocytes [14–16]. To enter macrophages and alveolar type II epithelial cells, AMDV has been reported to utilise antiviral antibodies using an Fc-receptor-mediated mechanism, which is known as an antibody-dependent enhancement of infection (ADE) [17,18]. In the case of other parvoviruses, many receptors are used to infect the target cells such as the transferrin receptor, bound by the canine and feline parvoviruses (CPV and FPV) and the erythrocyte P antigen for human parvovirus B19, whereas sialic acid is the main receptor for AMDV, minute virus of mice (MVM) and porcine parvovirus (PPV), respectively [19–22]. Upon receptor binding, parvoviruses are uptaken by clathrin-mediated endocytosis and slowly trafficked toward the cell nucleus, escaping the endosomal compartments with the help of the crucial phospholipase A_2 (PLA2) motifs present on the VP1 capsid protein [23,24]. The PLA2s are enzymes that produce lysophospholipids and fatty acids from phospholipid substrates. Amdoparvoviruses, such as AMDV, on the other hand, are the only viruses in the *Parvovirinae* subfamily that do not contain this particular motif in the VP1 protein, and so, the endolysosomal trafficking remains to be clarified in future studies [24–26].

Upon the delivery of the single-stranded DNA genome of parvoviruses to the nucleus via the nuclear pore complex, the host cell DNA polymerase attaches itself to the 3′ terminal hairpin telomere of the genome to convert it into a double-stranded DNA [27]. During the S-phase of the host cellular cycle, from the P3 promoter, a precursor mRNA (pre-mRNA) is produced which is followed by polyadenylation and splicing to generate six distinct mRNAs named R1, R1′, R2, R2′, Rx and Rx′. From these, R1 and R1′ mRNAs encode for the NS1 protein; R2, which is the most abundant mRNA produced, encodes for the viral capsid proteins VP1, VP2 and the NSP2. The R2′ mRNA encodes for NS2, while the NS3 protein, whose function is not yet clear, is encoded from the Rx and Rx′ mRNAs [28–31]. In a permissive infection, the NS1 protein is cleaved by caspase proteins at two particular sites being 224INTD↓S228 and 282DQTD↓S286, which facilitates the localization of the full-length NS1 protein to the nucleus to initiate viral replication. This localization has been proposed to be possible by the oligomerization of cleaved NS1 protein with full-length NS1 [30,32,33]. In some cases, AMDV infection can be persistent where this caspase-mediated cleavage event of NS1 does not take place, resulting in a lower amount of viral DNA replicated and a limited amplification of mature virions [34,35]. NS1, with its helicase, DNA binding, ATPase and endonuclease domains, binds covalently to specific ACCA motifs at 5′ end of one of the strands of the double-stranded DNA genome and then performs a single-strand nick. This enables a process known as rolling hairpin replication (RHR) in which the terminal hairpin is synthesized and rearranged many times allowing the replication fork to be reversed and to generate many copies of the single-stranded DNA genome [36,37]. The latter is then bound by the VP1 structural protein when it enters the nucleus and, together with VP2, AMDV virions are then assembled to, subsequently, exit out of the nuclear pore complex and from the cell [20]. In a permissive infection, where the NS1 protein was cleaved prior to replication, the mature virions exit the cell by inducing cell lysis [38]. A proposed AMDV viral cycle is summarized in Figure 1.

At both the nucleotide and amino acid sequence levels, AMDV has an unusually high genetic variability where several strains have been identified each having a varying degree of severity of the disease [39,40]. In the case of the Utah 1, Ontario, Danish -K and United strains, these are known to be highly pathogenic, whereas SL3 and Pullman strains are moderately virulent and AMDV-G strain is non-pathogenic [41–43]. In newborn farmed mink (*Mustela vison*), Aleutian disease manifests as a fatal acute interstitial pneumonia with a mortality rate higher than 90% for mink infected with highly virulent strains and ranging from 30% to 50% in low-virulence strains (in infected newborn mink). The most serious form of the disease, however, is persistent and occurs in adults, where it is mediated by the immune system, causing plasma cell proliferation (plasmacytosis), an abnormal amount of gamma globulins present in the blood (hypergammaglobulinemia), the formation of infectious immune complexes and inflammation of kidneys (glomerulonephritis) [3,44–46]. AMDV is transmitted horizontally by air and contact between farms and also vertically from mother to offspring [46–48]. It has been shown that AMDV can also infect different species such as ferrets, otters, polecats, foxes, racoons, skunks, dogs, cats, and mice, but the disease has only been reported to be manifested in mink. Since this virus can infect other animals, the latter can be regarded as possible reservoirs facilitating AMDV spread [49–53].

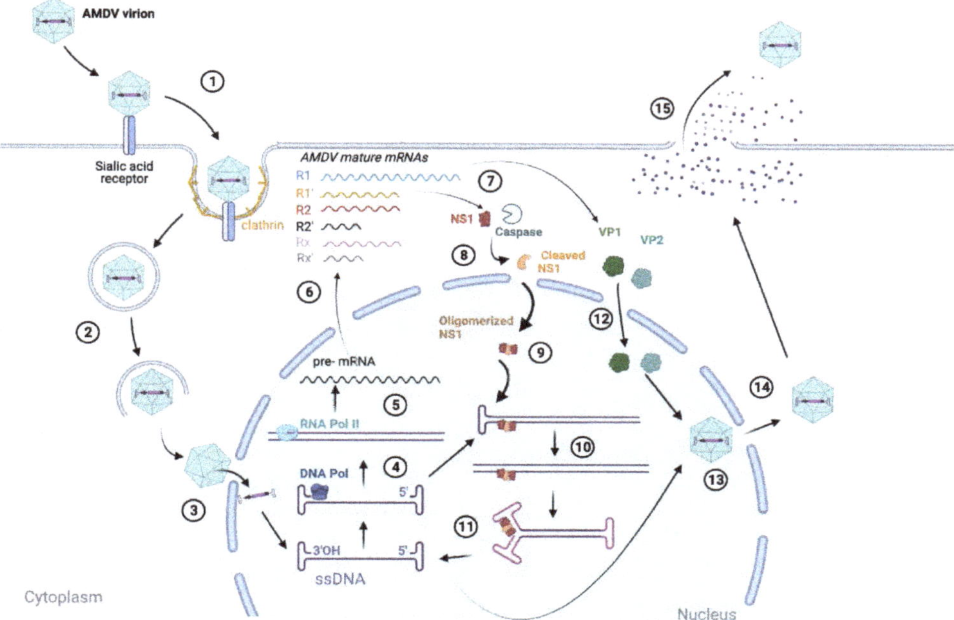

Figure 1. Proposed viral life cycle of Aleutian mink disease virus (AMDV). (1) The virus attaches and binds to the sialic acid receptor on host cells, triggering uptake by clathrin-mediated endocytosis. (2) Endolysosomal trafficking toward the nucleus. (3) Viral single-stranded DNA (ssDNA) genome is delivered into the nucleus. (4) Conversion of ssDNA into dsDNA by host cell DNA polymerase (DNA pol). (5) Transcription of double-stranded DNA (dsDNA) produces precursor mRNA (pre-mRNA) in the nucleus. (6) Splicing and polyadenylation of pre-mRNA yields six mature mRNAs, which exit the nucleus for (7) translation of viral proteins (only NS1, VP1 and VP2 shown). (8) Specifically, for AMDV: cleavage of NS1 protein by caspase proteins to facilitate localization of full-length NS1 to nucleus during permissive replication. (9) Full-length/cleaved NS1 oligomer enters the nucleus to covalently bind to the 5′ end of dsDNA, and (10) performs a nick. (11) Rolling hairpin replication (RHR) generates many copies of ssDNA genome. (12) Together, the VP1 and VP2 proteins enter the nucleus and (13) bind the ssDNA genome to form a mature AMDV virion. (14) Exit of AMDV virion from nucleus. (15) Release of AMDV virions through cell lysis. Created with BioRender.com (accessed on 15 August 2021).

Unfortunately, there are not any efficient methods to control the spread of this disease as AMDV is resistant to physical and chemical treatments, which is why farmers manually identify infected mink and kill those screened positive [18]. The diagnosis of AD mink takes place by looking for clinical signs and the detection of AMDV antibodies, which can be performed using non-specific methods such as iodine serum plate agglutination, where the excess of the gamma globulins levels in the serum can be visualized since certain gamma globulins can form complexes with iodine [54]. However, currently, more specific approaches are available such as counterimmunoelectrophoresis (CIEP), which involves using an electric current to assess binding resulting from the migration of unbound

AMDV antigens towards the specific AMDV antibodies [55,56]. In recent times, this approach has been shown to be successful in reducing the prevalence of infected mink in Nova Scotia Canada, but has failed in eradicating AMDV from most of the infected farms [57]. Despite its advantages, CIEP is not a very sensitive diagnostic technique, can yield false negative results and is very time consuming [58,59]. In contrast, the enzyme-linked immunosorbent assay (ELISA) is another approach that is frequently used, which is based on detecting anti-AMDV antibodies in blood with the advantage of being automated and more sensitive [60]. Both CIEP and ELISA are very comparable and in good agreement, according to a study by Dam-Tuxen et al., where both techniques were evaluated in Danish mink [61]. Although being a more sensitive approach, ELISA has been shown to be a less accurate technique than CIEP, as reported by Farid and Rupasinghe (2016) [62] Alternatively, the polymerase chain reaction (PCR) is another specific method that is used as a supplementary test, and this allows the detection of viral DNA by amplifying AMDV gene fragments [63]. A drawback of this method is that it does not detect all virus strains since AMDV has a great genetic diversity and is constantly mutating [64,65]. Another strategy to eliminate the harmful effects of AMDV is by selecting mink that are tolerant to this virus and this can be performed by identifying mink with a low antibody titer, which is performed in Europe and North America [58,66–69]. The selection for tolerance can also be performed by characterizing genomic regions of AD-tolerant mink [70]. Despite the different measures aimed to control the spread of AMDV and its harmful effects on mink the virus is, unfortunately, still persistent on farms, which increases the demand of more effective AD preventative measures, as well as effective therapies.

Throughout the years, vaccines have been proven to be ingenious tools to be used against highly infectious pathogens and, thus, have been successful in eradicating diseases such as rinderpest in cows [71]. Consequently, one can assume that a plausible approach to combat against the persistent and highly consequential AMDV is to immunize mink against this virus. In fact, there were many attempts of researchers trying to develop an AMDV-vaccine using different approaches with a wide range of results obtained. For this reason, in this article, we review the several attempts of AMDV-vaccine design and the challenges faced in the process to provide a thorough understanding for future studies in the goal of eradicating this deadly disease.

2. AMDV Preventative Measures

Aleutian disease is one of the most threatening diseases in modern day mink farming. Further, domesticated mink escaping farms either by accident or with the help of animal activists pose a threat to wild populations of mink which could act as reservoirs of AMDV [72]. For this reason, the development of a novel AMDV vaccine is of crucial importance. In the meantime, it is of great interest to use alternative strategies to limit the viral spread and avoid huge financial losses to farmers and protect feral mink populations from potential spillovers of the virus. One strategy that is used for temporarily controlling the pathogen in farms is by the mandatory testing of mink to have an early detection and surveillance of AMDV. Common approaches of AMDV diagnosis consist in using serological testing techniques such as ELISA and counterimmunoelectrophoresis (CIEP) or PCR [73,74]. PCR is not as widely used as a routine serological testing method for herd screening due to a short-lived viral replication and viremia which causes negative PCR, while AMDV is sequestered in organs, making viral detection more challenging [44,58,75–77]. An additional reason for this is the cost of an essential DNA extraction step, which requires the clearance of residual inhibitors that would lead to the loss of sample DNA if not removed [78,79]. Additionally, in a recent study, it was shown that CIEP turned out to be more accurate than conventional PCR testing in black American mink infected with AMDV [75]. When taken together, early diagnostic approaches of AMDV-positive mink could provide reinforcing tools to help limit viral spread. Indeed, these have been somewhat effective in several countries, including Spain, where farms with AD have been reduced from 100% in 1980 to around 25% in 2019 [68].

Despite this encouraging statistic, however, the eradication of the disease remains challenging and has failed in many countries, including Canada, Denmark, and others [57,63,68,80]. This can be explained by the reinfection of farms due to viral persistence and by contaminated human individuals visiting farms, leading to a failure of eradication systems [81]. For this reason, some researchers explored the idea of immunomodulation in AMDV-infected mink. In fact, this approach has been investigated in a variety of animals, where the host immune system is modulated to be enhanced against infectious diseases. This can be achieved with numerous substances such as cytokines, microbial products, traditional medicinal plants, nutraceuticals and pharmaceuticals [82]. An example of immunomodulators are β-glucans, which are readily used in sheep and swine production systems [83]. For Aleutian mink disease, this strategy has also been explored by some researchers. This is the case of a study by Kowalczyk et al. [84] where female brown AMDV-positive mink were given methisoprinol (isoprinosine; inosiplex; inosine pranobex), a drug with immunopotentiating properties for which antiviral and immunostimulatory effects on Aujeszky's and Newcastle disease have been reported by various groups [84]. Interestingly, the methisoprinol treatment of mink displayed a lower number of the AMDV DNA copies in the spleen and lymph node, as well as a higher fecundity compared to control mink [85]. More recently, Farid and Smith investigated the effect of kelp meal (*Ascophylum nodosum*) in mink infected with AMDV. From this study, it was shown that the supplementation of mink with kelp meal for 451 days did not improve the mink immune response to infection or virus replication. Interestingly, lower levels of blood, urea, nitrogen (BUN) and creatine were noticed, which suggested an improved kidney function [86].

Alternatively, another initiative for reducing the effects of AD on chronically infected farms is by passive immunity, which can provide immediate, but short-lived, protection to humans and animals [87]. In the context of Aleutian disease, the effect of the passive transfer of anti-ADV gamma globulins to newborn mink infected with a highly virulent strain of AMDV has already been investigated by Alexandersen et al. [88]. From their research, an acute AMDV infection was prevented by the passive antibodies, but mink still manifested a chronic infection [88]. The mechanism of the modulation from acute to chronic AMDV was then assessed in vitro, where infected alveolar type II cells treated with antibodies showed lower levels of AMDV replication and transcription intermediates. This was also noted in antibody-treated kits in vivo, which allowed the authors to suggest a role of antiviral antibodies in developing persistent infection in mink [88]. Furthermore, antiviral drugs are regarded as an important strategy to control the spread. Indeed, this strategy is very common and used for many infectious agents in humans and animals, such as human immunodeficiency virus (HIV) and Feline Herpes Virus (FHV) [89]. For AMDV, unfortunately, there is no approved antiviral treatments available for use; however, recently, Lu et al. [90] attempted a novel antiviral treatment using the "magnetic beads-based systemic evolution of ligands by exponential enrichment" (SELEX) strategy to generate aptamers for the AMDV VP2 protein. From their results, it was shown that, in vitro, the micromolar concentrations of the aptamers for AMDV VP2 protein were able to specifically inhibit by half the AMDV production by infected cells [90]. More investigation is, therefore, required in future studies in designing an adequate antiviral for AMDV.

Another strategy to prevent the harmful effects of AMDV is to select mink having low titers of anti-AMDV-protein antibodies in their blood, which is a common practice in many countries [58,66–69]. Besides this classical selection method, an alternative approach has been recently studied by Karimi et al. which is based on characterizing host-genomic patterns of AD-tolerant mink with the intention to select those that could be tolerant or resistant to AMDV. More specifically, the authors performed a sequencing approach to focus on genomic regions related to host specific immune responses. Using this strategy, two regions of the mink genome were shown to be strongly selected and these contained important genes involved in the immune response, viral–host interaction, reproduction and liver regeneration [70]. Additionally, since AMDV is a rapidly evolving pathogen, it is of great importance to be informed about current and past circulating strains to be able to develop epidemiological models, predictions and forecasting of the strains that may cause serious outbreaks. This knowledge would help us in devising an efficient and practical control strategy against this virus in the future, by setting up a database similar to the global initiative on sharing all influenza data (GISAID), which has been especially useful in the ongoing coronavirus disease-19 (COVID-19) pandemic in tracking novel variants [91]. Some studies have in fact performed phylogenetic analyses of the AMDV genomes deciphered by genetic sequencing. This was performed in order to develop new tools for outbreak investigation, the determination of virulence markers and development of more sensitive diagnostic tests in the future [92–99]. As an example of such work, Canuti et al. [93] studied the evolutionary dynamics of AMDV from 2004 to 2014 using sequences obtained from many regions of the world, including North America and Europe. From these, a very high viral genetic diversity as well as high rates of co-infection were noted in Newfoundland, Canada. Interestingly, a low level of diversifying selection was shown on structural proteins, which was explained to be due to the specific cell-entry mechanism of AMDV, which uses antibodies to enter cells. Finally, it was concluded that a great amount of circulating viruses on farms has the capacity to recombine and increase viral diversity by co-infections [93]. Therefore, summarizing, these methods could be of great use to control the spread of AMDV, while actively working on a development of effective vaccine candidates. A summary of these preventative measures is shown in Table 1.

Table 1. Some of the AMDV Preventative Measures Proposed: Diagnostic and Reduction Methods.

Measure	Results	Authors
Diagnostic method: CIEP [a]	Reduction in the prevalence of infected mink in Nova Scotia Canada [57] and in Spain [68]; more accurate than ELISA [75]	Farid et al. (2012) [57], Pietro et al. (2020) [68], Farid and Hussain [75]
Diagnostic method: ELISA [b]	Estimated sensitivity and specificity of 96.2% and 98.4%, respectively, for AMDV-VP2-recombinant antigen [60]; greater sensitivity than CIEP [61] but lower accuracy [61,62]	Knuuttila et al. (2014) [60], Dam Tuxen et al. (2014) [61], Farid and Rupasinghe (2016) [62] Chen et al. (2016) [74]
Diagnostic method: PCR [c]	Relative diagnostic sensitivity of 94.7%, and relative diagnostic specificity was 97.9% [63]; specificity and sensitivity of 97.9% and 97.3%, respectively, for VP2 332–452 ELISA [74]; estimated specificity of 88.9% for AMDV-G NS1 probe-based real-time PCR [64]; lower sensitivity than CIEP [75]	Jensen et al. (2011) [63], Virtanen et al. [64], Farid and Hussain [75]
Immunomodulator molecules	Methisoprinol-administered mink showed lower number of AMDV DNA copies in spleen and lymph node and higher fecundity compared to control mink [85]; significantly lower levels of blood, urea, nitrogen and creatine in kelp meal-administered mink [86]	Kowalczyk et al. (2019) [85], Farid and Smith (2020) [86]
Passive Antibody Therapy	Prevention of acute AMDV infection by passive antibodies, but mink still manifested a chronic infection; reduction in mortality by 50 to 75% [88]	Alexandersen et al. (1989) [88]
Antiviral molecules	Specific inhibition of AMDV production in infected cells by AMDV VP2 aptamers: reduction of 47% supernatant concentration of AMDV compared to controls [90]	Lu et al. (2021) [90]

Diagnostic methods are used for detection, surveillance, for early detection and to control the spread. Immunomodulator molecules can be used to enhance the host immune system against infectious diseases. [a]—counterimmunoelectrophoresis. [b]—enzyme-linked immunosorbent assay. [c]—polymerase chain reaction.

3. AMDV Vaccine Attempts

3.1. *Inactivated Vaccine*

The use of inactivated and live attenuated virus-based vaccines dates back to more than a century. The principle of an inactivated viral vaccine consists of exposing a virus to chemical or physical agents, which disable its infectivity while allowing it to stimulate an immune response [100]. An example of this is by using formaldehyde, a cross-linking agent of amino acids, or, more modern approaches such as ascorbic acid or hydrogen peroxide, have been shown in recent years [101]. Currently, there are many inactivated viral vaccines available and approved for usage against viruses, including Hepatitis A, Japanese encephalitis virus, poliovirus, rabies and SARS-CoV-2 [102–105]. A study by Karstad et al., in 1963 [106], was one of the first attempts to develop an AMDV vaccine using this approach. These Canadian researchers used the inoculations of formalin-treated suspensions of tissues of AMDV-infected mink as a candidate vaccine [106]. It was shown that upon inoculating mink with this vaccine candidate, no plasmacytosis was noted in contrast to those inoculated with untreated diseased tissue. However, formalin-treated tissue failed to protect inoculated mink from the challenge with a virulent inoculum. Further, it was also noted that mink receiving three doses of the formalin-treated tissue suspension subcutaneously were not able to stimulate the immune system, i.e., the mink experienced plasmacytosis upon challenge [106]. Less than 10 years later, Porter et al. [107] tried a similar approach, where formaldehyde was used to inactivate AMDV-infected spleen and liver tissue homogenates. The inactivated homogenate was then administered to pastel mink and performed in parallel with the administration of a control group. From this, it was shown that there was no change in gamma globulin levels due to AMDV or the control group 33 days after inoculation. However, upon the challenge with a 10^2 and 10^5 tissue culture infective dose ($TCID_{50}$), the gamma globulin levels in AMDV-vaccinated mink were slightly greater than the control group. Surprisingly, it was also reported that mink vaccinated against AMDV displayed more tissue lesions and enhanced plasmacytosis than the mink receiving the control vaccine, where both groups were challenged with the same dose of the virus. The authors then evaluated the effect of antibodies on this AD enhancement by giving 250 mg or 1 g of normal mink IgG antibodies to infected mink. A significantly enhanced disease was observed with acute necrotizing lesions. In contrast, this was not observed in infected mink receiving a lower dose of normal mink IgG, and in uninfected mink receiving both anti-AMDV IgG and normal mink IgG. Thus, it was suggested that antibody levels in the AMDV-vaccinated mink were not necessarily higher than in the control, since it was not noted as a significant number of lesions in the former compared to the latter. Moreover, based on immunofluorescence data, the authors suggested that observed enhanced pathology was due to antibody-complement-induced cytolysis followed by the complement-mediated attraction of polymorphonuclear leukocytes [107].

3.2. DNA Vaccine

Before AMDV was known to be the causative agent of Aleutian Disease, one study by Basrur and Karstad [108] investigated the effects of infection with DNA, extracted from the spleen tissue of mink with Aleutian disease. An in vitro inoculation of mink testis cells with viral DNA demonstrated a change in cellular morphology and abortive growth, which were absent in controls. This finding was also evaluated in vivo, on nine-month-old standard dark male mink, where plasmacytosis was observed in most of animals inoculated with the viral DNA. In contrast, the control group did not demonstrate such an effect, leading the authors to believe that the disease was of viral origin [108]. In more recent times, the use of DNA derived from viruses or other pathogens has been of great interest in the development of vaccines. This immunization method is referred to as DNA vaccination that consists of administering foreign viral protein-coding plasmid DNA into animals of interest [109,110]. Using the host cellular translation machinery, antigen proteins are then generated in situ followed by MHC-I and MHC-II pathways and the induction of CD8+ and CD4+ T cells, leading to an antigen-specific immunity [109–111]. DNA vaccines have been used in human clinical trials against infectious agents, for cancer immune therapies and for asthma and allergies as well as gene therapies for chronic diseases [112–114]. For animals, there are several DNA vaccines created against a variety of viruses. Some DNA vaccines have already been approved to be used, such as the first commercial DNA vaccine against H_5N_1 for chickens and the West Nile Virus vaccine for horses [115–117].

In the case of AMDV, there have been several attempts to develop a vaccine, but to date, none of them have been successful. One of those attempts was conducted by the group of Castelruiz et al. (2005), which designed a vaccine candidate based on an AMDV NS1 protein-coding plasmid (pNS1). This approach was first tested in vitro, where a high expression of the NS1 protein was revealed after transfecting Crandell feline kidney cells (CRFK) with the NS1 expressing plasmid. After confirming the NS1 protein expression in cell culture, the vaccine candidate was evaluated in vivo using three different groups of non-Aleutian female mink. In the first group, the animals were immunized with pNS1, followed by a booster inoculum containing recombinant NS1 protein (named "DNA + protein" group). The second group was identical to the first, with the exception of omitting the NS1-protein inoculation step (named "DNA only"). Finally, the third group received an empty plasmid, which did not contain the NS1 gene (named "control" group). After each of these groups were challenged, the "DNA + protein" vaccine cohort was shown to be the only group to exhibit antibodies against the NS1 protein at the time of challenge. With time, this was also observed in the other cohorts and, more specifically, at 12 weeks post infection the "DNA only" group displayed a significantly higher anti-NS1 titer than the "control" group. A similar result was reported for all groups, when assessing anti-VP1/2 antibodies, with the only difference of not observing any VP1/2 antibodies on the day of challenge with AMDV. Interestingly, after 1 month of challenging mink with the virus, the "DNA+ protein" group exhibited higher levels of CD8+ T lymphocytes than the two other groups. This effect was suggested to be due to a memory response after an increase in interferon γ-producing lymphocytes for the same group was observed between 4 and 8 weeks post-challenge. A mild vaccine effect was suggested when noting that the "control" group displayed 10% higher gamma globulin levels than the other groups 8 weeks post-challenge. In terms of post-challenge deaths, most of the "control" group succumbed, where two of the latter were not likely to be AD related. In contrast, in the "DNA only" and "DNA + protein" groups most of the animals survived. From the results of both AMDV NS1 DNA vaccinated groups, it was finally concluded that a mild vaccine effect was induced, but it was only partially protective, since there were still mink that died during the experiment [118]. Later another attempt was undertaken by Liu et al. [119] to achieve the AMDV DNA-based vaccine (2018). For their vaccine candidate, the AMDV-DL125 strain was selected as a template since it displayed the highest $TCID_{50}$ from virulent strains assessed in CRFK cells. From this strain, the whole genome was used to generate plasmid-vectored vaccines by truncating regions of VP2 and NS1 genes using an overlap extension PCR. Similar to the study of Castelruiz et al., these plasmids with truncated regions were then tested in vitro on CRFK cells, where the successful infection of the cells with vaccine plasmids was confirmed by immunofluorescence, in contrast to controls, where no immunofluorescence signal was observed. Next, groups of six-month-old female mink were used to conduct an in vivo study. Most of these groups were administered with distinct DNA vaccine candidates, whereas the rest served as control groups, receiving phosphate-buffered saline (PBS), empty vector and AMDV-inactivated virus, respectively. Similar to Castelruiz et al., the anti-AMDV antibody levels in each group were shown to increase with time after challenge with AMDV. Interestingly, at the 24-week timeframe post-challenge, all groups demonstrated similar antibody levels in the serum, except for the inactivated AMDV-administered control group. The latter also displayed rapidly increasing levels of circulating immune complexes (CIC), suggesting that the inactivated AMDV vaccine could help accelerate the development of AD. For other groups, the ones inoculated with VP2 gene carrying plasmid with deletions at nucleotides coding for residues 428–446 and 487–501 (pcDNA3.1-ADV-428–487) had the lowest serum gamma globulin and CIC levels. This was suggested to be the most protective cohort; nonetheless, dead mink in each vaccinated and control group were observed with the earliest death occurring in the AMDV-inactivated group, followed by the others. By the end of the experiment (36 weeks post-challenge), the pcDNA3.1-ADV-428–487 group had the lowest mortality, and this allowed Liu et al. to suggest that this vaccine could be potentially used in mink populations in the future [119].

3.3. Subunit Based Vaccine

In 1981, the Food and Drug Administration (FDA) approved, for the first time, the plasma-derived hepatitis B virus (HBV) Heptavax-B vaccine (Merck) [120]. This was based on a heat-inactivated HBV surface envelope protein (HBsAg) isolated from the blood of asymptomatic HBV-infected patients and was shown to provide good protection in immunized individuals [121]. However, with the imminent HIV/AIDS global epidemic, concerns of infecting vaccinated individuals by blood with other pathogens limited this initiative, leading it to be discontinued in 1990 [122,123]. This encouraged the development of recombinant DNA technology, where the HBsAgs were produced in a yeast culture and this approach was used for the first time by the RECOMBIVAX vaccine (Merck Sharp and Dohme) approved in 1986 [123,124]. Protein subunit-based vaccines such as these primarily induce humoral immunity and exclude pathogens from their production, which reduces the risk of an incomplete activation, pre-existing immunity or causing disease [125,126]. Despite these advantages, the subunit vaccination, on its own, stimulates weaker immunological responses, which is why additional components known as adjuvants are added in vaccine doses [127,128]. Adjuvants, such as aluminium salts, can enhance cell-entry and preserve the structural integrity of the antigen, stimulate macrophages that promote helper T-cell responses, induce CD8+ cytotoxic T-lymphocytes as well as help in slowly releasing the antigens to prolong the exposure time to the immune system [129,130]. Currently, alongside HBV vaccination, the subunit-based approach is actively being tested for use in a variety of human pathogenic viruses such as Ebola and HIV [131–133].

In the case of Aleutian mink disease virus, this approach was investigated by Aasted et al. in 1998, where recombinant VP1/2 and NS1 proteins of AMDV-G were used to inoculate nine-month-old black female mink. Mink vaccinated with VP1/2 were challenged with AMDV which resulted in a higher death rate, more extreme hypergammaglobulinemia, higher NS1 and VP1/2 titers and higher counts of CD8-positive lymphocytes in lower peripheral blood compared to unvaccinated mink. The authors suggested that the observed outcome was caused by the antibody-dependent enhancement (ADE) of infection, a key mechanism in the pathophysiological changes of the AD, arguing that AMDV replicating cells are likely to be Fc-receptor positive. In the case of the group of mink vaccinated with a 10-fold higher dose of NS1, a more positive immunization effect was noticed. Throughout 11 months after challenge, lower levels of serum gamma globulin and CD8 positive lymphocytes were noted in addition to the lower death rates compared to a group of the non-vaccinated mink. The researchers did not detect vaccine-induced antibodies for VP1/2 and NS1 using ELISA assays. It was suggested that it could have been due to the induction of antibodies with a low affinity to native AMDV proteins since this higher affinity could have been developed to denatured the antigen protein vaccine present in antigen preparations prior to inoculation [134].

Considering the different vaccination approaches designed throughout the years against AMDV, the most successful attempt can be concluded to be the DNA vaccines, more particularly, those engineered by the group of Liu et al. [119]. Given the limited number of approaches for a successful vaccine design, more testing and the use of recent technological advancements in vaccinology can be employed to reinforce the fight against AMDV. Such methods are further elaborated in the Discussion section. The different vaccination attempts covered in this section against AMDV are summarized in Table 2.

Table 2. AMDV Vaccination Attempts.

Vaccine Type	Approach	Disadvantage/Benefits	Authors
Inactivated virus	Formalin-treated infected kidney, liver and spleen suspension	No protection: challenged vaccinated mink developed plasmacytosis	Karstad et al. (1963) [106]
Inactivated virus	Formalin-treated infected spleen and liver tissue suspension	Enhancement of disease: challenged vaccinated mink displayed more tissue lesions/plasmacytosis than the non-vaccinated	Porter et al. (1972) [107]
DNA-based	NS1-coding plasmid	Partial Protection: majority of challenged vaccinated mink survived with the exception of a few deaths	Castelruiz et al. (2005) [118]
DNA-based	Whole gene-coding plasmid (pcDNA3.1-ADV) VP2-coding plasmid with deleted nucleotides coding for amino acids 428–446 (pcDNA3.1-ADV-428) VP2-coding plasmid with deleted nucleotides coding for amino acids 428–446 and 487–501 (pcDNA3.1-ADV-428–487) NS1-coding plasmid (pcDNA3.1-NS1) Truncated NS1-coding plasmid (pcDNA3.1-NS1-D) NS2-coding plasmid (pcDNA3.1-VP2) Truncated NS2-coding plasmid (pcDNA3.1-VP2-D)	Partial Protection: deaths observed in each category of challenged vaccinated mink, lowest number of deaths, the lowest serum gamma globulin and CIC levels for vaccinated cohort with pcDNA3.1-ADV-428–487	Liu et al. (2018) [119]
Subunit protein	VP1/2 and NS1 recombinant proteins ink aluminium hydroxide gel adjuvant	VP1/2—enhancement of disease: compared to control, a higher death rate and more extreme hypergammaglobulinemia in challenged vaccinated mink; NS1—partial protection: compared to control, lower death rates for challenged vaccinated mink	Aasted et al. (1998) [134]

4. Discussion

Throughout the years, many animal viruses have emerged and have caused serious financial and public health repercussions in many regions worldwide. A prime example of these is the Aleutian mink disease virus (AMDV), which poses a serious threat to mink since it is the most significant disease in their farming around the world [135]. Therefore, it is of great importance to produce effective treatments and vaccines to prevent significant losses caused by AMDV. As discussed in this review, there have been many attempts of producing vaccines against AMDV using different strategies. From these, three main techniques were used, including inactivated, subunit protein and DNA-based vaccines. In the case of inactivated vaccines, Karstad et al. showed an enhanced disease upon challenging mink that had been inoculated with formalin-treated AMDV-infected tissues, which was also shown by Porter et al. [106,107]. The latter additionally reported higher gamma globulin levels as well as more lesions in challenged vaccinated mink [107]. As for the design of subunit vaccines, Aasted et al. also demonstrated an enhanced disease after challenging VP1/2-vaccinated mink with higher death rates and extreme hyperglobulinemia [134]. In contrast, this effect was not noted in mink when a recombinant NS1 protein was used at a dose of 10-fold greater than that used for the VP1/2 inoculum. Instead, lower levels of gamma globulins and CD8 positive lymphocytes were present in the serum and less deaths were shown [134]. For DNA vaccine candidates, Castelruiz et al. obtained better results by using an AMDV NS1 coding plasmid either given alone or in combination with recombinant NS1 protein [118]. In fact, for both strategies, a mild vaccine effect as well as partial protection were found [118]. Similarly, a much promising result was reported by Liu et al., who designed seven plasmid-vectored vaccines, which were based on the genome of the infectious AMDV-DL125 strain [119]. From those, the best candidate showed partial protection and a higher efficacy based on lower levels of circulating immune complexes (CIC) gamma globulins in the serum [119]. This was based on the VP2 gene, which had deletions at nucleotides coding for residues 428–446 and 487–501 [119]. From all AMDV vaccination attempts, the most promising ones have been shown to be the DNA-based vaccines. Interestingly, for the other strategies, some authors have previously suggested an antibody-dependent enhancement of disease (ADE), which was used to explain the enhanced disease upon challenge of post-vaccinated mink [106,107,134]. Indeed, AMDV has already been reported to use the Fc-receptor-mediated mechanism to enter cells [17,18]. With this in hand, Bloom et al. investigated the molecular mechanism of ADE in the context of AMDV in vitro [18]. Namely, researchers evaluated effects of mono- and poly-clonal antibodies against short peptides designed from the immunoreactive VP2: 429–524 linear epitope using CRFK and K562_cells [18]. From their results, the VP2: 428–446 peptide-induced Fc-mediated ADE, the neutralization of AMDV and participated in an immune complex formation. A limited ADE was also noted in the case of the VP2: 487–501 peptide. It was finally concluded that this may be the mechanism by which capsid-based vaccines have the ability of inducing both neutralization and ADE, which agrees with the disease enhancement (vaccine-enhanced disease) shown in the inactivated and subunit protein approaches [18]. Further, this is also consistent with the study by Liu et al., where the exclusion of these sites led to a more efficient vaccine candidate [119]. Therefore, it might be suggested that in future vaccination attempts, sites from the immunoreactive VP2 segment should be excluded for a better response.

Another approach to this could be to analyze the full VP2 sequence to find epitopes that can further mediate ADE. One way to achieve this is by generating infectious clones of the virus and identifying capsid residues that are necessary for infecting cells in vitro. In fact, Xi et al. previously used this method to show that VP2 residues 92 and 94 are critical for AMDV replication in vitro [136].

Since there have been few vaccine technologies used to generate AMDV vaccine candidates, other vaccine development strategies or their combinations can be further used such as viral-vectored, virus-like particles and nanoparticle-based vaccines [137]. For example, for some diseases, viral vectors have been successfully used in combination with other vaccine approaches in a strategy called a heterologous prime-boost [138]. Lately, in the context of the current coronavirus-19 pandemic (COVID-19), one particular approach, based on the mRNA technologies, has been used in two licensed vaccines, which has helped significantly reduce infection numbers worldwide [139,140]. Both mRNA-1273 (Moderna) and BNT162b2 (Pfizer—BioNTech) rely on lipid-nanoparticle (LNP)-encapsulated mRNA expressing the pre-fusion stabilized spike glycoprotein, which is translated upon entering the cells [141,142]. With the great efficiency and success brought by these vaccines, a plausible approach could be to design a VP1/2 or an NS1-based mRNA vaccine for AMDV.

As the search for an efficient AMDV vaccine continues, novel methods are currently being explored to enhance vaccine (immunogen) delivery. This is the case of nanoparticle technology, which has been explored in the context of many viruses [143]. In fact, nano-vaccines (vaccines, where nanoparticles or nanomaterials are used as carriers) are created by displaying relevant antigenic sites on nanomaterials either by physical entrapment or by covalent binding [144]. Such vaccines have a similar size to the infectious pathogens, which makes it easier to enter the target cells, where they will be degraded and release the antigen over time [145]. In terms of their numerous advantages, these can induce immune responses, could be given intramuscularly or via mucosal sites, and they are stable at room temperature [144,146,147]. One interesting biodegradable nanovaccine approach that has been approved by the FDA is by using liposome-derived nanovesicles as carriers, since these can enhance the delivery of antigens to cells [148,149]. The use of such liposomal delivery systems has been previously reported to enhance immune responses against many viruses, including Newcastle disease virus [150,151]. Using this approach has also been shown to stimulate both Th1 and Th2 responses in response to an H3N2 influenza subunit vaccine candidate [152]. Another example of nano-vaccines is virus-like particles (VLPs), which mimic native viruses with the exception of not having the capacity to infect and replicate [153]. These are composed of self-assembling viral antigen proteins which can induce a stronger humoral immune response compared to single soluble antigens [154]. In recent years, there have been several approved vaccines using this technology, including Gardasil® and Gardasil9®, against human papilloma virus (HPV), and many others [155]

Alternatively, another new vaccination method that is actively being explored involves the use of cell-derived bi-layered extracellular vesicles (EVs) such as exosomes, microvesicles and apoptotic bodies [156,157]. Indeed, these have great advantages in terms of delivery and can increase overall immunogenicity since they have immunostimulatory molecules on their surfaces such as MHC class I or II molecules [156,158]. Further, exosomes have the capacity to carry pathogen antigens, as was shown by Montaner-Tarbes et al. in a study analyzing exosomes from pigs infected with porcine reproductive and respiratory virus (PRRSV) [159]. From this, a specific reaction was demonstrated in a PRRSV RNA-negative and seropositive pig upon testing the exosome-derived viral proteins [159]. The microvesicle approach has also been tested by many researchers, including Rappazzo et al., where mice were vaccinated with microvesicles expressing ClyA surface protein fused with influenza matrix protein 2 [157,160]. Upon challenge with a virulent mouse-adapted H1N1 influenza strain, vaccinated mice demonstrated a full protection and the passive transfer of their antibodies to non-vaccinated mice protected the latter when challenged [160]. Microvesical vaccines, such as liposomal nano-vaccine delivery methods, can also be given orally or nasally [161]. With the aforementioned advantages these novel approaches for designing effective vaccines could be used to vaccinate mink against AMDV. This could be performed by incorporating key AMDV antigens in aerosol based nanoparticles or EVs, which would help vaccinate a large number of mink rapidly from their cages by simply releasing these vaccines in the air. However, another important challenge faced, when designing AMDV vaccines, is that the virus has an unusually high genetic variability in both NS1 and VP2 proteins [39,40].

To enhance the design of future AMDV vaccines, one strategy can be the constant tracking of current and past circulating strains by phylogenetic analyses, which can help determine key virulence markers [92–97]. This is how Canuti et al. revealed that structural proteins were not under diversifying pressure in sequences analysed in Newfoundland probably due to the conserved cell-entry mechanism of AMDV [93]. Moreover, in the wait of effective vaccines, there are many methods used and still being developed to control AMDV spread. A commonly used approach is detecting infected mink earlier on by serological testing techniques and by, subsequently, eliminating them [73,74]. Other techniques that have been explored or that are in development include the usage of passive anti-AMDV antibodies, antivirals, immunomodulators and the selection of mink with AMDV-tolerant traits [70,85,86,88,90]. Unfortunately, despite some positive results, these are not enough to eradicate this deadly disease, which is why the design of an effective vaccine is of primordial importance.

To conclude, AMDV poses a great threat to a huge number of mink either in farms or in in the wild and represents a potential risk to other species. In this review, the attempts to design a much-needed vaccine against this virus were covered, as well as the challenges that researchers are faced in designing a novel vaccine. For future studies, novel vaccine technologies should be aimed at, as well as trying to design better preventive measures to control viral spread.

Author Contributions: Conceptualization, L.A.; methodology, L.A.; validation, L.A.; data curation, N.M.M. and L.A.; writing—original draft preparation, N.M.M. and L.A.; writing—review and editing, N.M.M. and L.A.; supervision, L.A.; project administration, L.A.; funding acquisition, L.A. All authors have read and agreed to the published version of the manuscript.

Funding: This work was partially supported by the Mink Research Partnership and Canada Mink Breeders Association grant (RY000238) and by the NSERC Discovery grant RGPIN/04897-2017 to Levon Abrahamyan (Université de Montréal). N. Markarian was a recipient of the NSERC Undergraduate Student Research Award, and he realized an internship under L. Abrahamyan's supervision.

Institutional Review Board Statement: Not applicable.

Informed Consent Statement: Not applicable.

Data Availability Statement: Not applicable.

Acknowledgments: We thank Hossain Farid (Dalhousie University) and John Easley (Mink Research Partnership) for their valuable comments that greatly improved the manuscript.

Conflicts of Interest: The authors declare no conflict of interest.

References

1. Gorham, J.R.; L, R.W.; Padgett, G.A.; Burger, D.; Henson, J.B. *Some Observations on the Natural Occurrence of Aleutian Disease in: Slow, Latent, and Temperate Virus Infections*; Gajdusek, D.C., Gibbs, C.J., Alpers, M., Eds.; NINDB Monograph No. 2; U.S. Government Printing Office: Washington, DC, USA, 1965; pp. 279–285.
2. Karstad, L.; Pridham, T.J. Aleutian Disease of Mink: I. Evidence of its Viral Etiology. *Can. J. Comp. Med. Vet. Sci.* **1962**, *26*, 97–102. [PubMed]
3. Bloom, M.E.; Kanno, H.; Mori, S.; Wolfinbarger, J.B. Aleutian mink disease: Puzzles and paradigms. *Infect. Agents Dis.* **1994**, *3*, 279–301.
4. Best, S.M.; Bloom, M.E. Aleutian mink disease parvovirus. In *The Parvoviruses*; Hodder Arnold: London, UK, 2006; pp. 457–471.
5. Alexandersen, S. Pathogenesis of disease caused by Aleutian mink disease parvovirus. *APMIS Suppl.* **1990**, *14*, 1–32. [PubMed]
6. Jensen, T.H.; Chriél, M.; Hansen, M.S. Progression of experimental chronic Aleutian mink disease virus infection. *Acta Vet. Scand.* **2016**, *58*, 35. [CrossRef] [PubMed]
7. Cotmore, S.F.; Agbandje-McKenna, M.; Canuti, M.; Chiorini, J.A.; Eis-Hubinger, A.-M.; Hughes, J.; Mietzsch, M.; Modha, S.; Ogliastro, M.; Pénzes, J.J.; et al. ICTV Virus Taxonomy Profile: Parvoviridae. *J. Gen. Virol.* **2019**, *100*, 367–368. [CrossRef] [PubMed]
8. Bloom, M.E.; Alexandersen, S.; Perryman, S.; Lechner, D.; Wolfinbarger, J.B. Nucleotide sequence and genomic organization of Aleutian mink disease parvovirus (ADV): Sequence comparisons between a nonpathogenic and a pathogenic strain of ADV. *J. Virol.* **1988**, *62*, 2903–2915. [CrossRef]
9. Alexandersen, S.; Bloom, M.E.; Perryman, S. Detailed transcription map of Aleutian mink disease parvovirus. *J. Virol.* **1988**, *62*, 3684–3694. [CrossRef] [PubMed]
10. Qiu, J.; Cheng, F.; Burger, L.R.; Pintel, D. The Transcription Profile of Aleutian Mink Disease Virus in CRFK Cells Is Generated by Alternative Processing of Pre-mRNAs Produced from a Single Promoter. *J. Virol.* **2006**, *80*, 654–662. [CrossRef]
11. Tattersall, P. Parvoviruses: General Features. In *Encyclopedia of Virology*, 3rd ed.; Mahy, B.W.J., Van Regenmortel, M.H.V., Eds.; Academic Press: Oxford, UK, 2008; pp. 90–97.
12. Cotmore, S.F.; Tattersall, P. A genome-linked copy of the NS-1 polypeptide is located on the outside of infectious parvovirus particles. *J. Virol.* **1989**, *63*, 3902–3911. [CrossRef] [PubMed]
13. Canuti, M.; Whitney, H.G.; Lang, A.S. Amdoparvoviruses in small mammals: Expanding our understanding of parvovirus diversity, distribution, and pathology. *Front. Microbiol.* **2015**, *6*, 1119. [CrossRef] [PubMed]
14. Viuff, B.; Aasted, B.; Alexandersen, S. Role of alveolar type II cells and of surfactant-associated protein C mRNA levels in the pathogenesis of respiratory distress in mink kits infected with Aleutian mink disease parvovirus. *J. Virol.* **1994**, *68*, 2720–2725. [CrossRef] [PubMed]
15. Kanno, H.; Wolfinbarger, J.B.; Bloom, M.E. Aleutian mink disease parvovirus infection of mink peritoneal macrophages and human macrophage cell lines. *J. Virol.* **1993**, *67*, 2075–2082. [CrossRef] [PubMed]
16. Kowalczyk, M.; Jakubczak, A.; Horecka, B.; Kostro, K. A comparative molecular characterization of AMDV strains isolated from cases of clinical and subclinical infection. *Virus Genes* **2018**, *54*, 561–569. [CrossRef] [PubMed]
17. Kanno, H.; Wolfinbarger, J.B.; Bloom, M.E. Aleutian mink disease parvovirus infection of mink macrophages and human macrophage cell line U937: Demonstration of antibody-dependent enhancement of infection. *J. Virol.* **1993**, *67*, 7017–7024. [CrossRef] [PubMed]
18. Bloom, M.E.; Best, S.M.; Hayes, S.F.; Wells, R.D.; Wolfinbarger, J.B.; McKenna, R.; Agbandje-McKenna, M. Identification of Aleutian Mink Disease Parvovirus Capsid Sequences Mediating Antibody-Dependent Enhancement of Infection, Virus Neutralization, and Immune Complex Formation. *J. Virol.* **2001**, *75*, 11116. [CrossRef] [PubMed]

19. Vihinen-Ranta, M.; Suikkanen, S.; Parrish, C.R. Pathways of Cell Infection by Parvoviruses and Adeno-Associated Viruses. *J. Virol.* **2004**, *78*, 6709–6714. [CrossRef]
20. Tu, M.; Liu, F.; Chen, S.; Wang, M.; Cheng, A. Role of capsid proteins in parvoviruses infection. *Virol. J.* **2015**, *12*, 114. [CrossRef]
21. López-Bueno, A.; Rubio, M.-P.; Bryant, N.; McKenna, R.; Agbandje-McKenna, M.; Almendral, J.M. Host-selected amino acid changes at the sialic acid binding pocket of the parvovirus capsid modulate cell binding affinity and determine virulence. *J. Virol.* **2006**, *80*, 1563–1573. [CrossRef] [PubMed]
22. Boisvert, M.; Fernandes, S.; Tijssen, P. Multiple pathways involved in porcine parvovirus cellular entry and trafficking toward the nucleus. *J. Virol.* **2010**, *84*, 7782–7792. [CrossRef]
23. Bilkova, E.; Forstova, J.; Abrahamyan, L. Coat as a dagger: The use of capsid proteins to perforate membranes during non-enveloped DNA viruses trafficking. *Viruses* **2014**, *6*, 2899–2937. [CrossRef]
24. Zádori, Z.; Szelei, J.; Lacoste, M.-C.; Li, Y.; Gariépy, S.; Raymond, P.; Allaire, M.; Nabi, I.R.; Tijssen, P. A Viral Phospholipase A2 Is Required for Parvovirus Infectivity. *Dev. Cell* **2001**, *1*, 291–302. [CrossRef]
25. Deng, F.; Ye, G.; Liu, Q.; Navid, M.; Zhong, X.; Li, Y.; Wan, C.; Xiao, S.; He, Q.; Fu, Z.; et al. Identification and Comparison of Receptor Binding Characteristics of the Spike Protein of Two Porcine Epidemic Diarrhea Virus Strains. *Viruses* **2016**, *8*, 55. [CrossRef] [PubMed]
26. Pénzes, J.J.; Marsile-Medun, S.; Agbandje-McKenna, M.; Gifford, R.J. Endogenous amdoparvovirus-related elements reveal insights into the biology and evolution of vertebrate parvoviruses. *Virus Evol* **2018**, *4*, vey026. [CrossRef] [PubMed]
27. Martin, D.P.; Biagini, P.; Lefeuvre, P.; Golden, M.; Roumagnac, P.; Varsani, A. Recombination in eukaryotic single stranded DNA viruses. *Viruses* **2011**, *3*, 1699–1738. [CrossRef]
28. King, A.M.Q.; Adams, M.J.; Carstens, E.B.; Lefkowitz, E.J. (Eds.) Family-Parvoviridae. In *Virus Taxonomy*; Elsevier: San Diego, CA, USA, 2012; pp. 405–425.
29. Guan, W.; Huang, Q.; Cheng, F.; Qiu, J. Internal polyadenylation of the parvovirus B19 precursor mRNA is regulated by alternative splicing. *J. Biol. Chem.* **2011**, *286*, 24793–24805. [CrossRef] [PubMed]
30. Huang, Q.; Luo, Y.; Cheng, F.; Best, S.M.; Bloom, M.E.; Qiu, J. Molecular characterization of the small nonstructural proteins of parvovirus Aleutian mink disease virus (AMDV) during infection. *Virology* **2014**, *452–453*, 23–31. [CrossRef]
31. Chen, A.Y.; Qiu, J. Parvovirus infection-induced cell death and cell cycle arrest. *Future Virol.* **2010**, *5*, 731–743. [CrossRef]
32. Best, S.M.; Shelton, J.F.; Pompey, J.M.; Wolfinbarger, J.B.; Bloom, M.E. Caspase cleavage of the nonstructural protein NS1 mediates replication of Aleutian mink disease parvovirus. *J. Virol.* **2003**, *77*, 5305–5312. [CrossRef]
33. Connolly, P.F.; Fearnhead, H.O. Viral hijacking of host caspases: An emerging category of pathogen-host interactions. *Cell Death Differ.* **2017**, *24*, 1401–1410. [CrossRef]
34. Richard, A.; Tulasne, D. Caspase cleavage of viral proteins, another way for viruses to make the best of apoptosis. *Cell Death Dis.* **2012**, *3*, e277. [CrossRef]
35. Cheng, F.; Chen, A.Y.; Best, S.M.; Bloom, M.E.; Pintel, D.; Qiu, J. The Capsid Proteins of Aleutian Mink Disease Virus Activate Caspases and Are Specifically Cleaved during Infection. *J. Virol.* **2010**, *84*, 2687–2696. [CrossRef] [PubMed]
36. Tattersall, P.; Ward, D.C. Rolling hairpin model for replication of parvovirus and linear chromosomal DNA. *Nature* **1976**, *263*, 106–109. [CrossRef] [PubMed]
37. Cotmore, S.F.; Tattersall, P. Resolution of Parvovirus Dimer Junctions Proceeds through a Novel Heterocruciform Intermediate. *J. Virol.* **2003**, *77*, 6245. [CrossRef] [PubMed]
38. Leng, X.; Liu, D.; Li, J.; Shi, K.; Zeng, F.; Zong, Y.; Liu, Y.; Sun, Z.; Zhang, S.; Liu, Y.; et al. Genetic diversity and phylogenetic analysis of Aleutian mink disease virus isolates in north-east China. *Arch. Virol.* **2018**, *163*, 1241–1251. [CrossRef] [PubMed]
39. Olofsson, A.; Mittelholzer, C.; Treiberg Berndtsson, L.; Lind, L.; Mejerland, T.; Belák, S. Unusual, high genetic diversity of Aleutian mink disease virus. *J. Clin. Microbiol.* **1999**, *37*, 4145–4149. [CrossRef]
40. Knuuttila, A.; Uzcátegui, N.; Kankkonen, J.; Vapalahti, O.; Kinnunen, P. Molecular epidemiology of Aleutian mink disease virus in Finland. *Vet. Microbiol.* **2009**, *133*, 229–238. [CrossRef]
41. Bloom, M.E.; Berry, B.D.; Wei, W.; Perryman, S.; Wolfinbarger, J.B. Characterization of chimeric full-length molecular clones of Aleutian mink disease parvovirus (ADV): Identification of a determinant governing replication of ADV in cell culture. *J. Virol.* **1993**, *67*, 5976–5988. [CrossRef]
42. Hadlow, W.J.; Race, R.E.; Kennedy, R.C. Comparative pathogenicity of four strains of Aleutian disease virus for pastel and sapphire mink. *Infect. Immun.* **1983**, *41*, 1016–1023. [CrossRef]
43. Li, Y.; Huang, J.; Jia, Y.; Du, Y.; Jiang, P.; Zhang, R. Genetic characterization of Aleutian mink disease viruses isolated in China. *Virus Genes* **2012**, *45*, 24–30. [CrossRef]
44. Persson, S.; Jensen, T.H.; Blomström, A.-L.; Appelberg, M.T.; Magnusson, U. Aleutian Mink Disease Virus in Free-Ranging Mink from Sweden. *PLoS ONE* **2015**, *10*, e0122194. [CrossRef]
45. Knuuttila, A.; Aronen, P.; Saarinen, A.; Vapalahti, O. Development and evaluation of an enzyme-linked immunosorbent assay based on recombinant VP2 capsids for the detection of antibodies to Aleutian mink disease virus. *Clin. Vaccine Immunol.* **2009**, *16*, 1360–1365. [CrossRef]
46. Kashtanov, S.N.; Salnikova, L.E. Aleutian Mink Disease: Epidemiological and Genetic Aspects. *Biol. Bull. Rev.* **2018**, *8*, 104–113. [CrossRef]

47. Broll, S.; Alexandersen, S. Investigation of the pathogenesis of transplacental transmission of Aleutian mink disease parvovirus in experimentally infected mink. *J. Virol.* **1996**, *70*, 1455–1466. [CrossRef] [PubMed]
48. Jackson, M.K.; Winslow, S.G.; Dockery, L.D.; Jones, J.K.; Sisson, D.V. Investigation of an outbreak of Aleutian disease on a commercial mink ranch. *Am. J. Vet. Res.* **1996**, *57*, 1706–1710. [PubMed]
49. Alexandersen, S.; Jensen, Å.U.; Hansen, M.; Aasted, B. Experimental Transmission of Aleutian Disease virus (ADV) to Different Animal Species. *Acta Pathol. Microbiol. Scand. Ser. B Microbiol.* **2009**, *93B*, 195–200. [CrossRef]
50. Oie, K.L.; Durrant, G.; Wolfinbarger, J.B.; Martin, D.; Costello, F.; Perryman, S.; Hogan, D.; Hadlow, W.J.; Bloom, M.E. The relationship between capsid protein (VP2) sequence and pathogenicity of Aleutian mink disease parvovirus (ADV): A possible role for raccoons in the transmission of ADV infections. *J. Virol.* **1996**, *70*, 852–861. [CrossRef]
51. Mañas, S.; Ceña, J.C.; Ruiz-Olmo, J.; Palazón, S.; Domingo, M.; Wolfinbarger, J.B.; Bloom, M.E. Aleutian mink disease parvovirus in wild riparian carnivores in Spain. *J. Wildl. Dis.* **2001**, *37*, 138–144. [CrossRef]
52. Ladouceur, E.E.B.; Anderson, M.; Ritchie, B.W.; Ciembor, P.; Rimoldi, G.; Piazza, M.; Pesti, D.; Clifford, D.L.; Giannitti, F. Aleutian Disease. *Vet. Pathol.* **2015**, *52*, 1250–1253. [CrossRef]
53. Porter, H.G.; Porter, D.D.; Larsen, A.E. Aleutian disease in ferrets. *Infect. Immun.* **1982**, *36*, 379–386. [CrossRef]
54. Greenfield, J.; Walton, R.; Macdonald, K.R. Detection of Aleutian Disease in Mink: Serum-plate Agglutination Using Iodine Compared with Precipitation by Agar-Gel Electrophoresis. *Res. Vet. Sci.* **1973**, *15*, 381–383. [CrossRef]
55. Cho, H.J.; Greenfield, J. Eradication of Aleutian disease of mink by eliminating positive counterimmunoelectrophoresis test reactors. *J. Clin. Microbiol.* **1978**, *7*, 18–22. [CrossRef]
56. Queiroz, L.H.; Moreira, W.C.; Moura, W.C.; Silva, M.V. Chapter Fifteen—Demonstration of Viral Antibodies by the Counterimmunoelectrophoresis Test. In *Current Laboratory Techniques in Rabies Diagnosis, Research and Prevention*; Rupprecht, C., Nagarajan, T., Eds.; Academic Press: Cambridge, MA, USA, 2015; Volume 2, pp. 175–186.
57. Farid, A.H.; Zillig, M.L.; Finley, G.G.; Smith, G.C. Prevalence of the Aleutian mink disease virus infection in Nova Scotia, Canada. *Prev. Vet. Med.* **2012**, *106*, 332–338. [CrossRef]
58. Farid, A.H.; Ferns, L.E. Reduced severity of histopathological lesions in mink selected for tolerance to Aleutian mink disease virus infection. *Res. Vet. Sci.* **2017**, *111*, 127–134. [CrossRef]
59. Ma, F.; Zhang, L.; Wang, Y.; Lu, R.; Hu, B.; Lv, S.; Xue, X.; Li, X.; Ling, M.; Fan, S.; et al. Development of a Peptide ELISA for the Diagnosis of Aleutian Mink Disease. *PLoS ONE* **2016**, *11*, e0165793. [CrossRef] [PubMed]
60. Knuuttila, A.; Aronen, P.; Eerola, M.; Gardner, I.A.; Virtala, A.-M.K.; Vapalahti, O. Validation of an automated ELISA system for detection of antibodies to Aleutian mink disease virus using blood samples collected in filter paper strips. *Virol. J.* **2014**, *11*, 141. [CrossRef] [PubMed]
61. Dam-Tuxen, R.; Dahl, J.; Jensen, T.H.; Dam-Tuxen, T.; Struve, T.; Bruun, L. Diagnosing Aleutian mink disease infection by a new fully automated ELISA or by counter current immunoelectrophoresis: A comparison of sensitivity and specificity. *J. Virol. Methods* **2014**, *199*, 53–60. [CrossRef]
62. Farid, A.H.; Rupasinghe, P.P. Accuracy of enzyme-linked immunosorbent assays for quantification of antibodies against Aleutian mink disease virus. *J. Virol. Methods* **2016**, *235*, 144–151. [CrossRef] [PubMed]
63. Jensen, T.H.; Christensen, L.S.; Chriél, M.; Uttenthal, Å.; Hammer, A.S. Implementation and validation of a sensitive PCR detection method in the eradication campaign against Aleutian mink disease virus. *J. Virol. Methods* **2011**, *171*, 81–85. [CrossRef]
64. Virtanen, J.; Aaltonen, K.; Vapalahti, O.; Sironen, T. Development and validation of nucleic acid tests to diagnose Aleutian mink disease virus. *J. Virol. Methods* **2020**, *279*, 113776. [CrossRef] [PubMed]
65. Kowalczyk, M.; Jakubczak, A.; Gryzińska, M. Duplex PCR for Detection of Aleutian Disease Virus from Biological and Environmental Samples. *Acta Vet.* **2019**, *69*, 402–413. [CrossRef]
66. Andersson, A.-M.; Nyman, A.-K.; Wallgren, P. A retrospective cohort study estimating the individual Aleutian disease progress in female mink using a VP2 ELISA and its association to reproductive performance. *Prev. Vet. Med.* **2017**, *140*, 60–66. [CrossRef] [PubMed]
67. Farid, A.H. Response of American mink to selection for tolerance to Aleutian mink disease virus. *EC Microbiol.* **2020**, *16*, 110–128.
68. Prieto, A.; Fernández-Antonio, R.; López-Lorenzo, G.; Díaz-Cao, J.M.; López-Novo, C.; Remesar, S.; Panadero, R.; Díaz, P.; Morrondo, P.; Díez-Baños, P.; et al. Molecular epidemiology of Aleutian mink disease virus causing outbreaks in mink farms from Southwestern Europe: A retrospective study from 2012 to 2019. *J. Vet. Sci.* **2020**, *21*, e65. [CrossRef]
69. Farid, A.H.; Daftarian, P.M.; Fatehi, J. Transmission dynamics of Aleutian mink disease virus on a farm under test and removal scheme. *J. Vet. Sci. Med. Diag.* **2018**, *7*, 2. [CrossRef]
70. Karimi, K.; Farid, A.H.; Myles, S.; Miar, Y. Detection of selection signatures for response to Aleutian mink disease virus infection in American mink. *Sci. Rep.* **2021**, *11*, 2944. [CrossRef]
71. Roeder, P.; Mariner, J.; Kock, R. Rinderpest: The veterinary perspective on eradication. *Philos. Trans. R. Soc. Lond. B Biol. Sci.* **2013**, *368*, 20120139. [CrossRef] [PubMed]
72. Nituch, L.A.; Bowman, J.; Beauclerc, K.B.; Schulte-Hostedde, A.I. Mink farms predict Aleutian disease exposure in wild American mink. *PLoS ONE* **2011**, *6*, e21693. [CrossRef] [PubMed]
73. Prieto, A.; Fernández-Antonio, R.; Díaz-Cao, J.M.; López, G.; Díaz, P.; Alonso, J.M.; Morrondo, P.; Fernández, G. Distribution of Aleutian mink disease virus contamination in the environment of infected mink farms. *Vet. Microbiol.* **2017**, *204*, 59–63. [CrossRef] [PubMed]

74. Chen, X.; Song, C.; Liu, Y.; Qu, L.; Liu, D.; Zhang, Y.; Liu, M. Development of an Enzyme-Linked Immunosorbent Assay Based on Fusion VP2332-452 Antigen for Detecting Antibodies against Aleutian Mink Disease Virus. *J. Clin. Microbiol.* **2016**, *54*, 439–442 [CrossRef]
75. Farid, A.H.; Hussain, I. Dose response of black American mink to Aleutian mink disease virus. *Immun. Inflamm. Dis.* **2020**, *8*, 150–164. [CrossRef]
76. Farid, A.H.; Hussain, I. A comparison between intraperitoneal injection and intranasal and oral inoculation of mink with Aleutian mink disease virus. *Res. Vet. Sci.* **2019**, *124*, 85–92. [CrossRef]
77. Hadlow, W.J.; Race, R.E.; Kennedy, R.C. Temporal replication of the Pullman strain of Aleutian disease virus in royal pastel mink *J. Virol.* **1985**, *55*, 853–856. [CrossRef]
78. Alaeddini, R. Forensic implications of PCR inhibition—A review. *Forensic Sci. Int. Genet.* **2012**, *6*, 297–305. [CrossRef]
79. Rådström, P.; Knutsson, R.; Wolffs, P.; Lövenklev, M.; Löfström, C. Pre-PCR processing. *Mol. Biotechnol.* **2004**, *26*, 133–146. [CrossRef]
80. Espregueira Themudo, G.; Houe, H.; Agger, J.F.; Østergaard, J.; Ersbøll, A.K. Identification of biosecurity measures and spatial variables as potential risk factors for Aleutian disease in Danish mink farms. *Prev. Vet. Med.* **2012**, *107*, 134–141. [CrossRef]
81. Díaz Cao, J.M.; Prieto, A.; López, G.; Fernández-Antonio, R.; Díaz, P.; López, C.; Remesar, S.; Díez-Baños, P.; Fernández, G Molecular assessment of visitor personal protective equipment contamination with the Aleutian mink disease virus and porcine circovirus-2 in mink and porcine farms. *PLoS ONE* **2018**, *13*, e0203144. [CrossRef] [PubMed]
82. Blecha, F. Immunomodulators for Prevention and Treatment of Infectious Diseases in Food-Producing Animals. *Vet. Clin. N. Am Food Anim. Pract.* **2001**, *17*, 621–633. [CrossRef]
83. Byrne, K.A.; Loving, C.L.; McGill, J.L. Innate Immunomodulation in Food Animals: Evidence for Trained Immunity? *Front Immunol.* **2020**, *11*, 1099. [CrossRef] [PubMed]
84. Siwicki, A.K.; Pozet, F.; Morand, M.; Kazuń, B.; Trapkowska, S. In vitro effect of methisoprinol on salmonid rhabdoviruses replication. *Bull.-Vet. Inst. Pulawy* **2002**, *46*, 53–58.
85. Kowalczyk, M.; Gąsiorek, B.; Kostro, K.; Borzym, E.; Jakubczak, A. Breeding parameters on a mink farm infected with Aleutian mink disease virus following the use of methisoprinol. *Arch. Virol.* **2019**, *164*, 2691–2698. [CrossRef]
86. Farid, A.H.; Smith, N.J. Dietary supplementation of Ascophylum nodosum improved kidney function of mink challenged with Aleutian mink disease virus. *BMC Vet. Res.* **2020**, *16*, 465. [CrossRef]
87. Marcotte, H.; Hammarström, L. Chapter 71—Passive Immunization: Toward Magic Bullets. In *Mucosal Immunology*, 4th ed. Mestecky, J., Strober, W., Russell, M.W., Kelsall, B.L., Cheroutre, H., Lambrecht, B.N., Eds.; Academic Press: Boston, MA, USA 2015; pp. 1403–1434.
88. Alexandersen, S.; Larsen, S.; Cohn, A.; Uttenthal, A.; Race, R.E.; Aasted, B.; Hansen, M.; Bloom, M.E. Passive transfer of antiviral antibodies restricts replication of Aleutian mink disease parvovirus in vivo. *J. Virol.* **1989**, *63*, 9–17. [CrossRef]
89. Villa, T.G.; Feijoo-Siota, L.; Rama, J.L.R.; Ageitos, J.M. Antivirals against animal viruses. *Biochem. Pharmacol.* **2017**, *133*, 97–116 [CrossRef]
90. Lu, T.; Zhang, H.; Zhou, J.; Ma, Q.; Yan, W.; Zhao, L.; Wu, S.; Chen, H. Aptamer-targeting of Aleutian mink disease virus (AMDV can be an effective strategy to inhibit virus replication. *Sci. Rep.* **2021**, *11*, 4649. [CrossRef]
91. Shu, Y.; McCauley, J. GISAID: Global initiative on sharing all influenza data—From vision to reality. *Eurosurveillance* **2017**, *22* 30494. [CrossRef]
92. Hagberg, E.E.; Pedersen, A.G.; Larsen, L.E.; Krarup, A. Evolutionary analysis of whole-genome sequences confirms inter-farm transmission of Aleutian mink disease virus. *J. Gen. Virol.* **2017**, *98*, 1360–1371. [CrossRef] [PubMed]
93. Canuti, M.; O'Leary, K.E.; Hunter, B.D.; Spearman, G.; Ojkic, D.; Whitney, H.G.; Lang, A.S. Driving forces behind the evolution o the Aleutian mink disease parvovirus in the context of intensive farming. *Virus Evol.* **2016**, *2*, vew004. [CrossRef]
94. Leimann, A.; Knuuttila, A.; Maran, T.; Vapalahti, O.; Saarma, U. Molecular epidemiology of Aleutian mink disease virus (AMDV in Estonia, and a global phylogeny of AMDV. *Virus Res.* **2015**, *199*, 56–61. [CrossRef] [PubMed]
95. Knuuttila, A.; Aaltonen, K.; Virtala, A.M.K.; Henttonen, H.; Isomursu, M.; Leimann, A.; Maran, T.; Saarma, U.; Timonen, P. Vapalahti, O.; et al. Aleutian mink disease virus in free-ranging mustelids in Finland—A cross-sectional epidemiological and phylogenetic study. *J. Gen. Virol.* **2015**, *96*, 1423–1435. [CrossRef] [PubMed]
96. Sang, Y.; Ma, J.; Hou, Z.; Zhang, Y. Phylogenetic analysis of the VP2 gene of Aleutian mink disease parvoviruses isolated from 2009 to 2011 in China. *Virus Genes* **2012**, *45*, 31–37. [CrossRef] [PubMed]
97. Jakubczak, A.; Kowalczyk, M.; Kostro, K.; Jezewska-Witkowska, G. Comparative molecular analysis of strains of the Aleutian Disease Virus isolated from farmed and wild mink. *Strain* **2017**, *3*, 4. [CrossRef] [PubMed]
98. Nituch, L.A.; Bowman, J.; Wilson, P.; Schulte-Hostedde, A.I. Molecular epidemiology of Aleutian disease virus in free-ranging domestic, hybrid, and wild mink. *Evol. Appl.* **2012**, *5*, 330–340. [CrossRef]
99. Xi, J.; Wang, J.; Yu, Y.; Zhang, X.; Mao, Y.; Hou, Q.; Liu, W. Genetic characterization of the complete genome of an Aleutian mink disease virus isolated in north China. *Virus Genes* **2016**, *52*, 463–473. [CrossRef] [PubMed]
100. Burrell, C.J.; Howard, C.R.; Murphy, F.A. Chapter 11—Vaccines and Vaccination. In *Fenner and White's Medical Virology*, 5th ed Burrell, C.J., Howard, C.R., Murphy, F.A., Eds.; Academic Press: London, UK, 2017; pp. 155–167.
101. Sanders, B.; Koldijk, M.; Schuitemaker, H. Inactivated Viral Vaccines. *Vaccine Anal. Strateg. Princ. Control* **2014**, 45–80. [CrossRef]
102. Hicks, D.J.; Fooks, A.R.; Johnson, N. Developments in rabies vaccines. *Clin. Exp. Immunol.* **2012**, *169*, 199–204. [CrossRef] [PubMed]

103. WHO. Japanese Encephalitis Vaccines: WHO position paper, February 2015–Recommendations. *Vaccine* **2016**, *34*, 302–303. [CrossRef] [PubMed]
104. Wallace, M.R.; Brandt, C.J.; Earhart, K.C.; Kuter, B.J.; Grosso, A.D.; Lakkis, H.; Tasker, S.A. Safety and Immunogenicity of an Inactivated Hepatitis A Vaccine among HIV-Infected Subjects. *Clin. Infect. Dis.* **2004**, *39*, 1207–1213. [CrossRef]
105. Zhang, Y.; Zeng, G.; Pan, H.; Li, C.; Hu, Y.; Chu, K.; Han, W.; Chen, Z.; Tang, R.; Yin, W.; et al. Safety, tolerability, and immunogenicity of an inactivated SARS-CoV-2 vaccine in healthy adults aged 18–59 years: A randomised, double-blind, placebo-controlled, phase 1/2 clinical trial. *Lancet Infect. Dis.* **2021**, *21*, 181–192. [CrossRef]
106. Karstad, L.; Pridham, T.J.; Gray, D.P. Aleutian Disease (Plasmacytosis) of Mink: II. Responses of Mink to Formalin-Treated Diseased Tissues and to Subsequent Challenge With Virulent Inoculum. *Can. J. Comp. Med. Vet. Sci.* **1963**, *27*, 124–128.
107. Porter, D.D.; Larsen, A.E.; Porter, H.G. The Pathogenesis of Aleutian Disease of Mink. *J. Immunol.* **1972**, *109*, 1. [CrossRef]
108. Basrur, P.K.; Karstad, L. Studies on viral plasmacytosis (Aleutian disease) of mink. VII. Infection of mink with DNA extracted from diseased spleens. *Can. J. Comp. Med. Vet. Sci.* **1966**, *30*, 295–300.
109. Pascolo, S. Messenger RNA-based vaccines. *Expert Opin. Biol. Ther.* **2004**, *4*, 1285–1294. [CrossRef]
110. Ulmer, J.B.; Sadoff, J.C.; Liu, M.A. DNA vaccines. *Curr. Opin. Immunol.* **1996**, *8*, 531–536. [CrossRef]
111. Leitner, W.W.; Ying, H.; Restifo, N.P. DNA and RNA-based vaccines: Principles, progress and prospects. *Vaccine* **1999**, *18*, 765–777. [CrossRef]
112. Bodles-Brakhop, A.M.; Draghia-Akli, R. DNA vaccination and gene therapy: Optimization and delivery for cancer therapy. *Expert Rev. Vaccines* **2008**, *7*, 1085–1101. [CrossRef] [PubMed]
113. Scheiblhofer, S.; Thalhamer, J.; Weiss, R. DNA and mRNA vaccination against allergies. *Pediatr. Allergy Immunol.* **2018**, *29*, 679–688. [CrossRef]
114. Whalen, R.G. DNA vaccines for emerging infectious diseases: What if? *Emerg. Infect. Dis.* **1996**, *2*, 168–175. [CrossRef]
115. Redding, L.; Weiner, D.B. DNA vaccines in veterinary use. *Expert Rev. Vaccines* **2009**, *8*, 1251–1276. [CrossRef] [PubMed]
116. Kutzler, M.A.; Weiner, D.B. DNA vaccines: Ready for prime time? *Nat. Rev. Genet.* **2008**, *9*, 776–788. [CrossRef] [PubMed]
117. Jazayeri, S.D.; Poh, C.L. Recent advances in delivery of veterinary DNA vaccines against avian pathogens. *Vet. Res.* **2019**, *50*, 78. [CrossRef]
118. Castelruiz, Y.; Blixenkrone-Møller, M.; Aasted, B. DNA vaccination with the Aleutian mink disease virus NS1 gene confers partial protection against disease. *Vaccine* **2005**, *23*, 1225–1231. [CrossRef] [PubMed]
119. Liu, D.; Li, J.; Shi, K.; Zeng, F.; Zong, Y.; Leng, X.; Lu, H.; Du, R. Construction and Immunogenicity Analysis of Whole-Gene Mutation DNA Vaccine of Aleutian Mink Virus Isolated Virulent Strain. *Viral Immunol.* **2018**, *31*, 69–77. [CrossRef]
120. Huzair, F.; Sturdy, S. Biotechnology and the transformation of vaccine innovation: The case of the hepatitis B vaccines 1968–2000. *Stud. Hist. Philos. Biol. Biomed. Sci.* **2017**, *64*, 11–21. [CrossRef]
121. Szmuness, W.; Stevens, C.E.; Zang, E.A.; Harley, E.J.; Kellner, A. A controlled clinical trial of the efficacy of the hepatitis B vaccine (heptavax B): A final report. *Hepatology* **1981**, *1*, 377–385. [CrossRef]
122. Tan, W.S.; Ho, K.L. Phage display creates innovative applications to combat hepatitis B virus. *World J. Gastroenterol.* **2014**, *20*, 11650–11670. [CrossRef]
123. Ho, J.K.-T.; Jeevan-Raj, B.; Netter, H.-J. Hepatitis B Virus (HBV) Subviral Particles as Protective Vaccines and Vaccine Platforms. *Viruses* **2020**, *12*, 126. [CrossRef]
124. Valenzuela, P.; Medina, A.; Rutter, W.J.; Ammerer, G.; Hall, B.D. Synthesis and assembly of hepatitis B virus surface antigen particles in yeast. *Nature* **1982**, *298*, 347–350. [CrossRef] [PubMed]
125. Wang, N.; Shang, J.; Jiang, S.; Du, L. Subunit Vaccines Against Emerging Pathogenic Human Coronaviruses. *Front. Microbiol.* **2020**, *11*, 298. [CrossRef] [PubMed]
126. Liljeqvist, S.; Ståhl, S. Production of recombinant subunit vaccines: Protein immunogens, live delivery systems and nucleic acid vaccines. *J. Biotechnol.* **1999**, *73*, 1–33. [CrossRef]
127. Lee, N.-H.; Lee, J.-A.; Park, S.-Y.; Song, C.-S.; Choi, I.-S.; Lee, J.-B. A review of vaccine development and research for industry animals in Korea. *Clin. Exp. Vaccine Res.* **2012**, *1*, 18–34. [CrossRef] [PubMed]
128. Pérez, O.; Romeu, B.; Cabrera, O.; González, E.; Batista-Duharte, A.; Labrada, A.; Pérez, R.; Reyes, L.M.; Ramírez, W.; Sifontes, S.; et al. Adjuvants are Key Factors for the Development of Future Vaccines: Lessons from the Finlay Adjuvant Platform. *Front. Immunol.* **2013**, *4*, 407. [CrossRef]
129. Awate, S.; Babiuk, L.A.; Mutwiri, G. Mechanisms of action of adjuvants. *Front. Immunol.* **2013**, *4*, 114. [CrossRef] [PubMed]
130. Beijnen, E.M.S.; van Haren, S.D. Vaccine-Induced CD8+ T Cell Responses in Children: A Review of Age-Specific Molecular Determinants Contributing to Antigen Cross-Presentation. *Front. Immunol.* **2020**, *11*, 3328. [CrossRef] [PubMed]
131. Kadam, A.; Sasidharan, S.; Saudagar, P. Computational design of a potential multi-epitope subunit vaccine using immunoinformatics to fight Ebola virus. *Infect. Genet. Evol.* **2020**, *85*, 104464. [CrossRef] [PubMed]
132. Liu, Y.; Ye, L.; Lin, F.; Gomaa, Y.; Flyer, D.; Carrion, R., Jr.; Patterson, J.L.; Prausnitz, M.R.; Smith, G.; Glenn, G.; et al. Intradermal Vaccination With Adjuvanted Ebola Virus Soluble Glycoprotein Subunit Vaccine by Microneedle Patches Protects Mice against Lethal Ebola Virus Challenge. *J. Infect. Dis.* **2018**, *218*, S545–S552. [CrossRef]
133. Pandey, R.K.; Ojha, R.; Aathmanathan, V.S.; Krishnan, M.; Prajapati, V.K. Immunoinformatics approaches to design a novel multi-epitope subunit vaccine against HIV infection. *Vaccine* **2018**, *36*, 2262–2272. [CrossRef]

134. Aasted, B.; Alexandersen, S.; Christensen, J. Vaccination with Aleutian mink disease parvovirus (AMDV) capsid proteins enhances disease, while vaccination with the major non-structural AMDV protein causes partial protection from disease. *Vaccine* **1998**, *16*, 1158–1165. [CrossRef]
135. Tong, M.; Sun, N.; Cao, Z.; Cheng, Y.; Zhang, M.; Cheng, S.; Yi, L. Molecular epidemiology of Aleutian mink disease virus from fecal swab of mink in northeast China. *BMC Microbiol.* **2020**, *20*, 234. [CrossRef]
136. Xi, J.; Zhang, Y.; Wang, J.; Yu, Y.; Zhang, X.; Li, Z.; Cui, S.; Liu, W. Generation of an infectious clone of AMDV and identification of capsid residues essential for infectivity in cell culture. *Virus Res.* **2017**, *242*, 58–65. [CrossRef]
137. Brisse, M.; Vrba, S.M.; Kirk, N.; Liang, Y.; Ly, H. Emerging Concepts and Technologies in Vaccine Development. *Front. Immunol.* **2020**, *11*, 2578. [CrossRef]
138. Nascimento, I.P.; Leite, L.C.C. Recombinant vaccines and the development of new vaccine strategies. *Braz. J. Med. Biol. Res.* **2012**, *45*, 1102–1111. [CrossRef] [PubMed]
139. Dagan, N.; Barda, N.; Kepten, E.; Miron, O.; Perchik, S.; Katz, M.A.; Hernán, M.A.; Lipsitch, M.; Reis, B.; Balicer, R.D. BNT162b2 mRNA Covid-19 Vaccine in a Nationwide Mass Vaccination Setting. *N. Engl. J. Med.* **2021**, *384*, 1412–1423. [CrossRef] [PubMed]
140. Tenforde, M.W.; Olson, S.M.; Self, W.H.; Talbot, H.K.; Lindsell, C.J.; Steingrub, J.S.; Shapiro, N.I.; Ginde, A.A.; Douin, D.J.; Prekker, M.E.; et al. Effectiveness of Pfizer-BioNTech and Moderna Vaccines against COVID-19 among Hospitalized Adults Aged ≥65 Years—United States, January–March 2021. *MMWR. Morb. Mortal. Wkly. Rep.* **2021**, *70*, 674–679. [CrossRef] [PubMed]
141. Baden, L.R.; El Sahly, H.M.; Essink, B.; Kotloff, K.; Frey, S.; Novak, R.; Diemert, D.; Spector, S.A.; Rouphael, N.; Creech, C.B.; et al. Efficacy and Safety of the mRNA-1273 SARS-CoV-2 Vaccine. *N. Engl. J. Med.* **2020**. [CrossRef]
142. Polack, F.P.; Thomas, S.J.; Kitchin, N.; Absalon, J.; Gurtman, A.; Lockhart, S.; Perez, J.L.; Pérez Marc, G.; Moreira, E.D.; Zerbini, C.; et al. Safety and Efficacy of the BNT162b2 mRNA Covid-19 Vaccine. *N. Engl. J. Med.* **2020**, *383*, 2603–2615. [CrossRef]
143. de Souza, G.A.; Rocha, R.P.; Gonçalves, R.L.; Ferreira, C.S.; de Mello Silva, B.; de Castro, R.F.; Rodrigues, J.F.; Júnior, J.C.; Malaquias, L.C.; Abrahão, J.S.; et al. Nanoparticles as Vaccines to Prevent Arbovirus Infection: A Long Road Ahead. *Pathogens* **2021**, *10*, 36. [CrossRef]
144. Al-Halifa, S.; Gauthier, L.; Arpin, D.; Bourgault, S.; Archambault, D. Nanoparticle-Based Vaccines Against Respiratory Viruses. *Front. Immunol.* **2019**, *10*, 22. [CrossRef] [PubMed]
145. Maina, T.W.; Grego, E.A.; Boggiatto, P.M.; Sacco, R.E.; Narasimhan, B.; McGill, J.L. Applications of Nanovaccines for Disease Prevention in Cattle. *Front. Bioeng. Biotechnol.* **2020**, *8*, 1424. [CrossRef]
146. Thakur, A.; Foged, C. Nanoparticles for mucosal vaccine delivery. *Nanoeng. Biomater. Adv. Drug Deliv.* **2020**, 603–646. [CrossRef]
147. Wagner-Muñiz, D.A.; Haughney, S.L.; Kelly, S.M.; Wannemuehler, M.J.; Narasimhan, B. Room Temperature Stable PspA-Based Nanovaccine Induces Protective Immunity. *Front. Immunol.* **2018**, *9*, 325. [CrossRef]
148. Luo, M.; Samandi, L.Z.; Wang, Z.; Chen, Z.J.; Gao, J. Synthetic nanovaccines for immunotherapy. *J. Control. Release* **2017**, *263*, 200–210. [CrossRef]
149. Wang, N.; Chen, M.; Wang, T. Liposomes used as a vaccine adjuvant-delivery system: From basics to clinical immunization. *J. Control. Release* **2019**, *303*, 130–150. [CrossRef]
150. Schwendener, R.A. Liposomes as vaccine delivery systems: A review of the recent advances. *Adv. Vaccines* **2014**, *2*, 159–182. [CrossRef] [PubMed]
151. Fan, Y.; Wang, D.; Hu, Y.; Liu, J.; Han, G.; Zhao, X.; Yuan, J.; Liu, C.; Liu, X.; Ni, X. Liposome and epimedium polysaccharide-propolis flavone can synergistically enhance immune effect of vaccine. *Int. J. Biol. Macromol.* **2012**, *50*, 125–130. [CrossRef]
152. Babai, I.; Samira, S.; Barenholz, Y.; Zakay-Rones, Z.; Kedar, E. A novel influenza subunit vaccine composed of liposome-encapsulated haemagglutinin/neuraminidase and IL-2 or GM-CSF. II. Induction of TH1 and TH2 responses in mice. *Vaccine* **1999**, *17*, 1239–1250. [CrossRef]
153. Perotti, M.; Perez, L. Virus-Like Particles and Nanoparticles for Vaccine Development against HCMV. *Viruses* **2019**, *12*, 35. [CrossRef] [PubMed]
154. Spice, A.J.; Aw, R.; Bracewell, D.G.; Polizzi, K.M. Synthesis and Assembly of Hepatitis B Virus-Like Particles in a Pichia pastoris Cell-Free System. *Front. Bioeng. Biotechnol.* **2020**, *8*, 72. [CrossRef] [PubMed]
155. Mohsen, M.O.; Zha, L.; Cabral-Miranda, G.; Bachmann, M.F. Major findings and recent advances in virus–like particle (VLP)-based vaccines. *Semin. Immunol.* **2017**, *34*, 123–132. [CrossRef]
156. Pironti, G.; Andersson, D.C.; Lund, L.H. Mechanistic and Therapeutic Implications of Extracellular Vesicles as a Potential Link Between Covid-19 and Cardiovascular Disease Manifestations. *Front. Cell Dev. Biol.* **2021**, *9*, 197. [CrossRef]
157. Shkair, L.; Garanina, E.E.; Stott, R.J.; Foster, T.L.; Rizvanov, A.A.; Khaiboullina, S.F. Membrane Microvesicles as Potential Vaccine Candidates. *Int. J. Mol. Sci.* **2021**, *22*, 1142. [CrossRef]
158. Jiang, L.; Luirink, J.; Kooijmans, S.A.A.; van Kessel, K.P.M.; Jong, W.; van Essen, M.; Seinen, C.W.; de Maat, S.; de Jong, O.G.; Gitz-François, J.F.F.; et al. A post-insertion strategy for surface functionalization of bacterial and mammalian cell-derived extracellular vesicles. *Biochim. Biophys. Acta (BBA)-Gen. Subj.* **2021**, *1865*, 129763. [CrossRef] [PubMed]
159. Montaner-Tarbes, S.; Borrás, F.E.; Montoya, M.; Fraile, L.; Del Portillo, H.A. Serum-derived exosomes from non-viremic animals previously exposed to the porcine respiratory and reproductive virus contain antigenic viral proteins. *Vet. Res.* **2016**, *47*, 59. [CrossRef]

160. Rappazzo, C.G.; Watkins, H.C.; Guarino, C.M.; Chau, A.; Lopez, J.L.; DeLisa, M.P.; Leifer, C.A.; Whittaker, G.R.; Putnam, D. Recombinant M2e outer membrane vesicle vaccines protect against lethal influenza A challenge in BALB/c mice. *Vaccine* **2016**, *34*, 1252–1258. [CrossRef] [PubMed]
161. Carvalho, A.L.; Miquel-Clopés, A.; Wegmann, U.; Jones, E.; Stentz, R.; Telatin, A.; Walker, N.J.; Butcher, W.A.; Brown, P.J.; Holmes, S.; et al. Use of bioengineered human commensal gut bacteria-derived microvesicles for mucosal plague vaccine delivery and immunization. *Clin. Exp. Immunol.* **2019**, *196*, 287–304. [CrossRef] [PubMed]

Communication

Construction and Immunogenicity of Virus-Like Particles of Feline Parvovirus from the Tiger

Cuicui Jiao [1,2,†], Hongliang Zhang [1,†], Wei Liu [2,†], Hongli Jin [3,4], Di Liu [2], Jian Zhao [2], Na Feng [4,*], Chuanmei Zhang [1,*] and Jing Shi [1,2,*]

1. Shandong Collaborative Innovation Center for Development of Veterinary Pharmaceuticals, College of Veterinary Medicine, Qingdao Agricultural University, Qingdao 266109, China; jcc19905l2@163.com (C.J.); zhanghongliang001@126.com (H.Z.)
2. Changchun Sino Biotechnology Co., Ltd., Changchun 130012, China; liuwei200314@163.com (W.L.); liudi8680@163.com (D.L.); zhj15948095541@163.com (J.Z.)
3. Key Laboratory of Zoonosis Research, Ministry of Education, College of Veterinary Medicine, Jilin University, Changchun 130062, China; jin8616771@163.com
4. Key Laboratory of Jilin Province for Zoonosis Prevention and Control, Military Veterinary Research Institute, Academy of Military Medical Sciences, Changchun 130122, China

* Correspondence: fengna0308@126.com (N.F.); zhangchuanmei100@163.com (C.Z.); 18604465777@163.com (J.S.)

† These authors contributed equally to this work.

Received: 9 February 2020; Accepted: 13 March 2020; Published: 16 March 2020

Abstract: Feline panleukopenia, caused by feline parvovirus (FPV), is a highly infectious disease characterized by leucopenia and hemorrhagic gastroenteritis that severely affects the health of large wild Felidae. In this study, tiger FPV virus-like particles (VLPs) were developed using the baculovirus expression system. The VP2 gene from an infected Siberian tiger (*Panthera tigris altaica*) was used as the target gene. The key amino acids of this gene were the same as those of FPV, whereas the 101st amino acid was the same as that of canine parvovirus. Indirect immunofluorescence assay (IFA) results demonstrated that the VP2 protein was successfully expressed. SDS-PAGE and Western blotting (WB) results showed that the target protein band was present at approximately 65 kDa. Electron micrograph analyses indicated that the tiger FPV VLPs were successfully assembled and were morphologically similar to natural parvovirus particles. The hemagglutination (HA) titer of the tiger FPV VLPs was as high as $1{:}2^{18}$. The necropsy and tissue sections at the cat injection site suggested that the tiger FPV VLPs vaccine was safe. Antibody production was induced in cats after subcutaneous immunization, with a $>1{:}2^{10}$ hemagglutination inhibition (HI) titer that persisted for at least 12 months. These results demonstrate that tiger FPV VLPs might provide a vaccine to prevent FPV-associated disease in the tiger.

Keywords: feline parvovirus; virus-like particles; VP2 protein; antibodies

---

Feline panleukopenia, caused by feline parvovirus (FPV), is an acute and highly contagious disease [1]. It clinically manifests as vomiting, high fever, leukopenia and enteritis in infected animals [2]. FPV has been known to induce disease in cats since the 1920s [3]. Since its discovery, FPV has been isolated from cats, raccoons, monkeys, and many different wild and captive carnivores [4,5]. FPV is very stable in the environment, with infectivity persisting for up to one year within contaminated material [6]. Consequently, FPV has a high risk of transmission, posing a great threat to rare wild animals. In addition, FPV is prevalent in the Chinese Siberian tiger (*Panthera tigris altaica*) population [7,8]. On June 6th 2016, an outbreak of fatal FPV infection among captive Siberian tigers in Zhengzhou Zoo in central China was reported [7]. An FPV strain was isolated and identified from a captive Siberian tiger in a wildlife park in the Jilin Province, China [8]. Wild Siberian tigers from the Russian Far East had a 68% FPV antibody-positivity rate [9]. Vaccination is an effective means to prevent FPV infection. Currently, attenuated vaccines and inactivated vaccines are approved for use against FPV in cats. Attenuated vaccines have been recommended for wildlife use in North American countries, Europe and Asia [10]. A live-attenuated vaccine (Virbac S.A., Carros 06511, France), which combines feline calicivirus (FCV), feline herpesvirus (FHV) and FPV strains, is used to prevent disease in tigers [11]. However, it has been reported that attenuated FPV vaccines with a risk of virulence reversion can cause cerebellar dysplasia in cat fetuses [12]. The antibody level produced by inactivated vaccines is low, the antibody duration is short, and there is a risk of incomplete inactivation. Furthermore, the safety of using inactivated vaccines in tigers remains unknown. Therefore, the development of a novel vaccine for feline parvovirus disease in tigers will provide material reserves and new products for immune protection against important diseases in tigers.

FPV, a single-stranded DNA virus, belongs to the family *Parvoviridae*, subfamily *Parvovirinae*, genus *Protoparvovirus* [13]. The virions have a symmetrical icosahedral structure [14]. The FPV genome contains two open reading frames encoding nonstructural proteins (NS1 and NS2) and structural proteins (VP1 and VP2). The VP2 protein contains the major epitopes that stimulate the production of neutralizing antibodies [15], and mutations in key amino acid sites of the VP2 protein can affect its antigenic characteristics and host range [16,17].

Virus-like particles (VLPs) are composed of multiple copies of one or more recombinant viral structural proteins and are spontaneously assembled into particles without the viral genome [18]. Due to being replication-incompetent and producing nonpathogenic effects, VLPs have been widely used in studies of human and veterinary candidate vaccines [19]. In this study, the parvovirus VP2 gene was identified from a dead Siberian tiger in a wildlife park in the Jilin Province, China [8]. The tiger had symptoms such as diarrhea and vomiting before it died. The VP2 sequence was analyzed and kindly provided by Dr. Wang [8]. We used the baculovirus expression system [20] to produce the VP2 protein of tiger FPV. The VP2 protein can be assembled into VLPs which resemble the natural virus in size and shape. Meanwhile, tigers and cats both belong to the family Felidae. Due to the rarity and importance of the tiger, cats were selected to conduct immunization experiments of the tiger FPV VLPs vaccine. Herein, the tiger FPV VLPs vaccine stimulated cats to produce hemagglutination inhibition (HI) antibody, providing a preliminary basis and technical support for the preparation of new vaccines against tiger (*Panthera tigris altaica*) feline parvovirus disease.

First, the amino acid sites of the parvovirus VP2 gene from a tiger were analyzed. The results showed that most of the amino acid sites were consistent with those of the FPV reference sequence, whereas the 101st amino acid was consistent with that of the canine parvovirus (CPV) reference sequence (Table 1). Therefore, the parvovirus VP2 gene from the tiger was named tiger FPV VP2.

Table 1. Amino acid variations in the VP2 protein.

Type	Amino Acid Site of VP2 [21]												
	80	**87**	**93**	**101**	**103**	**297**	**300**	**305**	**323**	**426**	**555**	**564**	**568**
Tiger FPV	K	M	K	T	V	S	A	D	D	N	V	N	A
FPV	K	M	K	I	V	S	A	D	D	N	V	N	A
CPV-2	R	M	N	I	V	S	A	D	N	N	V	S	G
New CPV-2a	R	L	N	T	A	A	G	Y	N	N	V	S	G
New CPV-2b	R	L	N	T	A	A	G	Y	N	D	V	S	G
CPV-2c	R	L	N	T	A	A	G	Y	N	E	V	S	G

The box shows the difference between Tiger FPV and FPV at 101st amino acid.

Next, we optimized the codons of tiger FPV VP2 for *Spodoptera frugiperda* 9 (Sf9) cells according to codon preference (shown in Supplementary Materials, Sequence 1). Sequence optimization and synthesis were completed by Shanghai Biological Engineering Co., Ltd. The VP2 gene was amplified using the primers shown in Table 2 and cloned into the pFastBac Dual vector (Invitrogen, Carlsbad, CA, USA). The pFBD-dVP2 vector, a donor plasmid containing two copies of the VP2 gene, was successfully constructed. The pFBD-dVP2 donor plasmid was transformed into *E. coli* DH10Bac competent cells (Invitrogen, Carlsbad, CA, USA) to obtain the rBacmid-dVP2 recombinant bacmid.

Table 2. Primer sequences.

Primer	Primer Sequence (5′-3′)	Product Length
PH-VP2-F	TAT**GCGGCCGC**ATGTCCGACGGTGCTGT (*Not* I)	1755 bp
PH-VP2-R	TAT**AAGCTT**TTAGTACAGCTTACGAGGA (*Hind* III)	
P10-VP2-F	TAT**CTCGAG**ATGTCCGACGGTGCTGTGC (*Xho* I)	1755 bp
P10-VP2-R	TAT**GGTACC**TTAGTACAGCTTACGAGGAG (*Kpn* I)	

The bold characters are the sequences of restriction enzyme sites.

The rBacmid-dVP2 was transfected into Sf9 cells and cultured in six-well plates at approximately 80% confluence with Cellfectin II Reagent Transfection Reagent (Invitrogen, Carlsbad, CA, USA) according to the manufacturer's instructions, and the cells were then cultured in sf-900 II SFM medium (Gibco, Grand island, NY, USA) at 27 °C. It took approximately 96 h to reach an 80% cytopathogenic effect (CPE). The supernatant was collected to obtain recombinant baculovirus, called rpFBD-dVP2 which was passage 1 (P1). The genome was extracted and determined to be correct by PCR. The baculovirus stock titer was measured using a Baculovirus Rapid Titer Kit (TaKaRa, Tokyo, Japan) according to the manufacturer's instructions. RpFBD-dVP2 was continuously passaged to P4 at a multiplicity of infection (MOI) of 0.5, and obvious CPE was observed at 48 h, with large, rounded and shed cells (Figure 1A).

The expression of VP2 protein was verified by an indirect immunofluorescence assay (IFA). Sf9 cells were infected with rpFBD-dVP2 at a multiplicity of infection (MOI) of 0.5 and fixed after two days. A monoclonal antibody against CPV (clone 8H7, HyTest Ltd., Turku, Finland) was used as a primary antibody (1:200), and fluorescein isothiocyanate (FITC)-labeled sheep anti-mouse IgG (Solarbio, Beijing, China) was used as the secondary antibody (1:200). The IFA results showed obvious green fluorescence (Figure 1Ba), and no specific fluorescence was observed in normal Sf9 cells (Figure 1Bb). This result indicated that VP2 protein was expressed in Sf9 cells and could be specifically recognized by the VP2 monoclonal antibody.

Sf9 suspension cells were cultured in triangular flasks with sf-900 II SFM (Gibco, Grand island, NY, USA) at a speed of 120 rpm. To assess the expression of VP2 protein, Sf9 suspension cells were harvested 4 days after infection. The mixture was separated into cells and supernatant at 1751× *g* for 30 min. The cells were lysed with filtered 25 mM $NaHCO_3$, after which the supernatant was harvested by centrifugation. The samples were characterized by SDS-PAGE, and a clear band was observed at 65 kDa (Figure 1C). The results showed that VP2 protein was successfully expressed in Sf9 cells and released into the supernatant after simple cell lysis. A Western blotting (WB) assay was carried out to verify whether the 65 kDa band was the target protein. Samples from SDS-PAGE were transferred to a nitrocellulose (NC) membrane (GE Healthcare, Dassel, Germany). In the WB assay, as the primary antibody, the monoclonal antibody against CPV (clone 8H7, Hytest, Turku, Finland) at a 1:500 dilution was incubated at room temperature (RT) for 1.5 h. Then, the horseradish peroxidase (HRP)-labeled sheep anti-mouse IgG at a 1:5000 dilution was incubated at RT for 1 h. As shown in Figure 1D, the WB analysis results showed that the target protein specifically bound the monoclonal antibody against parvovirus; a specific band was seen at the molecular weight of approximately 65 kDa in lane 2, whereas no specific band was seen in the control sample. These results indicated that the expressed VP2 protein had the capacity to elicit a specific antibody response.

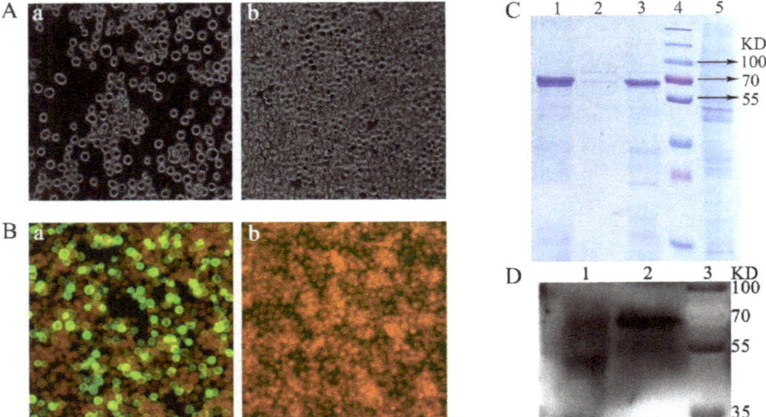

Figure 1. Recombinant baculovirus acquisition and the identification of VP2 protein expressed in Sf9 cells. (**A**) Cytopathogenic effect (CPE) of the recombinant rpFBD-dVP2 baculovirus (magnification, 200×). (Aa) Sf9 cells infected with rpFBD-dVP2 displayed typical CPE 48 h after infection, with large, rounded and shed cells, and (Ab) untreated Sf9 cells remained normal. (**B**) Indirect immune fluorescence antibody staining of VP2 protein in Sf9 cells infected with rpFBD-dVP2 (magnification, 200×). (Ba) Obvious green fluorescence was observed in Sf9 cells infected with rpFBD-dVP2, and (Bb) untreated Sf9 cells showed no fluorescence. (**C**) SDS-PAGE analysis of VP2 protein expression. Lane 1, untreated mixture; lane 2, supernatant after centrifugation; lane 3, supernatant after sodium bicarbonate treatment; lane 4, prestained protein markers; lane 5, cells infected with control baculovirus. (**D**) WB analysis of VP2 protein expression. Lane 1, cells infected with control baculovirus; lane 2, cells infected with recombinant baculovirus; lane 3, prestained protein markers.

HA tests were carried out as previously described to detect VP2 expression levels [22]. The HA titer was determined by the highest dilution of tiger FPV VLPs that caused pig erythrocytes agglutination. The results in Figure 2A showed that the HA titer of tiger FPV VLPs reached $1{:}2^{18}$. High yields of tiger FPV VLPs could be produced due to the high HA titer [23]. To inspect their morphology, the tiger FPV VLPs were observed by electron microscopy. The results showed that the tiger FPV VLPs were successfully assembled with a diameter of ~22 nm; their morphology was similar to that of natural parvovirus particles (Figure 2B).

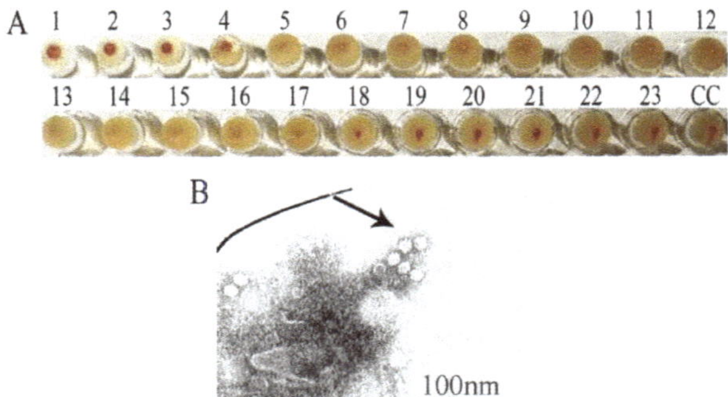

Figure 2. Determination of HA titer and observation of tiger FPV VLPs morphology. (**A**) HA activity of tiger FPV VLPs. (**B**) Electron micrograph analysis of tiger FPV VLPs. The arrow indicates VLPs, and the bar represents 100 nm.

The supernatant containing VLPs after cell lysis was used to prepare the vaccine. Tiger FPV VLPs were emulsified with Gel 02 adjuvant (Seppic, Paris, France) at a volume ratio of 7:1. To determine whether the vaccine was safe for cats, an overdose vaccination test was carried out on cats. Two 10-week old cats were subcutaneously immunized in the neck with 2 mL of tiger FPV VLPs vaccine and given an immunization boost with the same dose four weeks later. The feces production, behavior, and food and water consumption were observed continuously for 14 days after the last immunization. Fourteen days later, the immunized cats were euthanized. The injection site was dissected to observe whether there were side effects, such as nodules, and the tissue was sectioned to detect pathological changes. No anomalies in the daily observation parameters were observed in cats immunized with tiger FPV VLPs vaccine. There were no sarcomas in the injection site after necropsy (Figure 3Aa), which indicated that the absorption of the vaccine was good and that the tissue sections at the injection site were not abnormal (Figure 3Ab). These results demonstrated that the tiger FPV VLPs vaccine was safe for cats.

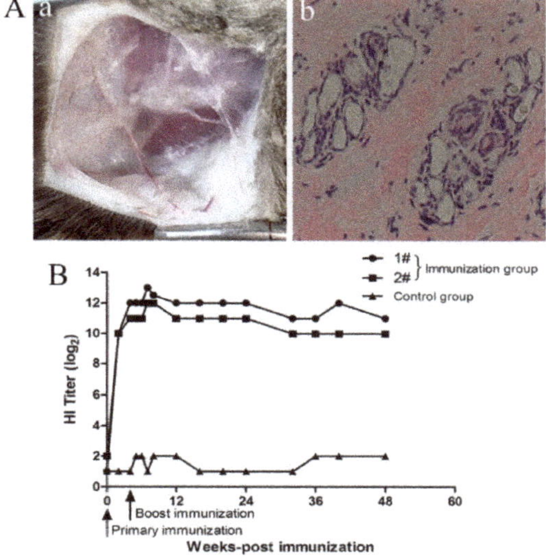

Figure 3. Safety and immunogenicity assay of tiger FPV VLPs. (**A**) Injection site necropsy and tissue section results. (Aa) There were no sarcomas in the injection site after necropsy. (Ab) The tissue sections were normal. (**B**) The immunogenicity of tiger FPV VLPs in cats was evaluated by HI assay after immunization. Cats were immunized twice via subcutaneous injection at four-week intervals. The immunization group was immunized with tiger FPV VLPs mixed with Gel 02 adjuvant. The control group was immunized with PBS.

To detect the immune effects, two 10-week old cats were subcutaneously immunized with 1 mL of tiger FPV VLPs vaccine, and one cat was given PBS as a control. Identical vaccinations were then repeated at four weeks after the primary immunization. Blood samples were collected at 2, 4, 5, 6, and 7 weeks and at 2, 3, 4, 5, 6, 7, 8, 9, 10, and 12 months after the primary immunization. Antibody levels of tiger FPV in the sera were measured by HI assay using pig erythrocytes, and the assay was performed as described previously [24]. The results, shown in Figure 3B, indicated that the anti-FPV HI titer was detected in the immunized cats two weeks after primary immunization, with a titer of approximately $1{:}2^{10}$. After boost immunization, the HI titer was essentially stable at $1{:}2^{10}$. The anti-FPV HI titer was still detected 12 months after the primary immunization, and the antibody titer was not less than $1{:}2^{10}$ (Figure 3B). The second batch of immunization was conducted in other independent tests, which are still under observation. The tendency of HI titer after immunization was the same as above, with

HI titers of 1:2^9–1:2^{10} six months after immunization (Figure S1). These results showed that in cats, tiger FPV VLPs effectively stimulated the immune response and induced the production of anti-FPV antibodies, which had a good immune effect.

VP2 is the main component of the capsid protein of FPV and determines its antigenicity [25]. The amino acid sites of the VP2 protein from tiger FPV were analyzed in this article. The 101st amino acid was consistent with that of CPV, and the rest of the amino acid variations in the VP2 protein [21] were consistent with those of FPV. It has been speculated that the strain might be the product of a recombination event or mutation in response to selection pressure. Therefore, the parvovirus VP2 gene from a tiger was selected to generate a vaccine for tiger parvovirus, which is more representative than other strains and provides a method to prevent and treat tiger parvovirus disease.

In this study, tiger FPV VLPs were developed using the baculovirus expression system, and the expressed protein was identified, confirming that VP2 protein was expressed in vitro and could self-assemble into VLPs. The recombinant baculovirus infected Sf9 cells, cultured in suspension, could express the VP2 protein at high yields and with a high HA titer, indicating their suitability for production on a large scale. VLPs are generally safer than inactivated and attenuated vaccines due to their lack of viral nucleic acids and infectivity. In addition, Gel 02 adjuvant, which is a water-soluble adjuvant, was used in this study. There were no sarcomas and site effects in cats immunized with the tiger FPV VLPs vaccine emulsified with Gel 02 adjuvant. Historically, HI is considered the gold standard for FPV antibody detection [26]. Generally, a HI titer of ≥1:40 is considered protective against FPV [27]. The highest HI titer in the sera of the cats in this experiment reached 1:2^{13}, and the HI titer remained stable at 1:2^{10} after 12 months. The successful preparation of tiger FPV VLPs provides a new approach for the prevention and control of tiger parvovirus disease and technical support for the prevention of rare wild animal viral diseases. Due to the rarity and importance of tigers, relevant institutions are particularly cautious about the inoculation of tigers. Therefore, cats of the same family as tigers were selected for the evaluation of the vaccine immunity effect in this study, with the aim of developing a new type of safe and effective vaccine against feline parvovirus disease in tigers and laying a foundation for the prevention, control and treatment of feline parvovirus disease. In the subsequent application, the tiger FPV VLPs vaccine can be used as a single vaccine or a multiple vaccine combined with other vaccines for routine immunizations of tigers in zoos.

In this study, animal experimentation was handled in compliance with the guidelines and protocols of the Welfare and Ethics of Laboratory Animals of China. The animal studies were approved by the Animal Welfare and Ethics Committee of Changchun Sino Biotechnology Co., Ltd.

Supplementary Materials: The following are available online at http://www.mdpi.com/1999-4915/12/3/315/s1, Figure S1: The immunogenicity of tiger FPV VLPs in cats was evaluated by HI assay, Sequence S1: The codon-optimized tiger parvovirus VP2 sequences.

Author Contributions: N.F., C.Z. and J.S. designed the experiments. C.J. and H.Z. performed the experiments. H.J., D.L. and J.Z. analyzed the data. C.J. and W.L. wrote the article. C.J., H.J. and J.S. reviewed the manuscript. All authors have read and agreed to the published version of the manuscript.

Funding: This work was supported by the National Key Research and Development Program of China—Research and Development of Technology for Prevention and Control of Important Diseases and Domestication and Breeding of Rare and Endangered Wild Animals under Grant No. 2017YFD0501704.

Conflicts of Interest: The authors declare no conflict of interest. The funders had no role in the design of the study; in the collection, analyses, or interpretation of data; in the writing of the manuscript, or in the decision to publish the results.

References

1. Stuetzer, B.; Hartmann, K. Feline parvovirus infection and associated diseases. *Vet. J.* **2014**, *201*, 150–155. [CrossRef] [PubMed]
2. Awad, R.A.; Khalil, W.K.B.; Attallah, A.G. Epidemiology and diagnosis of feline panleukopenia virus in Egypt: Clinical and molecular diagnosis in cats. *Vet. World* **2018**, *11*, 578–584. [CrossRef] [PubMed]

3. Steinel, A.; Munson, L.; van Vuuren, M.; Truyen, U. Genetic characterization of feline parvovirus sequences from various carnivores. *J. Gen. Virol.* **2000**, *81*, 345–350. [CrossRef] [PubMed]
4. Steinel, A.; Parrish, C.R.; Bloom, M.E.; Truyen, U. Parvovirus infections in wild carnivores. *J. Wildl. Dis.* **2001**, *37*, 594–607. [CrossRef] [PubMed]
5. Yang, S.T.; Wang, S.J.; Feng, H.; Zeng, L.; Xia, Z.P.; Zhang, R.Z.; Zou, X.H.; Wang, C.Y.; Liu, Q.; Xia, X.Z. Isolation and characterization of feline panleukopenia virus from a diarrheic monkey. *Vet. Microbiol.* **2010**, *143*, 155–159. [CrossRef]
6. Pfankuche, V.M.; Jo, W.K.; de Van, V.E.; Jungwirth, N.; Lorenzen, S.; Adme, O.; Baumgã Rtner, W.; Puff, C. Neuronal Vacuolization in Feline Panleukopenia Virus Infection. *Vet. Pathol.* **2018**, *55*, 294–297. [CrossRef]
7. Wang, X.G.; Li, T.Y.; Liu, H.Y.; Du, J.M.; Zhou, F.; Dong, Y.M.; He, X.Y.; Li, Y.T.; Wang, C.Q. Recombinant feline parvovirus infection of immunized tigers in central China. *Emerg. Microbes Infect.* **2017**, *6*, e42. [CrossRef]
8. Wang, K.; Du, S.S.; Wang, Y.Q.; Wang, S.Y.; Luo, X.Q.; Zhang, Y.Y.; Liu, C.F.; Wang, H.J.; Pei, Z.H.; Hu, G.X. Isolation and identification of tiger parvovirus in captive siberian tigers and phylogenetic analysis of VP2 gene. *Infect. Genet. Evol.* **2019**, *75*, 103957–103963. [CrossRef]
9. Goodrich, J.M.; Quigley, K.S.; Lewis, J.C.M.; Astafiev, A.A.; Slabi, E.V.; Miquelle, D.G.; Smirnov, E.N.; Kerley, L.L.; Armstrong, D.L.; Quigley, H.B. Serosurvey of Free-ranging Amur Tigers in the Russian Far East. *J. Wildl. Dis.* **2012**, *48*, 186–189. [CrossRef]
10. Yang, S.T.; Xia, X.Z.; Qiao, J.; Liu, Q.; Chang, S.; Xie, Z.J.; Ju, H.Y.; Zou, X.H.; Gao, Y.W. Complete protection of cats against feline panleukopenia virus challenge by a recombinant canine adenovirus type 2 expressing VP2 from FPV. *Vaccine* **2008**, *26*, 1482–1487. [CrossRef]
11. Risi, E.; Agoulon, A.; Allaire, F.; Le Dréan-Quénec'hdu, S.; Martin, V.; Mahl, P. Antibody response to vaccines for rhinotracheitis, caliciviral disease, panleukopenia, feline leukemia, and rabies in tigers (Panthera tigris) and lions (Panthera leo). *J. Zoo Wildl. Med.* **2012**, *43*, 248–255. [CrossRef] [PubMed]
12. Duenwald, J.C.; Holland, J.M.; Gorham, J.R.; Ott, R.L. Feline panleukopenia: Experimental cerebellar hypoplasia produced in neonatal ferrets with live virus vaccine. *Res. Vet. Sci.* **1971**, *12*, 394–396. [CrossRef]
13. Brussel, K.; Carrai, M.; Lin, C.; Kelman, M.; Setyo, L.; Aberdein, D.; Brailey, J.; Lawler, M.; Maher, S.; Plaganyi, I.; et al. Distinct lineages of feline parvovirus associated with epizootic outbreaks in Australia, New Zealand and the United Arab Emirates. *Viruses* **2019**, *11*, 1155. [CrossRef] [PubMed]
14. Truyen, U.; Parrish, C.R. Feline panleukopenia virus: Its interesting evolution and current problems in immunoprophylaxis against a serious pathogen. *Vet. Microbiol.* **2013**, *165*, 29–32. [CrossRef] [PubMed]
15. Chapman, M.S.; Rossmann, M.G. Structure, sequence, and function correlations among parvoviruses. *Virology* **1993**, *194*, 491–508. [CrossRef]
16. Diao, F.F.; Zhao, Y.F.; Wang, J.L.; Wei, X.H.; Cui, K.; Liu, C.Y.; Guo, S.Y.; Jiang, S.J.; Xie, Z.J. Molecular characterization of feline panleukopenia virus isolated from mink and its pathogenesis in mink. *Vet. Microbiol.* **2017**, *205*, 92–98.
17. Hueffer, K.; Parker, J.S.L.; Weichert, W.S.; Geisel, R.E.; Sgro, J.-Y.; Parrish, C.R. The natural host range shift and subsequent evolution of canine parvovirus resulted from virus-specific binding to the canine transferrin receptor. *J. Virol.* **2003**, *77*, 1718–1726. [CrossRef]
18. Xu, J.; Guo, H.C.; Wei, Y.Q.; Dong, H.; Han, S.C.; Ao, D.; Sun, D.H.; Wang, H.M.; Cao, S.Z.; Sun, S.Q. Self-assembly of virus-like particles of canine parvovirus capsid protein expressed from Escherichia coli and application as virus-like particle vaccine. *Appl. Microbiol. Biotechnol.* **2014**, *98*, 3529–3538. [CrossRef]
19. Antonis, A.F.G.; Bruschke, C.J.M.; Rueda, P.; Maranga, L.; Casal, J.I.; Vela, C.; Hilgers, L.A.T.; Belt, P.B.G.M.; Weerdmeester, K.; Carrondo, M.J.T. A novel recombinant virus-like particle vaccine for prevention of porcine parvovirus-induced reproductive failure. *Vaccine* **2006**, *24*, 5481–5490. [CrossRef]
20. Martínez-Solís, M.; Herrero, S.; Targovnik, A.M. Engineering of the baculovirus expression system for optimized protein production. *Appl. Microbiol. Biotechnol.* **2019**, *103*, 113–123. [CrossRef]
21. Decaro, N.; Buonavoglia, C. Canine parvovirus—A review of epidemiological and diagnostic aspects, with emphasis on type 2c. *Vet. Microbiol.* **2012**, *155*, 1–12. [CrossRef] [PubMed]
22. Feng, H.; Hu, G.Q.; Wang, H.L.; Liang, M.; Liang, H.R.; Guo, H.; Zhao, P.S.; Yang, Y.J.; Zheng, X.X.; Zhang, Z.F. Canine parvovirus VP2 protein expressed in silkworm pupae self-assembles into virus-like particles with high immunogenicity. *PLoS ONE* **2014**, *9*, e79575. [CrossRef] [PubMed]

23. Jin, H.L.; Xia, X.H.; Liu, B.; Fu, Y.; Chen, X.P.; Wang, H.L.; Xia, Z.Q. High-yield production of canine parvovirus virus-like particles in a baculovirus expression system. *Arch. Virol.* **2016**, *161*, 705–710. [CrossRef] [PubMed]
24. Carmichael, L.E.; Joubert, J.C.; Pollock, R.V. Hemagglutination by canine parvovirus: Serologic studies and diagnostic applications. *Am. J. Vet. Res.* **1980**, *41*, 784–791. [PubMed]
25. Cheng, N.; Zhao, Y.K.; Han, Q.X.; Zhang, W.J.; Xi, J.; Yu, Y.L.; Wang, H.L.; Li, G.H.; Gao, Y.W.; Yang, S.T.; et al. Development of a reverse genetics system for a feline panleukopenia virus. *Virus Genes* **2019**, *55*, 95–103. [CrossRef]
26. Bergmann, M.; Schwertler, S.; Reese, S.; Speck, S.; Truyen, U.; Katrin, H. Antibody response to feline panleukopenia virus vaccination in healthy adult cats. *J. Feline Med. Surg.* **2018**, *20*, 1087–1093. [CrossRef]
27. Mouzin, D.E.; Lorenzen, M.J.; Haworth, J.D.; King, V.L. Duration of serologic response to three viral antigens in cats. *J. Am. Vet. Med. Assoc.* **2004**, *224*, 61–66. [CrossRef]

Article

Changes in Mass Treatment of the Canine Parvovirus ICU Population in Relation to Public Policy Changes during the COVID-19 Pandemic

Kevin Horecka *, Nipuni Ratnayaka and Elizabeth A. Davis

Research Department, Austin Pets Alive!, Austin, TX 78703, USA;
nipuni.ratnayaka@americanpetsalive.org (N.R.); elizabethdavis7@gmail.com (E.A.D.)
* Correspondence: kevin.horecka@gmail.com

Academic Editor: Giorgio Gallinella
Received: 17 October 2020; Accepted: 8 December 2020; Published: 10 December 2020

Abstract: Previous work has indicated that canine parvovirus (CPV) prevalence in the Central Texas region may follow yearly, periodic patterns. The peak in CPV infection rates occurs during the summer months of May and June, marking a distinct "CPV season". We hypothesized that human activity contributes to these seasonal changes in CPV infections. The COVID-19 pandemic resulted in drastic changes in human behavior which happened to synchronize with the CPV season in Central Texas, providing a unique opportunity with which to assess whether these society-level behavioral changes result in appreciable changes in CPV patient populations in the largest CPV treatment facility in Texas. In this work, we examine the population of CPV-infected patients at a large, dedicated CPV treatment clinic in Texas (having treated more than 5000 CPV-positive dogs in the last decade) and demonstrate that societal–behavioral changes due to COVID-19 were associated with a drastic reduction in CPV infections. This reduction occurred precisely when CPV season would typically begin, during the period immediately following state-wide "reopening" of business and facilities, resulting in a change in the typical CPV season when compared with previous years. These results provide evidence that changes in human activity may, in some way, contribute to changes in rates of CPV infection in the Central Texas region.

Keywords: canine; canine parvovirus; COVID-19; veterinary epidemiology

---

1. Introduction

The canine parvovirus (CPV) is a highly contagious gastrointestinal virus which, without treatment, has a fatality rate as high as 90% among domestic dogs [1,2]. With treatment, the fatality rate improves to 14% [3], but is still among the most common infectious diseases in dogs with high morbidity and mortality [4]. The disease incubates for approximately five days [5]. Following symptom onset, CPV takes a median of 11 days to resolve [3] if the animal survives the infection. Because of the significant risk that CPV presents to domestic dog populations, understanding factors influencing CPV incidence is of critical importance to reduce disease transmission in unvaccinated populations.

Seasonal effects in CPV have been observed worldwide [6–10]. In Central Texas, seasonal trends in incidence have been observed during the past decade, specifically peaking in the late spring/early summer months of May and June. Although much is known about the mode of transmission (i.e., viral spread via the fecal–oral route [11]) and it is standard practice in Central Texas animal shelters to inoculate against for CPV using the highly effective and widely available CPV vaccine, significant natural disease reservoirs must persist to perpetuate the disease among domesticated dogs. These reservoirs may exist in the form of wildlife (big cats, racoons, racoon dogs, arctic foxes, and mink; though evidence of transmission via this route is unclear [12–14]) and/or unvaccinated feral animal populations. Additionally, surfaces in the environment such as grass or pavement may house infections, given CPV's ability to survive on surfaces for extended periods of time due to the hardy nature of this nonenveloped virus [4,15]. Previous work [3] has suggested that human activity may be, in some way, driving CPV infections. Thus, the drastic changes in human behavior during and immediately following the "stay-at-home" orders due to the coronavirus disease 2019 (COVID-19) pandemic provided a unique opportunity to gather further evidence with regard to this hypothesis.

The COVID-19 pandemic, caused by the severe acute respiratory syndrome coronavirus 2 (SARS-CoV-2) pathogen, resulted in major policy changes across the state of Texas, the United States, and much of the world. In Texas's Travis County, such policy changes included non-essential business closures [16–18], school closures [19,20], and stay-home orders [18,21], designed to mitigate COVID-19 transmission in an effort to avoid exceeding healthcare capacity. Mobility and navigation request data from cellular phones [22,23], consumer activity data [24], and transportation department traffic density maps [25] generally indicated that there were substantial reductions in extra-household human activities while stay-home orders were in effect (and immediately after orders were lifted). Although good quality data on how outdoor recreational and physical activities (e.g., hiking, camping, walking dogs, etc.) were impacted are sparse, it is possible that the initial panic or concern surrounding the pandemic, in conjunction with closures of some outdoor venues (i.e., parks, bars, dog parks, beaches, etc.), discouraged people from partaking in outdoor activities as frequently as usual, in particular earlier in the shutdown period. Changes in human behavior during and shortly after the "stay-at-home" period may have, subsequently, impacted the incidence of CPV infection compared with previous years by reducing the frequency, duration, and intensity of contact between owned dogs and CPV infection sources.

Austin Pets Alive! (APA!), a non-profit, closed admissions animal shelter in Austin, TX (Travis County) has maintained a dedicated intensive care unit (ICU) for the exclusive treatment of CPV-infected dogs since 2008, having treated over 5000 animals during that period [3]. Moreover, as the organization has evolved, data collection has substantially increased in resolution, from monthly aggregate population numbers only to individual animal care records beginning reliably in 2017. A previous investigation used these data sources to assess patient signalment, disease trajectory and prognosis, and seasonal trends in disease incidence [3]. In the present study, we use more recent data from APA! in order to examine changes in population in the ICU during the COVID-19 pandemic in conjunction with changes in policy which may have impacted human activities and behaviors contributing to the disease spread. We present several hypotheses that might further elucidate the relationship between human activity and CPV infection and, therefore, may lead to policy and protocol changes which can reduce the spread of CPV. Through this observational study, this work seeks to investigate whether human activity, marked by public policy-driven behavioral changes during the COVID-19 pandemic, is associated with CPV spread in the Central Texas region.

2. Materials and Methods

2.1. Data Sourcing

In order to assess population levels of CPV-infected animals over time, we examined data collected via our end-of-shift ICU report. These data include the number of animals in the ICU measured twice daily, once in the morning and once in the evening (see [3] for a detailed prior analysis of pre-COVID-19 elements of these data). Data on COVID-19 policies were collected from the Austin City Government website [18]. Information on Austin Animal Center and Austin Pets Alive! (the two primary shelters involved in this study) and their associated intake policies was collected directly from the shelter administrators at each site. Intake source location data were collected via the intake paperwork for each animal during the time periods in question (note that data of this granularity were only available from 2017 onward, while the population data being examined were available from 2013 and onward). Finally, monthly aggregate counts of at-home treatment for CPV patients (see [3] for details on when this protocol was implemented) were observed as a possible explanatory variable for changes in CPV populations in the ICU.

2.2. Shelter Intake Policy and Geographic Sourcing

Austin Pets Alive!'s Parvo ICU intake practices did not change during any of the periods in question (i.e., no artificial reduction in population due to policy was present). In addition, no changes in geographic distribution of animal intake sources at APA! were observed, as measured by a McNemar–Bowker test for multiple correlation proportions (2017–2019 vs. 2020; $\chi^2 = 9.48$, $\nu = 528$, $p \sim 1$).

2.3. Statistical Analyses

To analyze the difference in CPV infections between 2020 and prior years, statistical tests were performed on the difference between the nadir of CPV infections in 2020 compared with the remainder of CPV infections in prior years. This allowed us to assess the likelihood of such an extreme difference occurring by chance in the model. Because the distribution of differences in CPV infections between 2020 and prior years was not normal and not easily correctable to normal (via boxcox, sqrt, log, or other common transformations), we were unable to assess a z-score or other measures of likelihood for this observed difference using parametric methods. As a result, a non-parametric, Gaussian kernel density estimate over the distribution of differences was computed with a bandwidth parameter determined by Scott's factor ($w = 0.33$, in this case; for all comparisons, [26]).

Per previous methods of analyzing these types of data [3], a two-week moving average was used on all daily-reported time-series data. This was carried out primarily to reduce the presence of temporal autocorrelation due to animals being measured on recurring days, given that the disease time course is approximately 11 days. All statistical tests used significance thresholds of $p < 0.05$.

3. Results

Differences in CPV Infections during the COVID-19 Pandemic

In order to assess differences in CPV infections during the COVID-19 pandemic compared with previous years (raw data visualized in Figure 1), two primary comparisons were performed. First, a comparison of 2019 and 2020 CPV infections over time was conducted to assess the direct year-over-year difference which may have been due to the pandemic (Figure 2A). Second, a comparison of the average of the 2013–2019 CPV infections over time and the 2020 CPV infections over time was also conducted (Figure 2B).

A distinct dip in CPV ICU population can be visually observed during the "stay-at-home" period (defined as the time period between the City of Austin "stay-at-home" order and the state-wide "reopening" order; see Figures 1 and 2B). However, a nearly identical dip was present in 2019 as well (see Figure 1), suggesting that this dip cannot be easily associated with the pandemic. However, immediately following the state-wide "reopening" event, results from the kernel density estimate analysis show that the decrease in CPV ICU population after the "stay-at-home" period is statistically significant when compared with the differences throughout the rest of the year with prior years (Figure 2A [2019 vs. 2020; $p = 0.007$] and Figure 2B [2013–2019 vs. 2020; $p = 0.013$]).

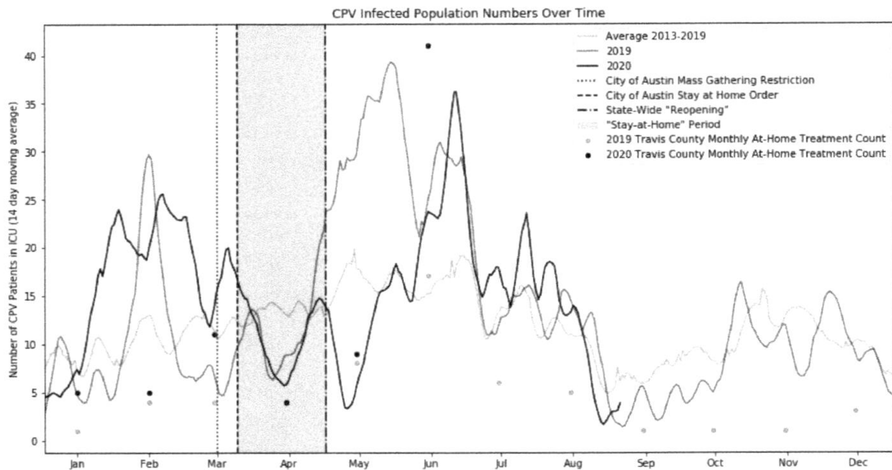

Figure 1. Average canine parvovirus (CPV)-infected animals over time. This figure, presented solely to illustrate the raw data under consideration in this work, shows the average population of CPV-infected animals in the intensive care unit (ICU) during 2019 and 2020 as well as the historical average ICU population from 2013–2019. The period of "stay-at-home" orders which lasted from 24 March 2020 until the Texas state-wide "reopening" on 01 May 2020 is highlighted (gray shading). Note the large difference between the 2019 and 2020 data which occurs immediately following the state-wide "reopening". Additionally, the monthly count of at-home treatment animals (all Travis County sourced) is shown for 2019 and 2020, and, although no difference is seen during the low period in May, a large upswing in at-home care patients is seen as CPV season begins. All data are presented in two-week moving averages.

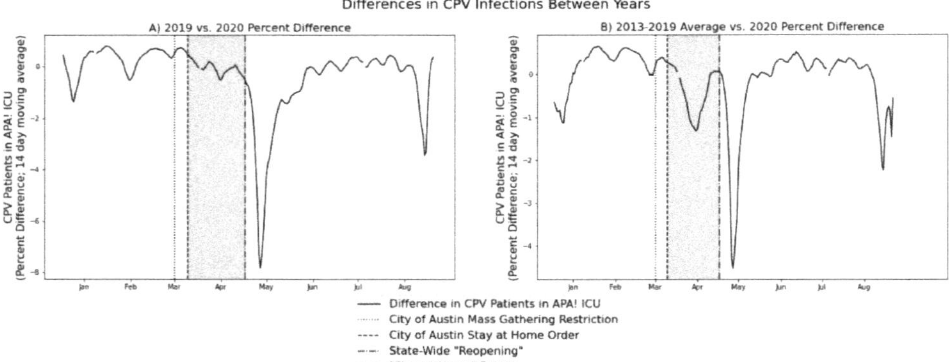

Figure 2. Differences in CPV infections between years. This figure contains the time series differences in CPV infections between 2019 and 2020 (**A**), as well as between the average of 2013–2019 and 2020 (**B**). Importantly, although a decrease in the CPV infection difference can be observed during the pandemic-related "stay-at-home" order (gray shading) when compared to the historical trend (**B**), this same decrease was indeed present in 2019 (**A**; also see Figure 1). Therefore, this decrease is unlikely to be related to changes in human behavior due to the "stay-at-home" order. The larger decrease, however, immediately after the "reopening" event, is not seen in any previous years. The associated date of maximum difference was 11 May 2020, 11 days after the state-wide "reopening". All data are presented in two-week moving averages.

4. Discussion

Although significant progress has been made in developing diagnostics [27–31], vaccines [29,32–34], and treatment protocols [3,35–38] for canine parvovirus (CPV) since its emergence, few methodologies have been developed to attempt to limit or slow the spread of this disease within vulnerable populations. This may be partially due to a belief that reductions like this are not possible, given the extensive period during which CPV can survive in the environment [4] and difficulty in disinfecting CPV-infected areas (i.e., dog parks, homes, shelters, etc.) [4,15]. However, if population-level human behavior can be linked to infection rates, preventative methods to reduce the risk of infection in vulnerable populations beyond the typical vaccination recommendations [39,40] might be developed and/or more widely adopted. In the present study, we take advantage of the extra-ordinary circumstances provided by the COVID-19 pandemic to examine changes in CPV population in a mass-treatment intensive care unit (ICU) dedicated exclusively to the treatment of CPV. We observed a substantial drop in CPV patients immediately after the state-wide "reopening" event on 01 May 2020, with a minimum on 11 May 2020 when viewed in a two-week moving average. This finding may indicate an overall reduction in infection rates or a delay in the typical CPV season (which was still observed in these data in June, though not in the typical peak month of May [3]). In either case, its coincidence with large-scale changes in human behavior associated with the COVID-19 pandemic suggest that this reduction in CPV infections is related to human behavior and, therefore, may be employed outside of the context of a pandemic for the purposes of reducing CPV infection.

In addition to the typical on-site treatment described in this work, Austin Pets Alive! offers an at-home protocol [3] to individuals who do not wish to give up their animal but cannot afford care by a veterinarian. This option is only given to owners whose animals are not critically ill at the time of evaluation (i.e., do not present with bloody diarrhea, excessive vomiting, inappetence, and lethargy). If increases in this at-home care were seen in May of 2020, this could explain the dip in CPV-infected patients that was observed. However, no such increase was observed (see Figure 1). It is interesting to note that as the 2020 CPV season began in the month of June, a large increase in at-home treatment cases was observed, which may represent a continuation of the effects of the pandemic.

Despite evidence suggesting that reduced CPV infections were related to widespread changes in human behavior during the "stay-at-home" mandate, several potential alternative explanations warrant evaluation and discussion. Although Austin Pets Alive! changed no policies regarding the intake of animals and did not reject animals with CPV infections at any point during the periods in question, we investigated the possibility that changes at other shelters in Central Texas were related to our results. We examined potential differences in geographical animal sourcing at APA! from other shelters across Texas during 2020 which might indicate changes in policy at these shelters; however, no differences were found (see "Shelter intake policy and geographical sourcing"). Of course, this does not rule out the possibility that a failure to seek care in general drove this difference, a possibility that is extraordinarily difficult to rule out in general. However, if failure to seek care drove the differences that were observed, it might be expected that the dip would persist into peak CPV season or start in a time-locked manner to the drastic increase in unemployment (or other macroeconomic factors) seen in April in Texas and the Austin area [41]. Thus, the transient nature of the observed effect reduces the likelihood that failure to seek care is driving the effect, though it does not eliminate this possibility.

Another alternative explanation for our results could be that changes in breeding practices or animal acquisition rates across the period analyzed led to an overall decrease in the vulnerable population available for CPV infection. This is also difficult to assess, though future studies may examine if there was a decrease in the number of dogs born in the first months of 2020. However, when accounting for the typical two-month gestation period [42] and the notion that maternal antibodies may provide protection for a short period after birth [43–48], this explanation leaves little time for the onset of COVID-19 policies in the United States (mid-March 2020) to produce a maximum effect of decreased CPV-vulnerable populations in mid-May 2020, as observed. Similarly, reductions in pet acquisitions in the early portions of the pandemic could potentially explain a decrease in CPV infections, though further studies would need to examine this more holistically. During the period of 01 March 2020 to 01 June 2020, on balance, Austin Pets Alive! and Austin Animal Center adopted out more animals than the previous year, making this explanation potentially less convincing [49].

Finally, it is possible that weather patterns could have contributed to the observed decrease in CPV infections. As far as the authors are aware, no significant weather events or natural disasters (such as hurricanes or fires) occurred during the period in question in Central Texas or immediately preceding it. The distribution of intakes from various geographic regions did not change over the period in question, indicating that if weather can be used to explain the effect, the weather pattern would have to span nearly the entirety of Central, South, and East Texas (a region approximately 430,000 square kilometers in area, or somewhere between the size of Sweden and the size of Spain). Future investigations may seek to determine the degree to which temperature and precipitation may relate to infection rates. However, it is noteworthy that even if this association can be established, this could still be due to changes these weather patterns cause in human activity.

The ecological study of CPV infection has many limitations, enumerated above, and cannot conclusively show that CPV infection rates are being modified by human activity. However, the present data analysis takes advantage of a generationally-defining event in the COVID-19 pandemic, combined with seven years of high-resolution historical data on CPV intakes at the largest CPV facility in Texas, to provide a potential correlation between human activity and CPV incidence. We hope the relationship uncovered herein inspires the pursuit of additional research which attempts to directly intervene in human behavior to reduce CPV infections. Future studies could attempt targeted interventions, potentially making use of the following practices: (1) enforcing age restrictions on high-risk CPV areas such as dog parks (to ultimately reduce CPV exposure for dogs too young to be fully vaccinated), (2) performing periodic testing of public spaces such as parks and waterways in which dogs are likely to be present, and/or (3) increased public education regarding regions at a high risk for propagating CPV infections and/or regions that have recently tested positive. In addition to these interventions, information campaigns educating the public on the importance of vaccinations and protecting unvaccinated, under-vaccinated, or incompletely vaccinated dogs may help owners make more informed choices on behalf of their pets, which could enable them to help avoid their pets contracting CPV.

Author Contributions: Conceptualization, K.H., E.A.D., and N.R.; methodology, K.H.; software, K.H.; validation, K.H.; formal analysis, K.H.; investigation, K.H.; resources, K.H.; data curation, K.H.; writing—original draft preparation, K.H.; writing—review and editing, K.H., E.A.D., and N.R.; visualization, K.H.; supervision, K.H.; project administration, K.H.; funding acquisition, K.H. All authors have read and agreed to the published version of the manuscript.

Funding: This research received no external funding.

Acknowledgments: The authors would like to thank all of the Austin Pets Alive! Parvo ICU volunteers and staff for all of their hard work and dedication over the past decade leading up to this publication. Additionally, all volunteers and staff at Austin Pets Alive!, our partners at Austin Animal Center, the City of Austin, and the greater Austin community should be congratulated for their great work in pushing the state of the art and ethics of animal welfare forward over the past decade. Additionally, special thanks go to Susan Amirian for her notes and insightful comments, Steve Porter for his advice on pulling data from internal data sources, Ellen Jefferson for her incredible leadership at Austin Pets Alive!, and Rory Adams for his contributions to the Research Committee goals.

Conflicts of Interest: The authors declare no conflict of interest. The funders of the treatments and operations of Austin Pets Alive! had no role in the design of the study; in the collection, analysis, or interpretation of data; in the writing of the manuscript, or in the decision to publish the results. Other components of this research were self-funded by volunteer researchers. Austin Pets Alive! and American Pets Alive! offer dogs who have survived CPV for adoption to the general public, but, as a nonprofit organization, are not financially incentivized to preferentially adopt these animals over others.

References

1. Mylonakis, M.; Kalli, I.; Rallis, T. Canine parvoviral enteritis: An update on the clinical diagnosis, treatment, and prevention. *Vet. Med. Res. Rep.* **2016**, *7*, 91–100. [CrossRef] [PubMed]
2. Otto, C.M.; Drobatz, K.J.; Soter, C. Endotoxemia and tumor necrosis factor activity in dogs with naturally occurring parvoviral enteritis. *J. Vet. Intern. Med.* **1997**, *11*, 65–70. [CrossRef] [PubMed]
3. Horecka, K.; Porter, S.; Amirian, E.S.; Jefferson, E. A Decade of Treatment of Canine Parvovirus in an Animal Shelter: A A Decade of Treatment of Canine Parvovirus in an Animal Shelter: A Retrospective Study. *Animals* **2020**, *10*, 939. [CrossRef] [PubMed]
4. Ettinger, S.; Feldman, E.; Côté, E. *Textbook of Veterinary Internal Medicine: Diseases of the Dog and the Cat*; Elsevier: St. Louis, MO, USA, 2017.
5. Potgieter, L.N.; Jones, J.B.; Patton, C.S.; Webb-Martin, T.A. Experimental parvovirus infection in dogs. *Can. J. Comp. Med.* **1981**, *45*, 212–216.
6. Kalli, I.; Leontides, L.S.; Mylonakis, M.E.; Adamama-Moraitou, K.; Rallis, T.; Koutinas, A.F. Factors affecting the occurrence, duration of hospitalization and final outcome in canine parvovirus infection. *Res. Vet. Sci.* **2010**, *89*, 174–178. [CrossRef]
7. Houston, D.M.; Ribble, C.S.; Head, L.L. Risk factors associated with parvovirus enteritis in dogs: 283 cases (1982–1991). *J. Am. Vet. Med. Assoc.* **1996**, *208*, 542–546.
8. Shima, F.K.; Apaa, T.T.; Mosugu, J.I.T. Epidemiology of Canine Parvovirus Enteritis among Hospitalized Dogs in Effurun/Warri Metropolitan Region of Delta State, Nigeria. *OALib* **2015**, *2*, 1–7. [CrossRef]
9. County, L. Canine Parvovirus in Dogs-Epidemiology Update 2015. Available online: https://admin.publichealth.lacounty.gov/vet/reports/2015CanineParvoEpiQ3.pdf (accessed on 27 May 2020).
10. Stann, S.E.; DiGiacomo, R.F.; Giddens, W.E.; Evermann, J.F. Clinical and pathologic features of parvoviral diarrhea in pound-source dogs. *J. Am. Vet. Med. Assoc.* **1984**, *185*, 651–655.
11. Goddard, A.; Leisewitz, A.L. Canine Parvovirus. *Vet. Clin. North Am. Small Anim. Pract.* **2010**, *40*, 1041–1053. [CrossRef]
12. Neuvonen, E.; Veijalainen, P.; Kangas, J. Canine parvovirus infection in housed raccoon dogs and foxes in Finland. *Vet. Rec.* **1982**, *110*, 448–449. [CrossRef]
13. Duarte, M.D.; Barros, S.C.; Henriques, M.; Fernandes, T.L.; Bernardino, R.; Monteiro, M.; Fevereiro, M. Fatal Infection with Feline Panleukopenia Virus in Two Captive Wild Carnivores (Panthera tigris and Panthera leo). *J. Zoo Wildl. Med.* **2009**, *40*, 354–359. [CrossRef] [PubMed]
14. Allison, A.B.; Kohler, D.J.; Fox, K.A.; Brown, J.D.; Gerhold, R.W.; Shearn-Bochsler, V.I. Frequent Cross-Species Transmission of Parvoviruses among Diverse Carnivore Hosts. *J. Virol.* **2013**, *87*, 2342–2347. [CrossRef] [PubMed]
15. Saknimit, M.; Inatsuki, I.; Sugiyama, Y.; Yagami, K. Virucidal Efficacy of Physico-chemical Treatments Against Coronaviruses and Parvoviruses of Laboratory Animals. *Exp. Anim.* **1988**, *37*, 341–345. [CrossRef] [PubMed]
16. Austin TM of the C of. Order of Control for Bars and Restraunts. United States. Available online: https://www.austintexas.gov/sites/default/files/files/Order%20-%20Prohibition%20of%20Community%20Gatherings%20of%2010%20persons%20or%20more%20and%20closure%20of%20restaurants%20and%20bars%203.17.2020.pdf (accessed on 8 December 2020).
17. County TCJ of T. Order of Control for Mass and Community Gatherings. United States. Available online: https://www.austintexas.gov/sites/default/files/files/County_Order__Gathering_of_250_(3_14_20_vSigned).pdf.pdf (accessed on 8 December 2020).
18. Austin C of. COVID-19 Information. AustintexasGov 2020. Available online: http://austintexas.gov/department/covid-19-information/orders-rules (accessed on 10 October 2020).
19. Bond, E.C.; Dibner, K.; Schweingruber, H. (Eds.) *Reopening K-12 Schools During the COVID-19 Pandemic*; National Academies Press: Washington, DC, USA, 2020. [CrossRef]
20. Austin Public Health. *Austin C of Austin Public Health Interim Guidance on Reopening for Austin-Travis County Schools*; Austin Public Health: Austin, TX, USA, 2020.
21. County TCJ of T. Stay Home Work Safe Order. United States. Available online: https://www.austintexas.gov/sites/default/files/files/20200324%20Judge\T1\textquoterights%20Order.pdf (accessed on 8 December 2020).

22. Olin, A. Mobility Data Shows some Staying at Home More than Others During COVID-19 Pandemic. Kinder Institute for Urban Reseach Rice University. COVID-19 and Cities, Health, Transportation. 2020. Available online: https://kinder.rice.edu/urbanedge/2020/04/08/mobility-data-shows-some-are-staying-home-more-others-houston-area-during-covid (accessed on 9 December 2020).
23. Valentino-DeVries, J.; Lu, D.; Dance, G.J. *Location Data Says It All: Staying at Home During Coronavirus Is a Luxury*; New York Times: New York, NY, USA, 2020.
24. Chetty, R.; Friedman, J.; Hendren, N.; Stepner, M. The Economic Impacts of COVID-19: Evidence from a New Public Database Built from Private Sector Data. *Oppor. Insights* **2020**. [CrossRef]
25. Rash, B. *Mobility Authority, TxDOT Make Adjustments Amid COVID 19 Situation*; Community Impact: Austin, TX, USA, 2020.
26. Scott, D.W. *Multivariate Density Estimation: Theory, Practice, and Visualization*, 2nd ed.; John Wiley & Sons: Chicester, NY, USA, 2015.
27. Lambe, U.; Guray, M.; Bansal, N.; Kumar, P.; Joshi, V.G.; Khatri, R.; Prasad, G. Canine parvovirus- an insight into diagnostic aspect. *J. Exp. Biol. Agric. Sci.* **2016**, *4*, 279–290. [CrossRef]
28. Yip HY, E.; Peaston, A.; Woolford, L.; Khuu, S.J.; Wallace, G.; Kumar, R.S.; Amanollahi, R. Diagnostic Challenges in Canine Parvovirus 2c in Vaccine Failure Cases. *Viruses* **2020**, *12*, 980. [CrossRef]
29. Decaro, N.; Buonavoglia, C. Canine parvovirus post-vaccination shedding: Interference with diagnostic assays and correlation with host immune status. *Vet. J.* **2017**, *221*, 23–24. [CrossRef]
30. Decaro, N.; Buonavoglia, C. Canine parvovirus—A review of epidemiological and diagnostic aspects, with emphasis on type 2c. *Vet. Microbiol.* **2012**, *155*, 1–12. [CrossRef]
31. Desario, C.; Decaro, N.; Campolo, M.; Cavalli, A.; Cirone, F.; Elia, G.; Buonavoglia, C. Canine parvovirus infection: Which diagnostic test for virus? *J. Virol. Methods* **2005**, *126*, 179–185. [CrossRef]
32. Decaro, N.; Buonavoglia, C.; Barrs, V.R. Canine parvovirus vaccination and immunisation failures: Are we far from disease eradication? *Vet. Microbiol.* **2020**, *247*, 108760. [CrossRef]
33. Bergmann, M.; Freisl, M.; Hartmann, K.; Speck, S.; Truyen, U.; Zablotski, Y.; Wehner, A. Antibody Response to Canine Parvovirus Vaccination in Dogs with Hyperadrenocorticism Treated with Trilostane. *Vaccines* **2020**, *8*, 547. [CrossRef] [PubMed]
34. Altman, K.D.; Kelman, M.; Ward, M.P. Are vaccine strain, type or administration protocol risk factors for canine parvovirus vaccine failure? *Vet. Microbiol.* **2017**, *210*, 8–16. [CrossRef] [PubMed]
35. Gerlach, M.; Proksch, A.L.; Unterer, S.; Speck, S.; Truyen, U.; Hartmann, K. Efficacy of feline anti-parvovirus antibodies in the treatment of canine parvovirus infection. *J. Small Anim. Pract.* **2017**, *58*, 408–415. [CrossRef] [PubMed]
36. Perley, K.; Burns, C.C.; Maguire, C.; Shen, V.; Joffe, E.; Stefanovski, D.; Watson, B. Retrospective evaluation of outpatient canine parvovirus treatment in a shelter-based low-cost urban clinic. *J. Vet. Emerg. Crit. Care* **2020**, *30*, 202–208. [CrossRef] [PubMed]
37. Venn, E.C.; Preisner, K.; Boscan, P.L.; Twedt, D.C.; Sullivan, L.A. Evaluation of an outpatient protocol in the treatment of canine parvoviral enteritis. *J. Vet. Emerg. Crit. Care* **2017**, *27*, 52–65. [CrossRef]
38. Yalcin, E.; Keser, G.O. Comparative efficacy of metoclopramide, ondansetron and maropitant in preventing parvoviral enteritis-induced emesis in dogs. *J. Vet. Pharmacol. Ther.* **2017**, *40*, 599–603. [CrossRef]
39. Association AVM. Canine Parvovirus. AvmaOrg. 2020. Available online: https://www.avma.org/resources-tools/pet-owners/petcare/canine-parvovirus#:~{}:text=Puppiesshouldreceiveadose,isup-to-date (accessed on 10 October 2020).
40. Ford, R.B.; Larson, L.J.; McClure, K.D.; Schultz, R.D.; Welborn, L.V. 2017 AAHA Canine Vaccination Guidelines. *J. Am. Anim. Hosp. Assoc.* **2017**, *53*, 243–251. [CrossRef]
41. Investments TCE and S. Unemployment Rates Summary & Interactive Data. Travis Cty Econ Dev Home, TraviscountytxGov. 2020. Available online: https://financialtransparency.traviscountytx.gov/EDSI_EcoRec_Economy/UnemploymentRates (accessed on 10 October 2020).
42. Okkens, A.C.; Teunissen, J.M.; Van Osch, W.; Van Den Brom, W.E.; Dieleman, S.J.; Kooistra, H.S. Influence of litter size and breed on the duration of gestation in dogs. *J. Reprod. Fertil. Suppl.* **2001**, *57*, 193–197.
43. Decaro, N.; Campolo, M.; Desario, C.; Elia, G.; Martella, V.; Lorusso, E.; Buonavoglia, C. Maternally-derived antibodies in pups and protection from canine parvovirus infection. *Biologicals* **2005**, *33*, 261–267. [CrossRef]
44. Pollock, R.V.; Carmichael, L.E. Maternally derived immunity to canine parvovirus infection: Transfer, decline, and interference with vaccination. *J. Am. Vet. Med. Assoc.* **1982**, *180*, 37–42.

45. Kelman, M.; Barrs, V.R.; Norris, J.M.; Ward, M.P. Canine parvovirus prevention and prevalence: Veterinarian perceptions and behaviors. *Prev. Vet. Med.* **2020**, *174*, 104817. [CrossRef] [PubMed]
46. Gooding, G.E.; Robinson, W.F. Maternal antibody, vaccination and reproductive failure in dogs with parvovirus infection. *Aust. Vet. J.* **1982**, *59*, 170–174. [CrossRef] [PubMed]
47. Hasan, K.; Rathnamma, D.; Narayanaswamy, H.D.; Malathi, V.; Tomar, N.; Gupta, S.; Singh, S.V. Current scenario and future perspectives of CPV-2 vaccines in India. *Adv. Anim. Vet. Sci.* **2017**, *5*, 446–448. [CrossRef]
48. Waner, T.; Naveh, A.; Wudovsky, I.; Carmichael, L.E. Assessment of Maternal Antibody Decay and Response to Canine Parvovirus Vaccination Using a Clinic-Based Enzyme-Linked Immunosorbent Assay. *J. Vet. Diagnostic Investig.* **1996**, *8*, 427–432. [CrossRef] [PubMed]
49. Dunning, S. *Austin Animal Center Reopens on-Site Adoptions*; Austin Monit: Austin, TX, USA, 2020.

Publisher's Note: MDPI stays neutral with regard to jurisdictional claims in published maps and institutional affiliations.

Article

Phylogenetic and Geospatial Evidence of Canine Parvovirus Transmission between Wild Dogs and Domestic Dogs at the Urban Fringe in Australia

Mark Kelman [1,*], Lana Harriott [2], Maura Carrai [1,3], Emily Kwan [1], Michael P. Ward [1] and Vanessa R. Barrs [1,3]

1 Sydney School of Veterinary Science, The University of Sydney, Sydney, NSW 2006, Australia; maura.carrai@sydney.edu.au (M.C.); emilypkwan2.7@gmail.com (E.K.); michael.ward@sydney.edu.au (M.P.W.); vanessa.barrs@cityu.edu.hk (V.R.B.)

2 Pest Animal Research Centre, Biosecurity Queensland, Department of Agriculture and Fisheries, Toowoomba, QLD 4350, Australia; Lana.Harriott@daf.qld.gov.au

3 Jockey Club College of Veterinary Medicine, City University of Hong Kong, Kowloon Tong, Hong Kong, China

* Correspondence: kelmanscientific@gmail.com

Received: 21 May 2020; Accepted: 17 June 2020; Published: 19 June 2020

Abstract: Canine parvovirus (CPV) is an important cause of disease in domestic dogs. Sporadic cases and outbreaks occur across Australia and worldwide and are associated with high morbidity and mortality. Whether transmission of CPV occurs between owned dogs and populations of wild dogs, including *Canis familiaris*, *Canis lupus dingo* and hybrids, is not known. To investigate the role of wild dogs in CPV epidemiology in Australia, PCR was used to detect CPV DNA in tissue from wild dogs culled in the peri-urban regions of two Australian states, between August 2012 and May 2015. CPV DNA was detected in 4.7% (8/170). There was a strong geospatial association between wild-dog CPV infections and domestic-dog CPV cases reported to a national disease surveillance system between 2009 and 2015. Postcodes in which wild dogs tested positive for CPV were 8.63 times more likely to also have domestic-dog cases reported than postcodes in which wild dogs tested negative ($p = 0.0332$). Phylogenetic analysis of CPV VP2 sequences from wild dogs showed they were all CPV-2a variants characterized by a novel amino acid mutation (21-Ala) recently identified in CPV isolates from owned dogs in Australia with parvoviral enteritis. Wild-dog CPV VP2 sequences were compared to those from owned domestic dogs in Australia. For one domestic-dog case located approximately 10 km from a wild-dog capture location, and reported 3.5 years after the nearest wild dog was sampled, the virus was demonstrated to have a closely related common ancestor. This study provides phylogenetic and geospatial evidence of CPV transmission between wild and domestic dogs in Australia.

Keywords: canine parvovirus; peri-urban; wild dogs; disease transmission; Australia

---

1. Introduction

Canine parvovirus (CPV) is a single-stranded DNA virus belonging to the family *Parvoviridae*, subfamily *Parvovirinae*, genus *Protoparvovirus* and species *Carnivore Protoparvovirus 1* (CPPV-1) [1]. CPV infection occurs worldwide and has been reported in a range of carnivores, including domestic dogs and various wildlife species [2–11]. Viruses closely related to CPV include Feline parvovirus (FPV), Mink enteritis virus (MEV), Blue fox parvovirus (BFPV) and Racoon parvovirus (RPV) [12,13], so named due to their initial detection in these species. Clinical disease caused by *Carnivore protoparvovirus 1* in different host species ranges from asymptomatic to severe gastroenteritis, leukopenia, dehydration and death [14].

Although the exact origin of CPV remains unknown, CPV is thought to have evolved from FPV, or, more likely, that both viruses share a common ancestor that is likely associated with a wild carnivore host [12,13,15–19]. CPV poses a continuing threat for the emergence of new variants with pathogenic potential for multiple carnivore species in the order *Carnivora*.

Across Australia, CPV disease is relatively common in owned domestic dogs, particularly in rural and remote regions. An estimated 20,000 cases occur annually in Australia, with a reported incidence of 4.12 cases per 1000 dogs [20,21]. CPV infection has also been reported in wild canids, including four dingoes (*Canis lupus dingo*) (including a litter of three dingo puppies) and a dingo–dog hybrid, based on fecal antigen testing [10], as well as a red fox (*Vulpes vulpes*), based on serological testing [22]. However, the presence of CPV in Australian wild dogs has not been investigated. Wild and feral dogs are distributed throughout Australia, ranging from remote, isolated regions to peri-urban areas [23], though wild-dog numbers are not well documented [24].

In Australia, the term "wild dog" includes dingoes, feral domestic dogs (*Canis familiaris)* and hybrids [25]. The only other wild canid in Australia is the red fox, *Vulpes vulpes*, which was introduced as a result of European settlement [26]. The other feral carnivore in Australia is the feral cat, *Felis catus* [27], which is a likely parvovirus reservoir, given that FPV and CPV have been demonstrated in the Australian feral cat population [28–30]. Domestic dogs are also suggested as likely reservoir hosts for CPV in Africa, China and Mexico [31–34]. Furthermore, molecular investigations of CPV have provided evidence of cross-species bidirectional transmission between domestic dogs and wild carnivores [15,32]. While an epidemiological association between CPV case-occurrence in domestic dogs and geographical proximity to wild dogs and foxes in Australia has been reported [10], evidence of infection with the same or related strains of CPV is lacking. Peri-urban regions in Australia are a likely location to detect the presence of CPV in wild dogs, as there is close contact between human settlements and wild carnivore populations, which has been associated with exposure to CPV [32,35,36].

As well as being present in blood and feces during active infection, parvoviruses persist in mononuclear cells in peripheral blood or tissue [29,37] after fecal shedding of virus has ceased. This allows tissue samples to be used to detect both active and resolved infections in asymptomatic hosts that are no longer shedding virus in feces, and which likely represent latent infection, as reported for human parvoviruses [12,38]. CPV has been detected in wild carnivores and domestic cats, in a range of tissue samples after viral shedding has ceased, including blood, bone marrow, mesenteric lymph nodes, tongue, spleen and myocardium [12,29,37].

Our goal was to investigate the epidemiology of parvoviruses at the wild–domestic dog interface. The specific objectives of this study were to determine whether wild-dog populations in Australia are exposed to CPV, and if so, to estimate its prevalence and identify circulating strains; and to assess the likelihood of transmission of CPV between wild dogs and domestic dogs in peri-urban regions.

2. Materials and Methods

2.1. Wild-Dog Sample Collection

The carcasses of 201 wild dogs captured and culled between August 2012 and May 2015 from peri-urban regions of South East Queensland (SEQ) and Northern New South Wales (NSW) were supplied by local councils, for a study investigating pathogens of public health or economic significance [24]. Ethics approval was granted by the University of Queensland AEC (Animal Ethics Committee) (SVS/145/13). Residual tissue samples (tongue), which had been stored frozen at −18 °C, were made available for CPV testing.

2.2. DNA Extraction and Conventional PCR

DNA was extracted from tongue tissue for PCR, using the Macherey-Nagel mini kit (Macherey-Nagel, Düren, Germany). To evaluate the presence and quality of canine DNA, a conventional PCR was performed, targeting a canine housekeeping gene, the ribosomal protein L32 (RPL32) gene, using the following primers: RPL32-F (5′- ACCTCTGGTGAAGCCCAAG-3′) and RPL32-R (5′-GGGATTGGTGACTCTGATGG-3′) [39]. The total reaction volume was 25 µL and contained 2.5 µL of template DNA, 0.5 µL each of forward- and reverse-primer, 0.5 µL of MyTaq™HS Red DNA Polymerase, 5 µL of MyTaq ™ Red Reaction Buffer and 16 µL of water. DNA amplification was performed by using an initial denaturation step at 95 °C for 1 min, followed by 35 cycles of 95 °C for 15 s, 60 °C for 15 s and 72 °C for 10 s, with a final extension step at 72 °C for 5 min. Following PCR, the samples were electrophoresed on a 1% agarose gel (Bio-Rad Laboratories, Hercules, CA, USA), using 1 × tris-borate EDTA running buffer, and visualized with SYBR safe DNA (Thermo Fisher Scientific, Waltham, MA, USA).

2.3. Real-Time PCR

Real-time PCR/quantitative PCR assay (qPCR) was performed to determine the CPV viral load in DNA extracts from tongue tissue of the wild-dog cadavers, as previously described, with minor modification [40]. In brief, real-time PCR was carried out in a 35 µL reaction containing 17.5 mL of IQ Supermix (Bio-Rad Laboratories Srl), 600 nM of primers CPV-For and CPV-Rev, 200 nM of probe CPV-Pb (Table 1), and 10 µL of template (diluted 1:10 in Tris EDTA buffer). Serial 8-fold dilutions (representing from 10^9 to 10^2 DNA copies/10 µL) of a plasmid, pFastBac™HTA, containing VP2 gene sequence were used to generate a standard curve. Each test sample and each dilution of standard DNA was tested in duplicate. An exogenous DNA internal control, Cal Orange 560 (Bioline, Meridian Bioscience, Cincinnati, OH, USA), was added to each sample, in order to control for PCR inhibition, according to the manufacturer's instructions. The thermal-cycle protocol used was the activation of Taq DNA polymerase at 95 °C for 10 min and 40 cycles consisting of denaturation at 95 °C for 15 s, primer annealing at 52 °C for 30 s and extension at 60 °C for 1 min. All reactions were conducted in an a CFX connect™Real Time PCR Detection System (Bio-Rad Laboratories Pty., Ltd. Gladesville, Australia), and the data were analyzed with the software CFX Maestro. Samples were considered positive only when results could be confirmed from paired testing in a single assay.

Table 1. Sequence and position of oligonucleotides used in the study [40] *.

Assay	Primer/Probe	Sequence 5′ - 3′	Polarity	Amplicon Size (bp)	Position †
Real Time Assay	CPV-For	AAACAGGAATTAACTATACTAATATATTTA	+	93	4104–4135
	CPV-Rev	AAATTTGACCATTTGGATAAACT	−		4176–4198
	CPV-Pb	FAM—TGGTCCTTTAACTGCATTAAATAATGTACC—TAMRA	+		4143–4172

* FAM 5 6-carboxyfluorescein; TAMRA 5 6-carboxytetramethylrhodamine. † Oligonucleotide position is referred to the sequence of strain CPV-b (accession M38245).

2.4. Conventional PCR and Sequence Analysis

Samples testing positive for CPV in the qPCR assay were subject to conventional PCR and sequencing of the complete VP2 gene, as previously described [30,41].

The VP2 sequences from wild dogs were assembled and aligned, using CLC Workbench (Qiagen, Hilden, Germany), and then compared to previously characterized VP2 sequences from domestic dogs in Australia [42] and other countries, as well as reference strains of CPV, including CPV-2a-like and related viruses obtained from the GenBank database. Sequences were aligned by using the Geneious prime software package (11.0.4) using the MAFT algorithm. Phylogenetic analysis was performed using Mega X version 10.0.5 and employing the Tamura 3-parameter model of nucleotide substitution using a discrete gamma distribution *(+G)* and assuming that a certain fraction of sites is evolutionarily invariable *(+I)*, with the Nearest-Neighbor Interchange heuristic method. The Tamura 3-parameter *+G +I* model provided the best maximum likelihood (ML) fit of 24 nucleotide substitution models, with a Bayesian Information Criterion Score of 9996.026. The analysis involved 78 nucleotide sequences.

Codon positions included were 1st+2nd+3rd+Noncoding. There were a total of 1719 positions in the final dataset.

2.5. Wild-Dog Sample Data and Owned-Dog CPV Case Occurrence Data

Data collected at the time each wild dog was culled included, sex, estimated age (from examination of dentition), date of capture, and latitude and longitude of capture location. These data were used in geospatial and statistical analysis. Eight samples missing data for date of capture or capture location were excluded from the analysis.

Retrospective data on owned-dog CPV cases occurring in the same postcodes in which wild-dog samples were collected were obtained to evaluate the potential for CPV transmission between wild- and owned-dog populations. Data from 2009 to 2015 were sourced from the Disease WatchDog® database [43], a collection of national disease surveillance data for companion animals in Australia (http://www.vetcompass.com.au). Only cases that had been confirmed by diagnostic testing (fecal antigen test "ELISA", PCR or immunofluorescence) were included in our analysis. For the purpose of geospatial and statistical analysis, only data relating to the case date (year) and postcode were used.

2.6. Geospatial Analysis of Wild-Dog Data and Owned-Dog Data

Mapping and geospatial analysis were performed, using ArcGIS® version 10.2 (ERSI, Redlands, CA). Wild dogs' positive and negative results were mapped by nearest postcode, which was identified from supplied latitude and longitude location data, in ArcGIS, using an ABS Postal Areas ASGS Ed 2016 Digital Boundaries Shapefile (ESRI Format) [44]. Owned-dog case occurrence was mapped by postcode. Maps were generated at the state level for SEQ and Northern NSW, and also at a regional level for Brisbane and south of Brisbane, to the border of Queensland (QLD) and NSW.

2.7. Statistical Analysis of Domestic-Dog CPV Cases and Association with Wild-Dog Infection

Data were analyzed by using Microsoft® Excel for Mac Version 16.16.15 and Statistix® version 10.0 (Analytical Software, Tallahassee, FL, USA). Odds ratios were calculated for the frequency of postcodes with wild dogs testing positive or negative for CPV and owned-dog CPV case occurrence (present or absent), for owned-dog cases occurring in the same year as wild-dog sampling, and across the entire owned dog sample period. Chi-squared ($\chi 2$) analysis was performed to test associations between wild-dog observations (present or absent) and categorical variables. For all statistical tests, a p-value of <0.05 was used to determine significance.

3. Results

3.1. Wild-Dog Sampling

Tissue samples from 171 wild dogs collected between 2012 and 2015 were available for PCR testing. Details of the cadavers sampled are reported in Table 2, and maps of the regions where wild dogs were trapped are depicted in Figures 1–3.

3.2. DNA Detection and Quantification

Canine DNA was identified in 170/171 wild-dog cadaver samples, and CPV DNA was amplified in 4.7% (8/170) of the remaining samples analyzed. The viral load of CPV DNA in the samples ranged from 3.41×10^1 to 1.95×10^7 copies/μL (Table 3).

Table 2. Australian wild-dog cadavers made available for PCR testing for canine parvovirus DNA, and test results, categorized by state of origin, sex, age and year trapped.

Category	Variable	No. Dogs Sampled	Percentage	No. Dogs Negative	No. Dogs Positive	Chi2	DF	*p*-Value
State	QLD	146	85.4	136	8	1.05	1	0.3051
	NSW	18	10.5	18	0			
	NR	7	4.1					
	Total	171						
Sex	Male	76	44.4	68	4	0.08	1	0.7833
	Female	87	50.9	83	4			
	NR	8	4.7					
	Total	171						
Age	<6 months	44	25.7	42	2	3.86	4	0.425
	6–12 months	44	25.7	40	3			
	1–2 years	34	19.9	33	0			
	2–5 years	14	8.2	13	1			
	>5 years	17	9.9	14	2			
	NR	18	10.5					
	Total	171						
Year captured	2012	2	1.2	2	0	0.91	3	0.8227
	2013	63	36.8	58	4			
	2014	90	52.6	85	4			
	2015	9	5.3	9	0			
	NR	7	4.1					
	Total	171						

NR = not recorded, QLD = Queensland, NSW = New South Wales, Chi2 = Chi squared statistic, DF= degrees of freedom. Seven dogs did not have corresponding location data, and sex, age and date trapped were not available for 8, 18 and 7 samples, respectively. While these were included in the PCR testing, they were excluded from the respective statistical analysis, for which their data were unavailable.

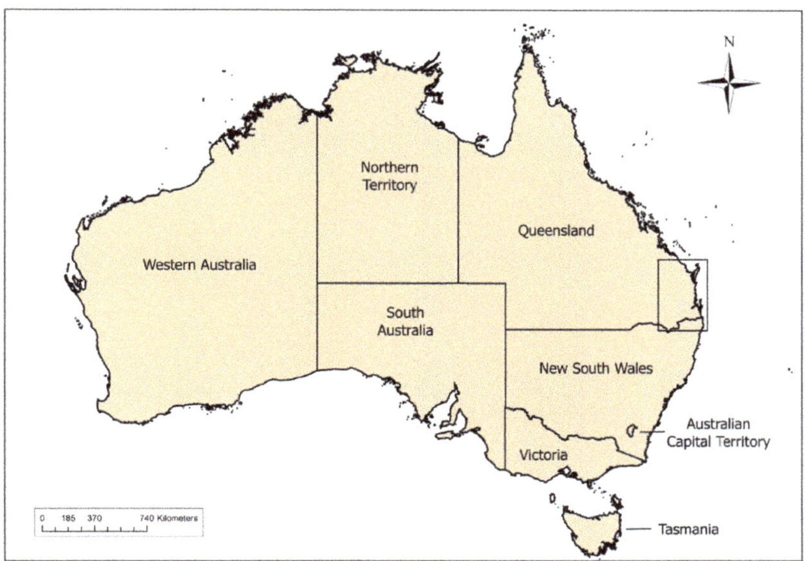

Figure 1. Region of wild-dog sampling (inset) between 2012 and 2015, in South East Queensland and Northeast New South Wales, Australia.

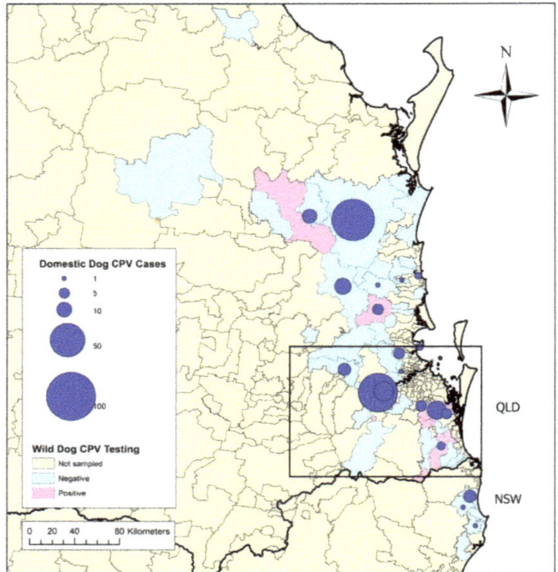

Figure 2. Distribution and frequency of canine parvovirus cases in domestic dogs reported by veterinarians on the Disease WatchDog® national disease surveillance system between 2009 and 2015, and postcodes where wild dogs, sampled between 2012 and 2015, tested positive or negative to canine parvovirus as determined by polymerase chain reaction. Map represents South East Queensland (QLD) and Northeast New South Wales (NSW), Australia. Inset depicts area represented in Figure 3.

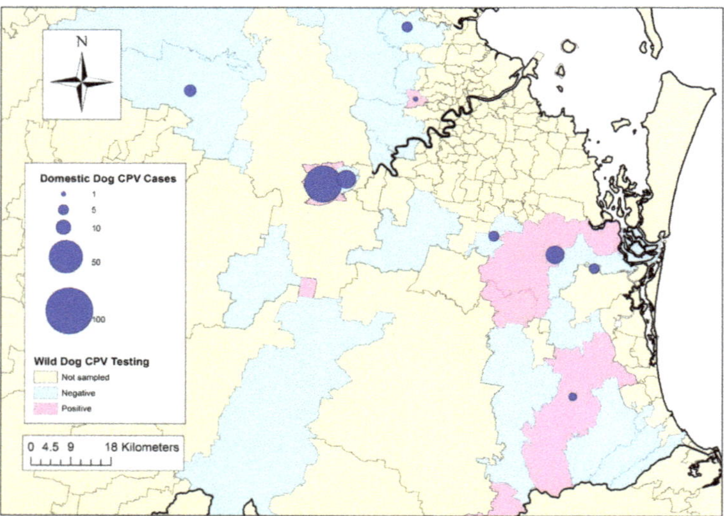

Figure 3. Distribution and frequency of canine parvovirus cases in domestic dogs reported by veterinarians on the Disease WatchDog® national disease surveillance system between 2009 and 2015, and postcodes where wild dogs, sampled between 2012 and 2015, tested positive or negative for canine parvovirus, as determined by polymerase chain reaction. Map represents Brisbane and south of Brisbane, to the border of Queensland (QLD) and New South Wales (NSW), Australia.

Table 3. Wild-dog samples testing positive for canine parvovirus DNA by quantitative PCR.

GenBank Accession No.	Sample ID	Sex	Age	Date Captured	Location Captured			Viral Copies per μL/DNA
					Region	Postcode	State Locality	
MT447094	WD29	M	<6 months	25/11/13	Woodford	4514	South East QLD	2.89×10^3
MT447095	WD46	M	6–12 months	28/2/14	The Gap	4061	Brisbane QLD	2.05×10^4
MT447096	WD48	F	>5 years	5/2/13	Tamborine	4270	South East QLD	1.95×10^7
MT447097	WD49	M	>5 years	15/5/14	Nerang	4211	Gold Coast QLD	1.57×10^6
MT447098	WD50	M	2–5 years	24/8/13	Beenleigh	4207	South East QLD	6.56×10^5
MT447099	WD51	F	<6 months	27/11/13	Ipswich	4305	South East QLD	2.19×10^4
NA	WD58	F	6–12 months	2/3/14	Goomeri	4601	South East QLD	4.08×10^1
NA	WD59	F	6–12 months	26/2/14	The Gap	4061	Brisbane QLD	3.41×10^1

QLD = Queensland, Australia; NA = not applicable.

3.3. *Geography and Prevalence of Wild-Dog Exposure to CPV*

CPV DNA was amplified from eight wild dogs from seven different postcodes of QLD, but from no dogs from NSW. However, CPV detection between the states of NSW and QLD was not statistically different ($\chi^2 = 1.05$, df = 1, $p = 0.3051$), nor was any difference based on sex ($\chi^2 = 0.08$, df = 1, $p = 0.7833$), age ($\chi^2 = 3.86$, df = 4, $p = 0.425$) or the year in which the dogs were captured ($\chi^2 = 0.91$, df = 3, $p = 0.8227$) (Table 2).

3.4. *Association between CPV Exposure in Wild Dogs and CPV Cases in Owned Dogs*

In total, wild dogs were sampled from 57 different postcodes. Postcodes with one or more wild dogs testing positive to CPV were 8.63 times more likely to have reported CPV cases in owned dogs in the same year ($p = 0.0332$) and 6.43 times more likely across the entire owned-dog case sampling period ($p = 0.0350$) (Table 4).

Table 4. Frequency of postcodes with wild dogs testing positive or negative for canine parvovirus (CPV) and domestic-dog CPV case occurrence (present or absent). Wild-dog sampling occurred between 2012 and 2015, and domestic-dog cases were reported between 2009 and 2015.

Domestic-Dog CPV Cases	Domestic-Dog CPV Case Occurrence	Wild-Dog CPV Status			Odds Ratio	*p*-value
		Positive	Negative	Total		
In all years between 2009 and 2015	Present	5	14	19	6.43	0.0350
	Absent	2	36	38		
	Total	7	50	57		
In same year as wild-dog sampling in postcode	Present	3	4	7	8.63	0.0332
	Absent	4	46	50		
	Total	7	50	57		

3.5. *Wild-Dog CPV VP2 Sequencing and Phylogenetic Analysis*

Of the eight dogs in which CPV DNA was amplified by qPCR, VP2 sequencing was successful for six dogs and unsuccessful in the two dogs with the lowest viral loads (Table 3). Phylogenetic analysis revealed that three of the six wild-dog sequences were identical and all six were closely related, belonging to a clade of CPV-2a viruses comprising those from wild dogs, as well as viruses from dogs with parvoviral enteritis, in three different states of Australia (Victoria, NSW and QLD) (Figure 4). Viruses in this clade were characterized by the VP2 mutation (Thr-21-Ala) that differed from all other variants analyzed (Supplementary Materials Table S1). One of the viruses in this clade (GenBank accession no. MN259063) was collected from an owned dog with parvoviral enteritis, approximately 10 km from the capture location of one of the wild dogs (WD50) 3.5 years after the wild dog was culled (Figure 4 and Supplementary Materials Table S1).

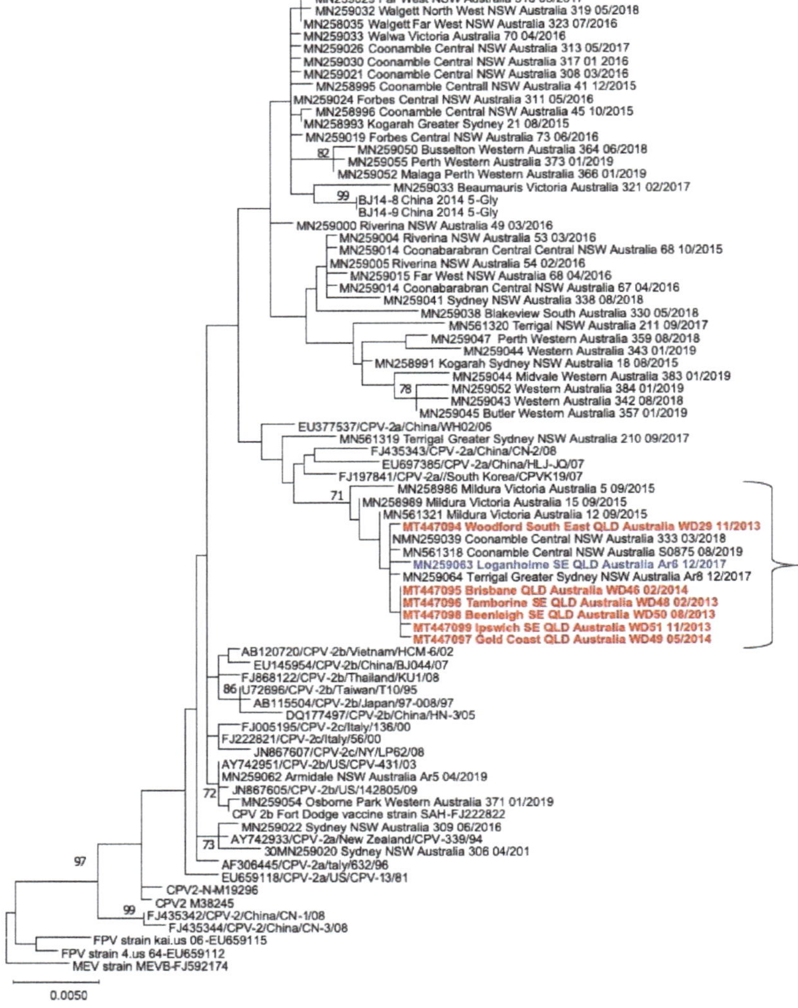

Figure 4. Phylogeny of CPV VP2 sequences from wild dogs in Australia (red) and a domestic dog (blue) within 10 km of the wild-dog sampling. The evolutionary history was inferred by using the Maximum Likelihood method and Tamura three-parameter model with 1000 bootstrap replicates [45]. Bootstrap values >70% are shown on the branches. The tree with the highest log likelihood (-4045.79) is shown. The tree is drawn to scale, with branch lengths measured in the number of substitutions per site.

4. Discussion

This study reports the first detection of CPV in wild-dog populations in Australia and provides phylogenetic evidence that a CPV strain from a wild-dog population was closely related to a strain from a domestic dog with parvoviral enteritis in close proximity. CPV strains from wild dogs were also closely related to viruses from other owned dogs collected over a large region of Eastern Australia. The viruses in this clade were characterized by the VP2 protein signature 21-Ala, 297-Ala, 324-Iso and 555-Val. Two of these residues (297-Ala and 555-Val) are common, well-characterized mutations among Australian CPV strains compared to the original CPV-2 virus that emerged in dogs in the late 1970s [46]. Another VP2 residue (324-Iso) in the wild-dog strains is common in Asian strains of CPV [47] and was recently identified in owned dogs in Australia [42]. The finding of this mutation in CPV strains from wild dogs further suggests introduction or circulation of Asian strains in Australia. The Thr-21-Ala mutation, present in the VP2 sequences from all of the wild dogs, was also recently identified as a novel mutation among CPVs from owned dogs in Australia [42]. The similarity of the wild-dog VP2 sequences to those from owned dogs suggests viral transmission between these two populations, although the directionality is uncertain.

Only wild dogs from SEQ and not Northern NSW tested positive for CPV, which is likely due to the smaller sample size for NSW (versus QLD), but may also reflect different risk factors for infection between these populations. Proximity to free-ranging owned dogs may be a risk factor for CPV infection in wild dogs [48]. However, our study did not investigate owned-dog population distribution or ranging behaviors, so such a transmission pathway cannot be directly assessed here. It is possible that a higher proportion of free-ranging owned dogs in an area could lead to increased infection rates of the domestic-dog population in those areas, but this requires further investigation. Close proximity to human settlements may also increase risk for CPV exposure in wild carnivores; however, previous studies have failed to demonstrate this to be statistically significant [32,36,49] or have shown no increased risk [50]. Conversely, wild-carnivore populations may have parvovirus infection cycles independent of domestic carnivores [51], and this may also be the case for some wild-dog populations in Australia; more research is needed to determine this.

No difference in age was observed between CPV-positive and -negative wild dogs, despite younger age being an identified risk factor for CPV infection and disease in unvaccinated domestic dogs [52,53] and wild canids (wolves) [54]. This may be because the method of CPV detection used in our study did not differentiate between active and recovered infections. Most (58%, 88/153) wild dogs trapped were <12 months old, which may have increased the likelihood of detecting recent or active infections and also likely reflects the shorter lifespan of wild dogs versus domestic dogs. Mortality rates from CPV in wild dogs have never been reported; however, CPV disease in young animals has been implicated in reduction of population renewal in gray wolves: the proportion of pups live-trapped each year (which had declined between 1984 and 2004) correlated with increasing CPV antibody prevalence ($r^2 = 0.51$; $p < 0.01$) [54]. CPV might also play a role in population reduction in wild dogs in Australia. Our finding of no difference between the year that wild dogs were sampled suggests that CPV may be endemic in these populations.

We identified a significantly increased risk for domestic-dog CPV cases in geographical regions where wild-dog CPV infection was also detected. The finding of a closely related CPV strain in an owned dog collected from the same geographic region several years after detection in wild dogs suggests both low viral diversity over time and that ongoing transmission of CPV might have been occurring between the wild-dog populations in this area and neighboring owned-dog populations. Interpopulation disease transmission could be unidirectional from either population, or bidirectional [32]. Interpopulation transmission could be facilitated by roaming dogs from either or both populations entering the others' territories and transmitting virus via fomites or directly from an infected animal through defecation. Peri-urban wild dogs in Australia have been found to have home ranges of around 17 km^2, travel an average of 7 km/day [55] and often spend time in urban habits [56], making CPV-transmission between wild and domestic dogs likely in these areas. Fomite transmission due to human movement or other domestic/wild/feral species traversing the two territories may also occur. The ability for CPV to survive for protracted periods in the environment makes indirect transmission more likely [57].

Our detected prevalence of CPV exposure (5.3%) is much lower than that reported by serological testing of a range of wild canids in the Yellowstone National Park, USA (98%); Canadian Rocky Mountains (95%); Montana, USA (65%); Chile (49%); Portugal (38%); and Spain (17.2%) [50,58–60]. The true level of CPV exposure in the wild-dog population we tested is likely to be higher. However, a limitation of our study was that we only had access to wild-dog cadaver tissue, and blood for serological testing was not available. Serological testing is able to detect prior CPV exposure in individuals, with high sensitivity, due to long-lasting seropositivity following natural infection [6], whereas PCR is very sensitive to detect CPV DNA, but successful amplification depends on the availability of infected monocytes being present in the sample tissue. The viral loads obtained from six of the eight positive samples were similar to the range reported in samples collected from lymph nodes of clinically affected FPV-positive cats [30], suggesting that these six dogs were undergoing active CPV infections, while the remaining two were latently infected. Given that the epidemiological features of CPV include fecal shedding in high viral loads, environmental persistence and a high degree of contagiousness, it is likely that a larger proportion of wild dogs are exposed than are reflected in our results. The ongoing culling programs from which our samples were sourced, and the resulting reduction of wild-dog population size, may have also reduced the transmission of CPV in these populations. Serological testing of wild dogs throughout Australia is therefore warranted to determine whether actual CPV prevalence is in fact higher than our findings reflect.

5. Conclusions

The detection of CPV in this small population of peri-urban wild dogs suggests that parvoviral infection might be widespread among sympatric populations of wild and owned dogs in Australia. The finding of related strains of CPV in wild and owned dogs in Australia suggests that viral transmission might be occurring between these two populations, although the directionality is uncertain. It is possible that wild dogs could be responsible for some outbreaks of disease among domestic-dog populations, particularly at the urban fringe, or that infection from domestic-dog populations can spill over to wild dogs. Further research is warranted to definitively determine if CPV transmission is occurring between these populations.

Supplementary Materials: The following are available online at http://www.mdpi.com/1999-4915/12/6/663/s1. Table S1: Amino acid substitutions in the VP2 region of strains of canine parvovirus (CPV) detected from tongue tissue of wild-dog cadavers from Northern New South Wales and South East Queensland, Australia (2013 to 2014), and in feces from dogs with confirmed parvoviral enteritis (2015 to 2019), compared to reference strains of CPV and feline parvovirus (FPV).

Author Contributions: Conceptualization, M.K., M.P.W. and V.R.B.; data curation, M.K., L.H., M.C. and E.K.; formal analysis, M.K. and V.R.B.; funding acquisition, M.K.; investigation, M.K., L.H., M.C. and E.K.; methodology, M.K., M.C., M.P.W. and V.R.B.; project administration, M.K.; resources, M.K. and V.R.B.; supervision, M.C., M.P.W. and V.R.B.; validation, M.K., M.P.W. and V.R.B.; visualization, M.K. and V.R.B.; writing—original draft, M.K. and V.R.B.; writing—review and editing, M.K., L.H., M.C., E.K., M.P.W. and V.R.B. All authors have read and agreed to the published version of the manuscript.

Funding: This research was funded by the Australian Companion Animal Health Fund, grant number 008/2019. The APC was funded by The University of Sydney.

Acknowledgments: The authors wish to thank the VetCompass consortium for access to data from the Disease WatchDog® database. Thanks are also given to the many veterinary clinic contributors of case reports to Disease WatchDog® and the local government officers and individuals involved in the wild-dog capture, without whom this research would not have been possible. The research in this paper was completed in partial fulfilment for the requirements of a PhD degree at The University of Sydney, by the lead author.

Conflicts of Interest: The authors declare no conflict of interest. The funders had no role in the design of the study; in the collection, analyses or interpretation of data; in the writing of the manuscript; or in the decision to publish the results.

References

1. Cotmore, S.F.; Agbandje-McKenna, M.; Canuti, M.; Chiorini, J.A.; Eis-Hubinger, A.-M.; Hughes, J.; Mietzsch, M.; Modha, S.; Ogliastro, M.; Pénzes, J.J.; et al. ICTV Virus Taxonomy Profile: Parvoviridae. *J. Gen. Virol.* **2019**, *100*, 367–368. [CrossRef] [PubMed]
2. Behdenna, A.; Lembo, T.; Calatayud, O.; Cleaveland, S.; Halliday, J.E.B.; Packer, C.; Lankester, F.; Hampson, K.; Craft, M.E.; Czupryna, A.; et al. Transmission ecology of canine parvovirus in a multi-host, multi-pathogen system. *Proc. R. Soc. B Biol. Sci.* **2019**, *286*, 20182772. [CrossRef] [PubMed]

3. Calatayud, O.; Esperón, F.; Cleaveland, S.; Biek, R.; Keyyu, J.; Eblate, E.; Neves, E.; Lembo, T.; Lankester, F. Carnivore Parvovirus Ecology in the Serengeti Ecosystem: Vaccine Strains Circulating and New Host Species Identified. *J. Virol.* **2019**, *93*, 1–18. [CrossRef] [PubMed]
4. Calatayud, O.; Esperón, F.; Velarde, R.; Oleaga, Á.; Llaneza, L.; Ribas, A.; Negre, N.; Torre, A.; Rodríguez, A.; Millán, J. Genetic characterization of Carnivore Parvoviruses in Spanish wildlife reveals domestic dog and cat-related sequences. *Transbound. Emerg. Dis.* **2019**, 1–9. [CrossRef] [PubMed]
5. Chen, C.-C.; Chang, A.-M.; Wada, T.; Chen, M.-T.; Tu, Y.-S. Distribution of Carnivore protoparvovirus 1 in free-living leopard cats (Prionailurus bengalensis chinensis) and its association with domestic carnivores in Taiwan. *PLoS ONE* **2019**, *14*, e0221990. [CrossRef]
6. Gese, E.M.; Schultz, R.D.; Rongstad, O.J.; Andersen, D.E. Prevalence of Antibodies against Canine Parvovirus and Canine Distemper Virus in Wild Coyotes in Southeastern Colorado. *J. Wildl. Dis.* **1991**, *27*, 320–323. [CrossRef]
7. Ikeda, Y.; Mochizuki, M.; Naito, R.; Nakamura, K.; Miyazawa, T.; Mikami, T.; Takahashi, E. Predominance of Canine Parvovirus (CPV) in Unvaccinated Cat Populations and Emergence of New Antigenic Types of CPVs in Cats. *Virology* **2000**, *278*, 13–19. [CrossRef]
8. Steinel, A.; Munson, L.; van Vuuren, M.; Truyen, U. Genetic characterization of feline parvovirus sequences from various carnivores. *J. Gen. Virol.* **2000**, *81*, 345–350. [CrossRef]
9. Truyen, U.; Platzer, G.; Parrish, C.R. Antigenic type distribution among canine parvoviruses in dogs and cats in Germany. *Vet. Rec.* **1996**, *138*, 365–366. [CrossRef]
10. Van Arkel, A.; Kelman, M.; West, P.; Ward, M.P. The relationship between reported domestic canine parvovirus cases and wild canid distribution. *Heliyon* **2019**, *5*, e02511. [CrossRef]
11. Zarnke, R.L.; Ballard, W.B. Serologic survey for selected microbial pathogens of wolves in Alaska, 1975–1982. *J. Wildl. Dis.* **1987**, *23*, 77–85. [CrossRef] [PubMed]
12. Allison, A.B.; Kohler, D.J.; Fox, K.A.; Brown, J.D.; Gerhold, R.W.; Shearn-Bochsler, V.I.; Dubovi, E.J.; Parrish, C.R.; Holmes, E.C. Frequent Cross-Species Transmission of Parvoviruses among Diverse Carnivore Hosts. *J. Virol.* **2013**, *87*, 2342–2347. [CrossRef] [PubMed]
13. Shackelton, L.A.; Parrish, C.R.; Truyen, U.; Holmes, E.C. High rate of viral evolution associated with the emergence of carnivore parvovirus. *Proc. Natl. Acad. Sci. USA* **2005**, *102*, 379–384. [CrossRef]
14. Mylonakis, M.; Kalli, I.; Rallis, T. Canine parvoviral enteritis: an update on the clinical diagnosis, treatment, and prevention. *Vet. Med. Res. Rep.* **2016**, *7*, 91–100. [CrossRef] [PubMed]
15. Allison, A.B.; Kohler, D.J.; Ortega, A.; Hoover, E.A.; Grove, D.M.; Holmes, E.C.; Parrish, C.R. Host-Specific Parvovirus Evolution in Nature Is Recapitulated by In Vitro Adaptation to Different Carnivore Species. *Plos Pathog.* **2014**, *10*, e1004475. [CrossRef]
16. Allison, A.B.; Harbison, C.E.; Pagan, I.; Stucker, K.M.; Kaelber, J.T.; Brown, J.D.; Ruder, M.G.; Keel, M.K.; Dubovi, E.J.; Holmes, E.C.; et al. Role of Multiple Hosts in the Cross-Species Transmission and Emergence of a Pandemic Parvovirus. *J. Virol.* **2012**, *86*, 865–872. [CrossRef]
17. Steinel, A.; Parrish, C.R.; Bloom, M.E.; Truyen, U. Parvovirus infections in wild carnivores. *J. Wildl. Dis.* **2001**, *37*, 594–607. [CrossRef]
18. Truyen, U.; Gruenberg, A.; Chang, S.-F.; Obermaier, B.; Veijalainen, P.; Parrish, C.R. Evolution of the Feline-Subgroup Parvoviruses and the Control of Canine Host Range In Vivo. *J. Virol.* **1995**, *69*, 9. [CrossRef]
19. Truyen, U.; Evermann, J.F.; Vieler, E.; Parrish, C.R. Evolution of canine parvovirus involved loss and gain of feline host range. *Virology* **1996**, *215*, 186–189. [CrossRef]
20. Kelman, M.; Ward, M.P.; Barrs, V.R.; Norris, J.M. The geographic distribution and financial impact of canine parvovirus in Australia. *Transbound. Emerg. Dis.* **2019**, *66*, 299–311. [CrossRef]
21. Zourkas, E.; Ward, M.P.; Kelman, M. Canine parvovirus in Australia: A comparative study of reported rural and urban cases. *Vet. Microbiol.* **2015**, *181*, 198–203. [CrossRef] [PubMed]
22. Mulley, R.C.; Claxton, P.D.; Feilen, C.P. A serological survey of some infectious diseases in the red fox (Vulpes vulpes) in New South Wales. *Proc. 4th Int. Conf. Wildl. Dis. Assoc.* **1981**, 44–46.
23. Gabriele-Rivet, V.; Arsenault, J.; Wilhelm, B.; Brookes, V.J.; Newsome, T.M.; Ward, M.P. A Scoping Review of Dingo and Wild-Living Dog Ecology and Biology in Australia to Inform Parameterisation for Disease Spread Modelling. *Front. Vet. Sci.* **2019**, *6*, 1–14. [CrossRef] [PubMed]
24. Harriott, L. Prevalence, Risk Factors, and Geographical Distribution of Zoonotic Pathogens Carried by Peri-urban Wild Dogs. Ph.D. Thesis, The University of Queensland, Brisbane, Australia, 2018.

25. Jackson, S.M.; Groves, C.P.; Fleming, P.J.S.; Aplin, K.P.; Eldridge, M.D.B.; Gonzalez, A.; Helgen, K.M. The Wayward Dog: Is the Australian native dog or Dingo a distinct species? *Zootaxa* **2017**, *4317*, 201. [CrossRef]
26. Kinnear, J.E.; Sumner, N.R.; Onus, M.L. The red fox in Australia—an exotic predator turned biocontrol agent. *Biol. Conserv.* **2002**, *108*, 335–359. [CrossRef]
27. Glen, A.S.; Dickman, C.R. Complex interactions among mammalian carnivores in Australia, and their implications for wildlife management. *Biol. Rev.* **2005**, *80*, 387. [CrossRef]
28. Brailey, J.; Jenkins, D.; Aghazadeh, M.; Slapeta, J.; Beatty, J.A.; Barrs, V.R. Feral carnivores are reservoirs of Carnivore protoparvovirus 1 in Australia. *Ecvim-Ca Congr. St Julian'smalta* **2017**.
29. Haynes, S.; Holloway, S. Identification of parvovirus in the bone marrow of eight cats. *Aust. Vet. J.* **2012**, *90*, 136–139. [CrossRef]
30. Van Brussel, K.; Carrai, M.; Lin, C.; Kelman, M.; Setyo, L.; Aberdein, D.; Brailey, J.; Lawler, M.; Maher, S.; Plaganyi, I.; et al. Distinct Lineages of Feline Parvovirus Associated with Epizootic Outbreaks in Australia, New Zealand and the United Arab Emirates. *Viruses* **2019**, *11*, 1155. [CrossRef]
31. Laurenson, K.; Sillero-Zubiri, C.; Thompson, H.; Shiferaw, F.; Thirgood, S.; Malcolm, J. Disease as a threat to endangered species: Ethiopian wolves, domestic dogs and canine pathogens. *Anim. Conserv.* **1998**, *1*, 273–280. [CrossRef]
32. López-Pérez, A.M.; Moreno, K.; Chaves, A.; Ibarra-Cerdeña, C.N.; Rubio, A.; Foley, J.; List, R.; Suzán, G.; Sarmiento, R.E. Carnivore Protoparvovirus 1 at the Wild–Domestic Carnivore Interface in Northwestern Mexico. *EcoHealth* **2019**, *16*, 502–511. [CrossRef]
33. Mainka, S.A.; Xianmeng, Q.; Tingmei, H.; Appel, M.J. Serologic Survey of Giant Pandas (Ailuropoda melanoleuca), and Domestic Dogs and Cats in the Wolong Reserve, China. *J. Wildl. Dis.* **1994**, *30*, 86–89. [CrossRef] [PubMed]
34. Woodroffe, R.; Prager, K.C.; Munson, L.; Conrad, P.A.; Dubovi, E.J.; Mazet, J.A.K. Contact with Domestic Dogs Increases Pathogen Exposure in Endangered African Wild Dogs (Lycaon pictus). *PLoS ONE* **2012**, *7*, e30099. [CrossRef] [PubMed]
35. Acosta-Jamett, G.; Cunningham, A.A.; Bronsvoort, B.D.; Cleaveland, S. Serosurvey of canine distemper virus and canine parvovirus in wild canids and domestic dogs at the rural interface in the Coquimbo Region, Chile. *Eur. J. Wildl. Res.* **2015**, *61*, 329–332. [CrossRef]
36. Truyen, U.; Müller, T.; Heidrich, R.; Tackmann, K.; Carmichael, L.E. Survey on viral pathogens in wild red foxes (Vulpes vulpes) in Germany with emphasis on parvoviruses and analysis of a DNA sequence from a red fox parvovirus. *Epidemiol. Infect.* **1998**, *121*, 433–440. [CrossRef] [PubMed]
37. Miyazawa, T.; Ikeda, Y.; Nakamura, K.; Naito, R.; Mochizuki, M.; Tohya, Y.; Vu, D.; Mikami, T.; Takahashi, E. Isolation of Feline Parvovirus from Peripheral Blood Mononuclear Cells of Cats in Northern Vietnam. *Microbiol. Immunol.* **1999**, *43*, 609–612. [CrossRef] [PubMed]
38. Norja, P.; Hokynar, K.; Aaltonen, L.-M.; Chen, R.; Ranki, A.; Partio, E.K.; Kiviluoto, O.; Davidkin, I.; Leivo, T.; Eis-Hubinger, A.M.; et al. Bioportfolio: Lifelong persistence of variant and prototypic erythrovirus DNA genomes in human tissue. *Proc. Natl. Acad. Sci. USA* **2006**, *103*, 7450–7453. [CrossRef]
39. Peters, I.R.; Peeters, D.; Helps, C.R.; Day, M.J. Development and application of multiple internal reference (housekeeper) gene assays for accurate normalisation of canine gene expression studies. *Vet. Immunol. Immunopathol.* **2007**, *117*, 55–66. [CrossRef] [PubMed]
40. Decaro, N.; Elia, G.; Martella, V.; Desario, C.; Campolo, M.; Trani, L.D.; Tarsitano, E.; Tempesta, M.; Buonavoglia, C. A real-time PCR assay for rapid detection and quantitation of canine parvovirus type 2 in the feces of dogs. *Vet. Microbiol.* **2005**, *105*, 19–28. [CrossRef]
41. Decaro, N.; Desario, C.; Elia, G.; Martella, V.; Mari, V.; Lavazza, A.; Nardi, M.; Buonavoglia, C. Evidence for immunisation failure in vaccinated adult dogs infected with canine parvovirus type 2c. *Microbiol. Q. J. Microbiol. Sci.* **2008**, *31*, 125–130.

42. Kwan, E.; Carrai, M.; Lanave, G.; Hill, J.; Parry, K.; Kelman, M.; Meers, J.; Decaro, N.; Beatty, J.; Martella, V.; et al. Analysis of canine parvoviruses circulating in Australia reveals predominance of variant 2b strains and identifies feline parvovirus-like mutations in the capsid proteins. *Transbound. Emerg. Dis.* **2020**. under review.
43. Ward, M.P.; Kelman, M. Companion animal disease surveillance: A new solution to an old problem? *Spat. Spatio-Temporal Epidemiol.* **2011**, *2*, 147–157. [CrossRef] [PubMed]
44. Australian Bureau of Statistics 1270.0.55.003—Australian Statistical Geography Standard (ASGS): Volume 3—Non ABS Structures, July 2016. Available online: https://www.abs.gov.au/AUSSTATS/abs@.nsf/DetailsPage/1270.0.55.003July%202016?OpenDocument (accessed on 18 June 2020).
45. Tamura, K. Estimation of the number of nucleotide substitutions when there are strong transition-transversion and G+C-content biases. *Mol. Biol. Evol.* **1992**, *9*, 678–687. [CrossRef] [PubMed]
46. Meers, J.; Kyaw-Tanner, M.; Bensink, Z.; Zwijnenberg, R. Genetic analysis of canine parvovirus from dogs in Australia. *Aust. Vet. J.* **2007**, *85*, 392–396. [CrossRef]
47. Wang, J.; Lin, P.; Zhao, H.; Cheng, Y.; Jiang, Z.; Zhu, H.; Wu, H.; Cheng, S. Continuing evolution of canine parvovirus in China: Isolation of novel variants with an Ala5Gly mutation in the VP2 protein. *Infect. Genet. Evol.* **2016**, *38*, 73–78. [CrossRef]
48. Frölich, K.; Streich, W.J.; Fickel, J.; Jung, S.; Truyen, U.; Hentschke, J.; Dedek, J.; Prager, D.; Latz, N. Epizootiologic Investigations of Parvovirus Infections in Free-ranging Carnivores from Germany. *J. Wildl. Dis.* **2005**, *41*, 231–235. [CrossRef]
49. Knobel, D.L.; Butler, J.R.A.; Lembo, T.; Critchlow, R.; Gompper, M.E. Dogs, disease, and wildlife. In *Free-Ranging Dogs and Wildlife Conservation*; Gompper, M.E., Ed.; Oxford University Press: Oxford, UK, 2013; pp. 144–169. ISBN 978-0-19-966321-7.
50. Nelson, B.; Hebblewhite, M.; Ezenwa, V.; Shury, T.; Merrill, E.H.; Paquet, P.C.; Schmiegelow, F.; Seip, D.; Skinner, G.; Webb, N. Prevalence of antibodies to canine parvovirus and distemper virus in wolves in the Canadian Rocky Mountains. *J. Wildl. Dis.* **2012**, *48*, 68–76. [CrossRef]
51. Courtenay, O.; Quinnell, R.J.; Chalmers, W.S.K. Contact rates between wild and domestic canids: no evidence of parvovirus or canine distemper virus in crab-eating foxes. *Vet. Microbiol.* **2001**, *81*, 9–19. [CrossRef]
52. Houston, D.M.; Ribble, C.S.; Head, L.L. Risk factors associated with parvovirus enteritis in dogs: 283 cases (1982-1991). *J. Am. Vet. Med. Assoc.* **1996**, *208*, 542.
53. Ling, M.; Norris, J.M.; Kelman, M.; Ward, M.P. Risk factors for death from canine parvoviral-related disease in Australia. *Vet. Microbiol.* **2012**, *158*, 280–290. [CrossRef]
54. Mech, L.D.; Goyal, S.M.; Paul, W.J.; Newton, W.E. Demographic effects of canine parvovirus on a free-ranging wolf population over 30 years. *J. Wildl. Dis.* **2008**, *44*, 824–836. [CrossRef] [PubMed]
55. McNeill, A.; Leung, L.; Goullet, M.; Gentle, M.; Allen, B. Dingoes at the Doorstep: Home Range Sizes and Activity Patterns of Dingoes and Other Wild Dogs around Urban Areas of North-Eastern Australia. *Animals* **2016**, *6*, 48. [CrossRef] [PubMed]
56. Allen, B.L.; Goullet, M.; Allen, L.R.; Lisle, A.; Leung, L.K.-P. Dingoes at the doorstep: Preliminary data on the ecology of dingoes in urban areas. *Landsc. Urban Plan.* **2013**, *119*, 131–135. [CrossRef]
57. Gordon, J.C.; Angrick, E.J. Canine parvovirus: environmental effects on infectivity. *Am. J. Vet. Res.* **1986**, *47*, 1464.
58. Almberg, E.S.; Mech, L.D.; Smith, D.W.; Sheldon, J.W.; Crabtree, R.L. A Serological Survey of Infectious Disease in Yellowstone National Park's Canid Community. *PLoS ONE* **2009**, *4*, e7042. [CrossRef]
59. Santos, N.; Almendra, C.; Tavares, L. Serologic Survey for Canine Distemper Virus and Canine Parvovirus in Free-ranging Wild Carnivores from Portugal. *J. Wildl. Dis.* **2009**, *45*, 221–226. [CrossRef] [PubMed]
60. Sobrino, R.; Arnal, M.C.; Luco, D.F.; Gortázar, C. Prevalence of antibodies against canine distemper virus and canine parvovirus among foxes and wolves from Spain. *Vet. Microbiol.* **2008**, *126*, 251–256. [CrossRef] [PubMed]

Communication

Phylogenetic, Evolutionary and Structural Analysis of Canine Parvovirus (CPV-2) Antigenic Variants Circulating in Colombia

Sebastián Giraldo-Ramirez, Santiago Rendon-Marin and Julián Ruiz-Saenz *

Grupo de Investigación en Ciencias Animales—GRICA, Facultad de Medicina Veterinaria y Zootecnia, Universidad Cooperativa de Colombia, Sede Bucaramanga 680005, Colombia; sebastian.giraldor2@udea.edu.co (S.G.-R.); santiago.rendonm@udea.edu.co (S.R.-M.)

* Correspondence: julianruizsaenz@gmail.com or julian.ruizs@campusucc.edu.co

Received: 10 April 2020; Accepted: 24 April 2020; Published: 30 April 2020

Abstract: Canine parvovirus (CPV-2) is the causative agent of haemorrhagic gastroenteritis in canids. Three antigenic variants—CPV-2a, CPV-2b and CPV-2c—have been described, which are determined by variations at residue 426 of the VP2 capsid protein. In Colombia, the CPV-2a and CPV-2b antigenic variants have previously been reported through partial VP2 sequencing. Mutations at residues Asn428Asp and Ala514Ser of variant CPV-2a were detected, implying the appearance of a possible new CPV-2a variant in Colombia. The purpose of the present study was to characterise the full VP2 capsid protein in samples from Antioquia, Colombia. We conducted a cross-sectional study with 56 stool samples from dogs showing clinical symptoms of parvoviral disease. Following DNA extraction from the samples, VP2 amplification was performed using PCR and positive samples were sequenced. Sequence and phylogenetic analyses were performed by comparison with the VP2 gene sequences of the different CPV-2 worldwide. VP2 was amplified in 51.8% of the analysed samples. Sequencing and sequence alignment showed that 93.1% of the amplified samples belonged to the new CPV-2a antigenic variant previously. Analysing the amino acid sequences revealed that all CPV-2a contain Ala297Asn mutations, which are related to the South America I clade, and the Ala514Ser mutation, which allows characterization as a new CPV-2a sub-variant. The Colombian CPV-2b variant presented Phe267Tyr, Tyr324Ile and Thr440Ala, which are related to the Asia-I clade variants. The CPV-2c was not detected in the samples. In conclusion, two antigenic CPV-2 variants of two geographically distant origins are circulating in Colombia. It is crucial to continue characterising CPV-2 to elucidate the molecular dynamics of the virus and to detect new CPV-2 variants that could be becoming highly prevalent in the region.

Keywords: antigenicity; sequencing; virus

---

1. Introduction

One of the main infectious agents that affect the canine population is the canine parvovirus type 2 (CPV-2), which causes acute haemorrhagic diarrhoea, primarily in puppies [1]. CPV-2 is a single-stranded DNA virus that belongs to the genus *Protoparvovirus* [2], with a genome of approximately 5200 nucleotides that contain two open reading frames (ORFs). The ORF 3′ codes for non-structural proteins NS1 and NS2 are important in controlling viral replication and assembly. The ORF 5′ codes for structural proteins are termed viral protein 1 (VP1) and 2 (VP2). VP2 is the major component of the viral capsid, which has an icosahedral structure of 60 subunits with a T = 1 symmetry, comprising approximately 4–5 copies of VP1 and 54–55 copies of VP2 [3].

VP2 subunits interact with transferrin receptors (TfRs) present on the outer surface of the cell membrane [4]. Subsequently, the absorption to the host cell is facilitated by clathrin-mediated endocytosis [5]. The VP2 capsid protein plays a crucial role because its mutations determine the antigenic changes that originate in the different antigenic CPV-2 variants [4].

It is believed that CPV-2 is derived from the feline panleukopenia virus (FPLV) [6], where specific mutations at Lys80Arg, Lys93Asn, Val103Ala, Asp323Asn, Asn564Ser and Ala568Gly capsid protein VP2 residues facilitated a change of host, thereby allowing the virus to infect canines and losing the ability to infect felines [7,8]. These mutations originated in the CPV-2 variant, firstly reported in the 1970s, and spread to countries in Europe, America, Asia and Oceania by 1978 [6].

By 1982, CPV-2 was replaced by a variant of the virus that genetically and antigenically differed [9]. This variant was called CPV-2a and differs from CPV-2 in 6 amino acids: Met87Leu, Ile101Thr, Ser297Ala, Ala300Gly, Asp305Tyr and Val555Ile residues [10].

Residue 426 of VP2 is located in the outermost part of the threefold axis, where 3 VP2 subunits converge. It is the site of greatest antigenicity of the virus [11]; therefore, the amino acid variation causing the antigenic changes that led to the origin of the CPV-2b antigenic variants (Asn426Asp) was reported in 1984 [10] and that for CPV-2c (Asp426Glu) was reported in Italy in 2001 [12]. Unlike CPV-2, CPV-2a, CPV-2b and CPV-2c variants infect canines as well as regaining the ability to infect felines and other wild carnivores [8,13]

In 2017, the antigenic CPV-2a and CPV-2b variants were reported via the study of a partial region of VP2 in Colombia [14]. The CPV-2c variant was not detected in Colombia, despite being the most prevalent antigenic variant in South America [15] and despite Colombia sharing borders with countries such as Peru, Ecuador and Brazil, where the variant has previously been reported [16]. Additionally, two mutations were observed in the Colombian CPV-2a variants (Asn428Asp and Ala514Ser) that suggest the emergence of a new CPV-2a variant unique in Colombia [14].

The aim of the present study was to characterise the coding region of the entire VP2 capsid protein of the circulating parvovirus variants in Antioquia (north-western Colombia) for determining the total amino acid variations in VP2 and to obtain information regarding the molecular evolution of the CPV-2 in Colombia.

2. Materials and Methods

2.1. Patient Selection and Sampling

A cross-sectional study was conducted with a convenience sampling of canine patients attending different veterinary clinics of the department of Antioquia and reporting clinical symptoms compatible with canine parvovirus, such as: haemorrhagic diarrhoea, dehydration, vomiting, loss of appetite and weakness. The samples were collected from February 2018 to March 2019. Gender, breed, age, and vaccination status of each patient was registered. No previous confirmation nor positive faecal antigen ELISA were required for any patient. Diagnosis was based on typical clinical signs. Faecal samples of each animal were collected (approximately 5 g) and stored under freezing conditions (−80 °C) until further use; prior authorisation was obtained from the owner of the animals that met the inclusion criteria. This study was approved by the Bioethics Committee of the Universidad Cooperativa de Colombia (project INV1473–bioethics Acta 002-2016). An informed consent was obtained from all owners.

2.2. DNA Extraction and Quantification

Viral DNA extraction from the collected stool samples was performed using QIAamp DNA fast stool mini kit (Qiagen®, Hilden, Germany), in accordance with the manufacturer's instructions. DNA obtained was quantified using 1 µL of the product with NanoDrop 2000 (Thermo Fisher Scientific®, Waltham, MA, USA).

2.3. VP2 Amplification Using PCR

To identify CPV-2-positive samples, PCR amplification of VP2 was conducted using the conventional method. For each PCR reaction, 25 μL of DreamTaqTM PCR Master Mix (2×) (Thermo Fisher Scientific®) was used, with 4 μL of the Forward Ext1F primer (5′-ATGAGTGATGGAGCAGTTCA-3′) and 4 μL of the Ext3R Reverse primer (5′-AGGTGCTAGTTGAGATTTTTCATATAC-3′) described by [17] and 17 μL of a mixture of DNA and molecular grade water, reaching an amount of 500 ng DNA for each reaction. Molecular grade water was used as a negative control for amplifications. PCR protocol used was as follows: an initial denaturation cycle at 94 °C for 5 min, 35 denaturation cycles at 94 °C for 30 s, alignment at 50 °C for 45 s, extension at 72 °C for 1 min and a final extension cycle at 72 °C for 5 min.

PCR amplification results were visualised using 1.5% horizontal agarose gel electrophoresis. Gels were stained with the Invitrogen ™ SYBR®Safe DNA Gel Stain (Thermo Fisher Scientific®). In each well, 4.2 μL of each sample obtained after amplification and 0.8 μL of the 6× DNA loading buffer were used, and the GeneRuler™ 100-bp DNA Plus Ladder (Thermo Fisher Scientific®) was used as a molecular weight marker. Gels were developed in the ultraviolet light Gel Doc™ XR+ imaging system (Bio-Rad, Molecular imager®, USA) and were visualised using ImageLab™ software.

2.4. Sequencing and Sequence Analysis

Samples positive for VP2 amplification after electrophoresis visualisation were purified and sequenced at Macrogen Inc. (Seoul, Korea) using Forward Ext1F and Reverse Ext 3R primers, along with a set of internal sequencing primers to amplify the entire VP2 [18]—F1: 5′-AGATAGTAATAATACTATGCCATTT-3′, F2: 5′-ACAGGAGAAACACCTGAGAGATTTA-3′, R1: 5′-TGGTTGGTTTCCATGGATA-3′, and R2: 5′-TTTTGAATCCAATCTCCTTCTGGAT-3′. The resulting electropherograms from the sequencing were analysed using Chormas™ v. 2.6 software. Contig generation, resulting from the overlapping of the sequences amplified by the primers, was performed on the SeqMan Pro platform with Lasergene™. After constructing the complete nucleotide sequences for each sample, alignment was performed using the ClustalW method, following which these sequences were compared with the DNA sequences of CPV-2a, CPV-2b and CPV-2c obtained from GenBank. All analyses were performed in the MEGA™ 7.0 software for Windows®.

2.5. Phylogenetic Analysis

For the phylogenetic analysis, we calculated the best nucleotide substitution model for the dataset generated with the sequences of FPLV, CPV-2, CPV-2a, CPV-2b and CPV-2c obtained from GenBank. Phylogenetic analysis was inferred using distance-based (neighbor-joining) and character based (maximum likelihood—Bayesian) approaches implemented in MEGA™ 7.0 and Mr Bayes™ software for Windows®. The nucleotide replacement model selected was Tamura-3 parameter with Gamma distributed rate and invariant sites (T92+G+I) and then the Markov chain Monte Carlo (MCMC) Bayesian analysis. We ran the MCMC searches for 1,000,000 generations. The TRACER™ v1.7.1 software was used to confirm all the parameters generated in the Bayesian analysis. The effective sample size was up to 200. The FigTree v1.4.3 software was used to display the consensus phylogenetic tree generated after Bayesian analysis.

2.6. Evolutionary Analysis

The construction of the phylogenetic evolutionary tree required the generation of a dataset containing the sequences obtained from GenBank and the Colombian CPV-2a and CPV-2b samples. The evolutionary rates, time to the most recent common ancestor (tMRCA) and geographic movements of CPV-2 were performed using the BEAST v1.8.4 software package. The phylogenetic evolutionary tree was generated according to the Hasegawa, Kishino and Yano nucleotide substitution model + gamma distribution + invariable sites (HKY + G + I) and a strict molecular clock. The Bayesian stochastic search variable selection was used to determine links between sequences. The length of the Markov chain Monte Carlo chain was 15 million. All parameters generated in the analysis were confirmed by verifying the effective sample size of >200 using the TRACER v1.7.1 software. With TreeAnnotator v1.8.4 software, 10% of the steps (1.5 million burn-in) were eliminated to obtain the tree with the most credible clades. FigTree v1.4.3 software was used to display the generated tree.

2.7. Structural Analysis

The VP2 tertiary structure construction and surface analysis was performed using the sequences of sample 1 (CPV-2a) and sample 50 (CPV-2b) as references. Homology modeling of the CPV-2a and CPV-2b sequences under study was generated using MODELLER software based on PDB: 1C8D structure. The three-dimensional models of the VP2 tertiary structure were created using the PyMOL™ software. VP2 surface analysis was performed using the RIVEM (Radial Interpretation of Viral Electron density Maps) software that facilitates the generation of a 'road map' of the viral surface and determination of the locations of the amino acids that constitute the analysed structure [19].

2.8. Analysis of Selection Pressure

The relationship of non-synonymous (dN) to synonymous (dS) substitutions was calculated using ML phylogenetic reconstruction and the general reversible nucleotide substitution model available through the web program Datamonkey. To detect non-neutral selection, the Fast, Unconstrained Bayesian AppRoximation (FUBAR) was implemented in the Datamonkey program. The values dN/dS > 1, dN/dS = 1, and dN/dS <1 were used to define positive selection (adaptive molecular evolution), neutral mutations, and negative selection (purification selection), respectively. Also, we implemented FEL (Fixed Effects Likelihood) which uses a ML approach to infer nonsynonymous (dN) and synonymous (dS) substitution rates on a per-site basis for a given coding alignment and corresponding phylogeny and permits the identification of positive selection at specific sites along particular clades.

2.9. Statistical Analyses

A descriptive analysis of the collected data was performed representing the qualitative and quantitative variables in tables and graphs.

3. Results

For this study, a total of 56 faecal matter samples were collected from dogs that presented clinical symptoms compatible with canine parvovirus, from different veterinary centres located in the department of Antioquia, Colombia.

PCR amplification of VP2 showed that a total of 29 samples (51.8%) were CPV-2-positive. Of these samples, a total of 41.4% (n = 12) belonged to females and 58.6% (n = 17) to males. Approximately 55.1% of the positive samples belonged to mixed-breed animals. According to the data provided at the time of sample collection, when visiting the veterinary centres with symptoms compatible with those of parvovirus infection, 20.7% (n = 6) of the animals had undergone a complete vaccination schedule as required by age, whereas 44.8% (n = 13) did not comply with the vaccination schedule. The remaining 34.5% (n = 10) of the animals presented incomplete vaccination schedules (Table 1).

Table 1. Information on Canine Parvovirus (CPV-2)-positive samples included in the present study.

Sample	Variant	Age	Sex	Race	Vaccination
1	CPV-2a	4 months	Male	Mixed-breed	Without vaccination
5	CPV-2a	3 months	Female	French Bull Dog	Incomplete
7	CPV-2a	1 month	Female	Mixed-breed	Without vaccination
17	CPV-2a	3 months	Female	Golden Retriever	Incomplete
19	CPV-2a	2 months	Female	French Bulldog	Incomplete
20	CPV-2a	2 months	Female	French Bulldog	Incomplete

Table 1. *Cont.*

Sample	Variant	Age	Sex	Race	Vaccination
21	CPV-2a	2 months	Male	French Bulldog	Incomplete
24	CPV-2a	2 months	Female	Beagle	Without vaccination
26	CPV-2a	3 months	Female	Mixed-breed	Without vaccination
29	CPV-2a	2 months	Male	Mixed-breed	Without vaccination
32	CPV-2a	9 months	Female	Pinscher	Without vaccination
33	CPV-2a	2 months	Male	Cocker Spaniel	Incomplete
36	CPV-2a	2 months	Male	Siberian Husky	Without vaccination
37	CPV-2a	7 months	Male	Mixed-breed	Incomplete
38	CPV-2a	6 months	Male	Cocker Spaniel	Complete
40	CPV-2a	4 months	Female	Mixed-breed	Incomplete
41	CPV-2a	4 months	Female	Mixed-breed	Incomplete
43	CPV-2a	9 months	Male	Mixed-breed	Incomplete
44	CPV-2a	3 months	Male	French Bulldog	Incomplete
45	CPV-2a	12 months	Female	Mixed-breed	Without vaccination
47	CPV-2a	7 months	Male	Mixed-breed	Without vaccination
48	CPV-2a	6 months	Male	Mixed-breed	Complete
50	CPV-2b	9 months	Male	Mixed-breed	Incomplete
51	CPV-2a	2 months	Male	Mixed-breed	Without vaccination
52	CPV-2a	2 months	Female	Mixed-breed	Incomplete
53	CPV-2a	4 months	Male	Mixed-breed	Without vaccination
54	CPV-2a	24 months	Male	Mixed-breed	Without vaccination
55	CPV-2b	3 months	Male	German Shepherd	Without vaccination
56	CPV-2a	7 months	Male	Shih Tzu	Incomplete

According to age distribution, 15 positive samples (51.7%) belonged to animals aged 1–3 months, which was the most prevalent age group for the virus, whereas only 1 sample (3.4%) belonged to an animal aged >1 year (Figure 1).

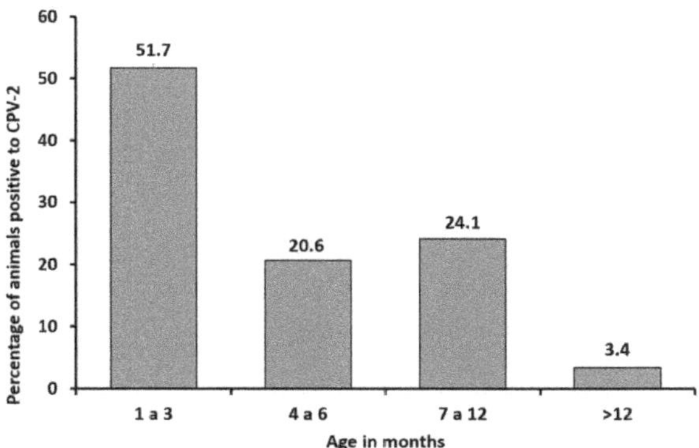

Figure 1. Distribution of CPV-2-positive samples by animal's age (in months).

3.1. *Sequence Analysis*

An almost full sequence of 1711 nucleotides of the VP2 gene was achieved. Sequencing analysis showed that the antigenic variants present in the study are CPV-2a and CPV-2b. The CPV-2a variant (Asn426) was present in 27 from the positive samples (93.1%) and was the most prevalent variant in the study, whereas the CPV-2b variant (Asp426) was detected in two samples (6.9%). No evidence was obtained regarding the presence of the CPV-2c variant (Glu426) in any sample (Table 1).

Sequence analysis revealed specific mutations with respect to the reference sequences of CPV-2a, CPV-2b and CPV-2c obtained from GenBank. The Colombian CPV-2a variants showed mutations at the amino acid residues Ala297Asn, Tyr324Ile and Ala514Ser, with the latter being reported for the first time in Colombia in 2017 [14]. Regarding the CPV-2b variants in the present study, the amino acid variations detected were Phe267Tyr and Thr440Ala. Similar to the CPV-2a variants, the CPV-2b sequenced from the sampling showed the Tyr324Ile mutation; however, a single Ala514Ser mutation was not detected in these samples (Table 2).

Table 2. Amino acid variations in the samples identified as CPV-2a and CPV-2b in relation to reference variants (in bold).

Sample	Variant	Amino Acid Position					
		267	297	324	426	440	514
MF177231	**CPV-2a**	**F**	**A**	**Y**	**N**	**T**	**A**
EU659119	**CPV-2b**	**F**	**A**	**Y**	**D**	**T**	**A**
MF177238	**CPV-2c**	**F**	**A**	**Y**	**E**	**T**	**A**
1	CPV-2a	F	N	I	N	T	S
5	CPV-2a	F	N	I	N	T	S
7	CPV-2a	F	N	I	N	T	S
17	CPV-2a	F	N	I	N	T	S
19	CPV-2a	F	N	I	N	T	S
20	CPV-2a	F	N	I	N	T	S
21	CPV-2a	F	N	I	N	T	S
24	CPV-2a	F	N	I	N	T	S
26	CPV-2a	F	N	I	N	T	S
29	CPV-2a	F	N	I	N	T	S
32	CPV-2a	F	N	I	N	T	S
33	CPV-2a	F	N	I	N	T	S
36	CPV-2a	F	N	I	N	T	S
37	CPV-2a	F	N	I	N	T	S
38	CPV-2a	F	N	I	N	T	S
40	CPV-2a	F	N	I	N	T	S
41	CPV-2a	F	N	I	N	T	S
43	CPV-2a	F	N	I	N	T	S
44	CPV-2a	F	N	I	N	T	S
45	CPV-2a	F	N	I	N	T	S
47	CPV-2a	F	N	I	N	T	S
48	CPV-2a	F	N	I	N	T	S
50	CPV-2b	Y	A	I	D	A	A
51	CPV-2a	F	N	I	N	T	S
52	CPV-2a	F	N	I	N	T	S
53	CPV-2a	F	N	I	N	T	S
54	CPV-2a	F	N	I	N	T	S
55	CPV-2b	Y	A	I	D	A	A
56	CPV-2a	F	N	I	N	T	S

3.2. Phylogenetic Analysis

The phylogenetic relationships based on the nucleotide alignment of VP2 Sequences inferred by distance (neighbor joining) and character approaches (maximum likelihood and Bayesian inference) resulted in trees with a similar topology. The phylogenetic tree generated by Bayesian inference included the CPV-2 positive samples representing the antigenic variants detected in the study (CPV-2a and CPV-2b). GenBank accession numbers for Colombian nucleotide sequences are: MT152347 to MT152375. Additionally, these samples exhibited differences in the pairwise distances. Representative samples of FLPV, CPV-2, CPV2a, CPV-2b and CPV-2c from different countries reported in GenBank were used for phylogenetic evolutionary tree construction. The Colombian CPV-2a antigenic variant constituted a monophyletic clade that substantially differs from the European CPV-2a antigenic variants as well as from Uruguayan variants and only shared distribution with an Ecuadorian CPV-2a variant (MG264075). The two samples of the study from the antigenic CPV-2b variant were located within a clade that includes the Uruguayan CPV-2a variants and Asian CPV-2b variants (Figure 2).

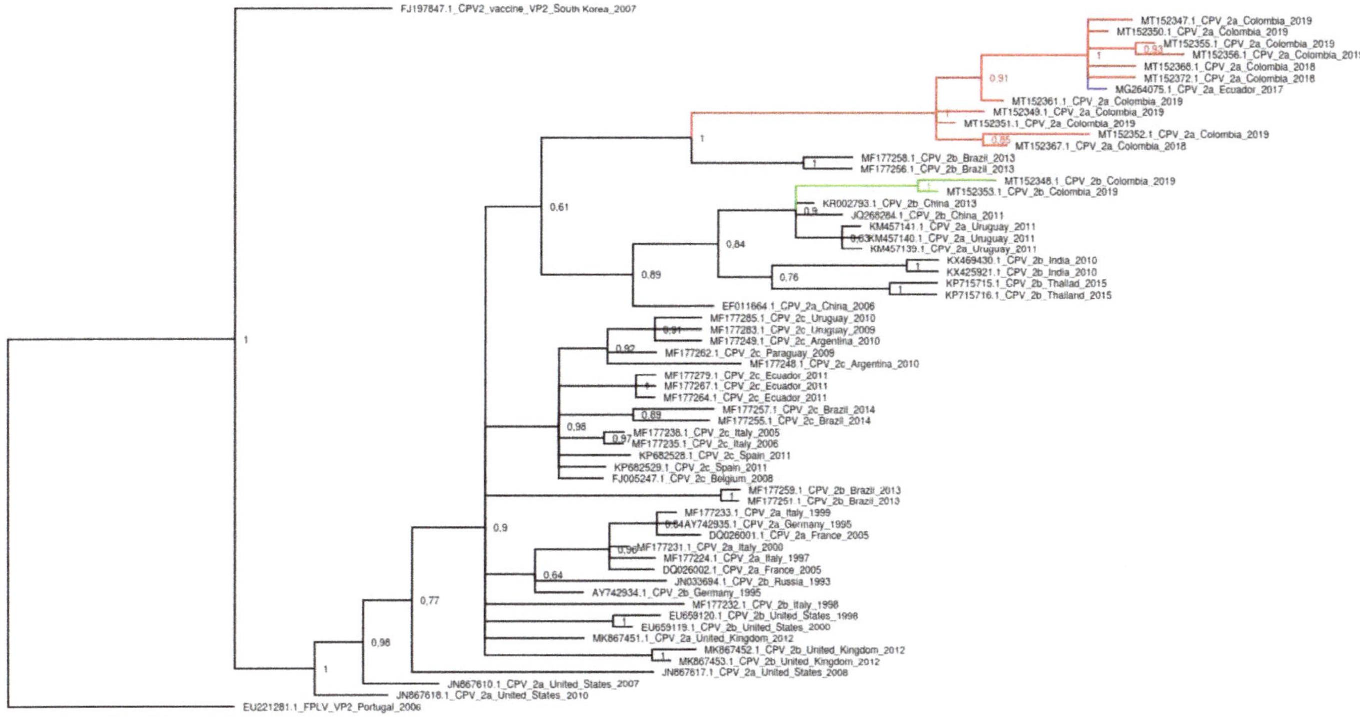

Figure 2. Phylogenetic Bayesian analysis of CPV-2 sequences. The phylogenetic analysis was performed using the nucleotide sequences of VP2 of the Colombian CPV-2a and CPV 2b variants and other sequences belonging to the three antigenic CPV-2a, CPV-2b and CPV-2c variants from different countries, as well as feline panleukopenia virus (FPLV) and original CPV-2 variants. The sequences are identified with the accession number, country, and date of collection. Tree was constructed with Bayes and every clade was supported by Bayesian posterior probabilities (BPP). The tree was rooted with the sequence of FPLV (EU221281.1). Red lines indicate the sequences belonging to the Colombian CPV-2a variant and the blue line indicates that belonging to the Ecuadorian CPV-2a variant related to Colombian samples. Green lines indicate the Colombian CPV-2b variant.

In 2017, by using partial VP-2 sequences, the presence of "new" CPV-2 variants in Colombia was shown [14]. To confirm this, we performed maximum likelihood phylogenetic analysis using VP-2 partial sequences, including those that showed the change Ala514Ser in CPV-2a and the sequences belonging to CPV-2b (Supplementary Figure S1). The analysis reveals that the CPV-2a variants of both studies have strong similar sequences. The CPV-2a sequences from 2017 and current analysis showed minor nucleotide differences; however, they turn out to be synonymous variations that do not represent amino acid changes, and for this reason they are all in the same clade. Regarding the CPV-2b sequences, it can also be seen that they are very similar and are located in a single well differentiated clade. Clade 2b sequences show variations at the amino acid level. CPV-2b 2019 sequences (M50 and M55) are located on a separate branch. This subdivision is due to the fact that only the CPV-2b 2019 sequences present the Thr440Ala mutation, one of the mutations reported in the present work (Supplementary Figure S1).

3.3. Evolutionary Analysis

For the evolutionary analysis, the same sequences of samples used in the phylogenetic analysis and a dataset of the sequences of CPV-2a, CPV-2b and CPV-2c variants from different countries were used. The estimated tMRCA for the phylogenetic evolutionary tree was generated 40 years ago (1979), based on the most recent analysed sequence (2019), with a 95% highest probability density and a range of 38–44 years. The analysis revealed that the Colombian CPV-2a and CPV-2b variants have their common ancestor in sequences in Italy (Figure 3), from which a branch was identified that originated a clade with South American sequences (Uruguay, Argentina and Brazil) derived ca.1990; the Colombian CPV-2a variant is located in this clade. Another branch deriving from the sequences from Italy originated a clade with sequences from Asia (China and India) and Uruguay around 1994, and the CPV-2b variant is located in this clade.

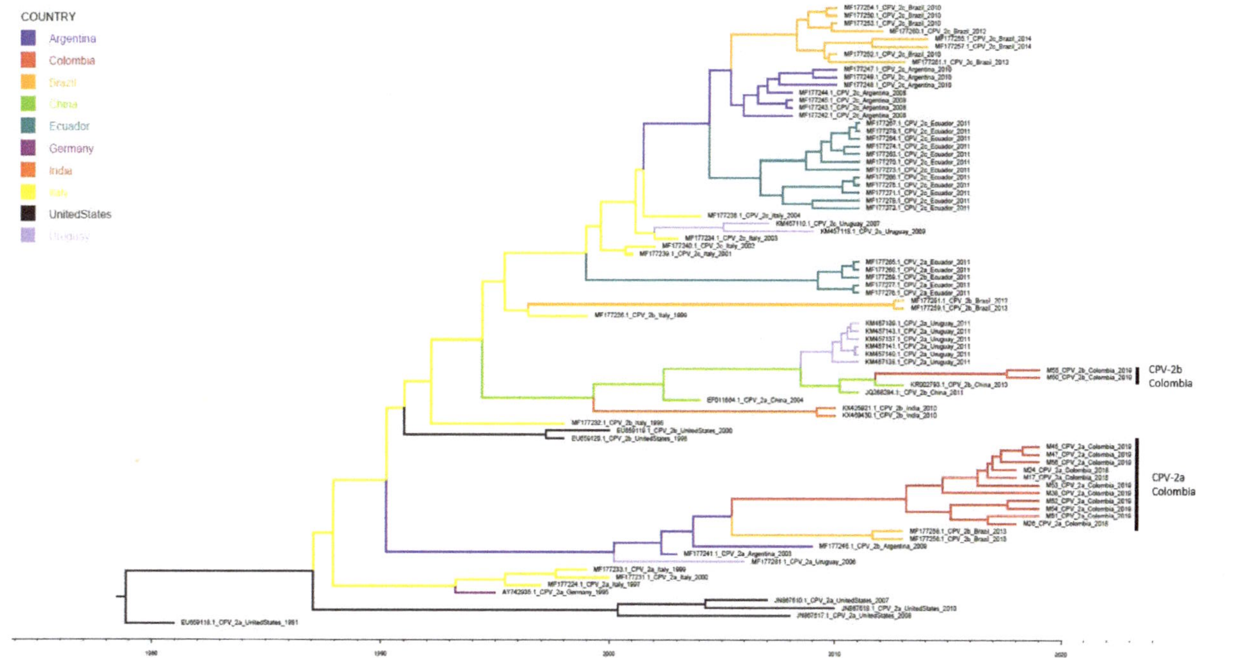

Figure 3. Phylogenetic evolutionary tree of Colombian CPV-2 variants. The tree was generated using CPV-2a, CPV-2b and CPV-2c variants from different countries. The sequences are identified with the accession number, country, and date of collection. The timeline shows the moment of evolutionary divergence of the sequences of each country. The common ancestor for both the Colombian CPV-2a and CPV-2b variants, in red, are sequences of Italian origin, represented in yellow. The Italian variants originated a clade with South American variants in 1990. In 2006, there is a divergence in the Argentine variants that originated the Brazilian variants and the Colombian CPV-2a variant. In 1994, a divergent branch appears that originated the Asian variants. In this clade, there are Chinese and Indian variants. Additionally, Uruguayan sequences related to the Chinese variants are found. In 2012, there is a divergence from the Chinese origin variants that originated the Colombian CPV-2b variant.

3.4. Structural Analysis

The three-dimensional reconstruction of the VP2 tertiary structure reveals the spatial location of the mutations detected by sequencing for the Colombian CPV-2a and CPV-2b variants. The CPV-2a variant exhibited the Ala297Asn, Tyr324Ile and Ala514Ser mutations, with all mutations being detected in the exposed VP2 regions. The amino acids 297Asn and 324Ile were located in the loop 3, whereas 514Ser was located in a less prominent region. The mutations found in the Colombian CPV-2a sequences are located at the region that divides the barrel and the depression of the 2-fold axis, an important site for the interaction between VP2 and the TfR receptor. The CPV-2b variant exhibited the Phe267Tyr, Tyr324Ile and Thr440Ala mutations. In the three-dimensional VP2 reconstruction, 267Tyr has been found to be located in an unexposed area in the internal structure of the protein. By contrast, 324Ile and 440Ala were detected in more prominent regions of the protein (Supplementary Figure S2).

3.5. Sites under Positive Selection

The positive selection sites were evaluated using the FUBAR and FEL methods. In CPV-2a variant, we found two sites under positive selection: 297 and 324 with a posterior probability of 0.9 and a Bayes factor of 48.8 and 39.4, respectively. In CPV-2b variant we found two positive selection sites 324 and 440 with a posterior probability of 0.9 and a Bayes factor of 39.4 and 53.9, respectively. As previously reported, both variants showed that the 426 amino acid also had a positive selection (Supplementary Table S1). By FEL, we also analysed the positive selection between clades and we found that Ala514Ser in the Colombian new CPV-2a had strong positive selection ($p = 0.033$).

4. Discussion

Despite the wide distribution of vaccination strategies for CPV-2 and the knowledge obtained from its evolution, CPV-2 remains a viral pathogen that has the greatest of impacts on animal health [20]. In our study, 51.7% of the samples were CPV-2-positive demonstrating that canine parvovirus is the causative agent for most cases of haemorrhagic gastroenteritis in canines [21]. The new CPV-2a variant was detected in 93.1% of the samples, indicating that this new CPV-2a variant previously reported [14] has gradually become the most prevalent virus in Colombia and this site is under positive selection.

Although the youngest animals—aged between 1 and 3 months—represent the population most affected by CPV-2 infection, consistent with the usual presentation of this infection [21], the animal population aged between 7 and 12 months showed a high rate of CPV-2-positive cases. However, animals belonging to the group of mature puppies (aged 7–12 months) showed an incomplete vaccination schedule or absence of any vaccination history (Table 1), rendering them susceptible to CPV-2 because they lack acquired immunity. Only one animal in the sample was aged >12 months; however, this animal had no vaccination history. Despite this unconventional finding, this serves as a starting point to demonstrate the importance and infectious potential of CPV-2 in immunologically mature animals, even in animals that have received a complete vaccination schedule, indicating that the mutational ability of the virus can result in immune response evasion [22].

Amino acid sequence analysis revealed important changes in the Colombian CPV-2b (Table 2). The Phe267Tyr change has been reported since the early 2000s in Asian CPV-2b variants [23] and in the Uruguayan CPV-2a antigenic variant [24]. Since its detection, this mutation has consistently appeared in the sequences reported in different studies and has become predominant in the CPV-2 population since 2014, suggesting that this mutation has a positive selection. Although this amino acid was not detected in an exposed area of antigenic change site, the fixation of this mutation reflects an evolutionary advantage for CPV-2 that has not yet been elucidated [6,15].

The Tyr324Ile mutation observed in our CPV-2b variant (Table 2) could be related to the ability of the virus to infect different hosts. The residue 324 is adjacent to the residue 323 and, in combination with the amino acid residue 93 of VP2, these residues reportedly exert effects on the host change of the CPV-2 because they are involved in the binding of the virus with TfR [4,25,26]. This amino acid is found in the loop 3 [15], a moderately exposed VP2 region, which is a part of the 'shoulder' of the structure that forms the threefold axis, a site of greater antigenic importance [11]. This mutation was first reported in Asia [27,28], and similar to residue 267, there exists a relationship with Uruguayan variants [24], where this mutation has been reported.

Another mutation observed in our samples regarding the reference variants was Thr440Ala (Table 2). This residue is found in the loop 4 in the most prominent region of the viral capsid—the threefold axis [29]. Therefore, mutations in this region could greatly impact antigenic changes that have implications on the host immune response [24]. Considering that this mutation has undergone a strong positive selection, identifying this mutation in different populations is possible as they underwent an independent evolution [30]. This mutation was first reported in 1993, and it has consistently occurred in the viral populations of CPV-2 since 2005 [15].

It is evident that our findings regarding the Colombian CPV-2b variant are related to the Uruguayan CPV-2a variant. In both cases, these mutations are observed in the same amino acid residues (267, 324 and 440), although they differ in 426. According to the phylogenetic analysis, both the Colombian CPV-2b and Uruguayan CPV-2a variants are present in the same clade as the Asian CPV-2b variants (Figure 2). Based on the study by Grecco et al. [16], the Uruguayan CPV-2a variant originated from the Asian variants, and they named this group the Asia I clade. This clade originated in Asia in the late 80s and arrived in South America between 2009 and 2010, when the dissemination of a divergent CPV-2a variant was reported in Uruguay, where the predominant population at that time was CPV-2c [24].

In the phylogenetic evolutionary analysis (Figure 3), the Colombian CPV-2b variant, although appearing in the same clade as the Uruguayan CPV-2a variant, has a direct relationship with the Chinese variants rather than with the Uruguayan ones. This is possibly attributable to the Uruguayan variants being CPV-2a, whereas in Asia, CPV-2b variants have been reported to have the same mutations as the Colombian CPV-2b sequences [31]. This relationship with the Asiatic variants has been well described recently in Italy, where a case of transcontinental spread of CPV-2c variant was identified with mutation Tyr324Ile and Phe267Tyr characteristically from Asiatic variants [32]. Further, in recent years a local spread of Asiatic-like CPV-2c variant in Italy has been reported [33]. Just as in Uruguay, this spread in Italy demonstrates the intercontinental dissemination of Asiatic CPV-2 variants being able to reach Colombia in a similar way. Structural analysis of CPV-2b has shown that the 440Ala mutation is located in a zone of antibody neutralization. Similarly, it is contiguous to the capsid interaction domain with the TfR cell receptor [34].

Regarding the Colombian CPV-2a, an Ala297Asn mutation was shown that has been reported in South American countries, such as Brazil, Uruguay and Argentina [16]. Residue 297 is located at medium exposure zone in the structure of the viral capsid [11] at a site of lower antigenicity and is not located in the most prominent place in the structure [6]. The Colombian CPV-2a-positive and CPV-2b samples contain the Tyr324Ile mutation in the present study, indicating that equivalent changes can occur in different antigenic variants.

The amino acid residue 514 showed the Ala → Ser mutation, which was first reported in Colombia [14] in CPV-2a variants and occurred in 66% of the samples evaluated. However, in our study, 100% of the samples belonging to the new antigenic CPV-2a variant were reported. Surface structure analysis of the virus revealed the spatial position of these mutations. Amino acids 297, 324 and 514 are located in a central region of VP2 (between the depression of the two-fold axis and the canyon). This area is critical for the interaction between the virus and the TfR receptor. The mutations in 297 and 324 are immediately adjacent to the region determined for the coupling of the receptor on the surface of the virus [35]. The mutation in 514 is distant from the virus-receptor interaction zone (Supplementary Figure S2); however, it is located at the antibody neutralization areas [34]. It is possible that these mutations favour the union between the virus with TfR receptor or may represent a change in the structure of VP2 that allows avoidance of neutralization by antibodies (514Ser) which, added to the fact of being under positive selection, can help explain the reason for this mutation becoming predominant in Antioquia by replacing the viruses that lack these mutations. Interestingly, these same mutations have been reported in Ecuador [36] in a single sample belonging to the CPV-2a antigenic variant, which is closely related to the Colombian CPV-2a variant, highlighting the need to continue the genotypic surveillance of CPV-2 variants circulating in the region due to the possible appearance of new genotypic variants with different possible pathogenic or antigenic potentials.

In agreement with these amino acid variations in the CPV-2a variants of the present study and based on phylogenetic analysis, it is possible that both the Colombian and Ecuadorian CPV-2a samples belong to the clade called South America I. This clade has the peculiarity of containing both South American CPV-2a and CPV-2b variants with the Ala297Asn mutation [16]. The phylogenetic tree presented in Figure 2 shows a well differentiated large clade where all our CPV-2a samples are grouped along with the Ecuadorian CPV-2a. These samples share an identical amino acid sequence, although they differ in some nucleotide level changes that are synonymous variations.

Similarly, in our phylogenetic evolutionary tree construction, Brazilian CPV-2b variants, which contains the Ala297Asn mutation and is present within the clade known as South America I, are most closely related to CPV-2a variants. Although the mutation in 297 groups our CPV-2a variants with variants belonging to the South America I clade, the unique mutations in 324 and 514 differentiate our samples from the others belonging to this clade. This suggests that the Colombian CPV-2 sequences configure a Colombian CPV-2a sub-variant within the CPV-2a variants belonging to the South American clade I, as postulated by Duque-García et al. [14].

To understand the evolutionary importance of the mutations found in CPV-2a and CPV-2b variants, an analysis was performed to determine the positive selection sites (Supplementary Table S1). Our results confirmed for CPV-2b, that sites 324 and 440 have a strong positive selection, as previously reported in Asian CPV-2 variants [37,38], which are related to the Colombian CPV-2b variant. In the case of CPV-2a, the positive selection sites were 297 and 324. Both sites are adjacent to the virus-receptor interaction domain [34], which demonstrates the importance of these residues in the evolutionary adaptation of the virus and therefore how these mutations have been established in viral populations. Also, by FEL analysis we confirmed that 514Ser in Colombian CPV-2a have strong positive selection, highlighting the importance of our results and the need for active genomic surveillance programs that help to early detect new CPV-2 variants that could be becoming highly prevalent in the region.

According to the phylogenetic analyses performed in the present study, it can be inferred that there are two antigenic variants (CPV-2a and CPV-2b) with different origins currently circulating in Antioquia, Colombia. However, the Colombian CPV-2a variants contain the Tyr324Ile mutation, which is characteristic of the variants belonging to the Asia I clade [16]. To clarify the relationship and origin of the two variants found in Colombia, the phylogenetic evolutionary analysis was performed. Our analysis revealed that two variants have different origins or present specific differential mutations while sharing a common origin.

The phylogenetic evolutionary analysis revealed that the Colombian CPV-2b variant shows a direct relationship with Asian variants that comprise the Asia I clade. This clade arrives in South America, initially in Uruguay in 2009, followed by Colombia in 2012. According to these results, the CPV-2b variant in Antioquia arrived in Colombia in a manner similar to that in Uruguay, i.e., it arrived directly from Asia and not as a migratory process within the continent from Uruguay to Colombia (Figure 3). On the contrary, the Colombian CPV-2a variant was related to variants of the South American clade I, which originated from the European variants that subsequently underwent evolutionary and migratory processes into the interior of the continent to finally configure a clade with characteristics of variants present only in South America [16]. According to the evolutionary analysis, the CPV-2a variant arrived in Colombia between 2005 and 2006.

As previously reported [16], our evolutionary analysis shows that the most ancestral sequences of CPV-2 are the viral sequences reported in the United States at the end of the 1970s. The subsequent arrive of CPV-2 to Europe can be observed, mainly represented by Italian sequences, which give rise to the clade South America-I, in which the Colombian CPV-2a variant is located (Figure 3). Additionally, the Colombian CPV-2b variant iw related to variants of Asian origin, in a similar way to the Uruguayan variants. The evolutionary rate in VP-2 was shown to be similar to that reported by others [16]. However, the resulting trees diverge in their topology without modifying the common ancestry of the clades. These differences in topology in comparison to previously published results may be due to the addition of new and updated sequences to the dataset. It is clear that variations in spatiotemporal sampling added different bias to the analysis, as has been previously reported [39].

Both the CPV-2a and the CPV-2b variant found in our study present the 324Ile mutation (Table 2; Supplementary Figure S2). This change has been previously reported in cats infected with CPV-2a [38], evidencing that it is an important amino acid in determining host range. Additionally, it has been recently shown that this amino acid residue is close to the virus-receptor interaction domain and that few structural changes are required in CPV-2 to be able to adapt and interact with TfR receptors from different species [34]. This result supports the fact that this mutation has a strong positive selection and is present in all our samples (CPV-2a and CPV-2b). It is possible that other carnivorous species different to canines could be involved in determining the changes in VP2 supporting the adaptation of the virus to carry out an efficient replicative cycle.

5. Conclusions

The antigenic variants, CPV-2a and CPV-2b, circulating in Antioquia, Colombia, originated in the South America I clade and Asia I clade, respectively. The mutations detected in CPV-2a variant have gradually undergone positive selection that appears to favour the virus–receptor interaction, rendering this Colombian CPV-2a sub-variant the most predominant in the region. The antigenic implications of the 440Ala mutation in CPV-2b and those related to the virus–receptor interaction should be elucidated in future investigations, with special emphasis on the proximity to sites of virus-receptor and virus-antibody interactions; results that will allow us to understand the functional impact of these genomic changes on the biology, immunology and pathogenesis of CPV-2.

Supplementary Materials: The following are available online at http://www.mdpi.com/1999-4915/12/5/500/s1, Supplementary Figure S1. Partial VP-2 Maximum likelihood phylogenetic analysis of Colombian CPV-2 sequences (Bootstrap 1000—Tamura 3 parameter model). Current sequences denote as Colombia 2019. Supplementary Figure S2. Three-dimensional reconstruction and structural analysis of the VP2 surface of the Colombian CPV-2a and CPV-2b variants. Supplementary Table S1. Sites under positive selection in Colombian CPV-2a and CPV-2b variants.

Funding: This study was supported by CONADI—Universidad Cooperativa de Colombia, by a grant to JRS. The funding source had no role in study design, analysis, and interpretation of data, writing of the report and in the decision to submit the article for publication.

Acknowledgments: The authors are grateful to the veterinary clinics and hospitals that participated in the sampling. Also, authors want to thank to July Duque-Valencia for her helpful suggestions in phylogenetic analysis.

Conflicts of Interest: The authors of this work do not have any personal or financial conflict that may improperly influence the content of this document.

References

1. Goddard, A.; Leisewitz, A.L. Canine parvovirus. *Vet. Clin. N. Am. Small Anim. Pract.* **2010**, *40*, 1041–1053. [CrossRef] [PubMed]
2. Cotmore, S.F.; Agbandje-McKenna, M.; Chiorini, J.A.; Mukha, D.V.; Pintel, D.J.; Qiu, J.; Soderlund-Venermo, M.; Tattersall, P.; Tijssen, P.; Gatherer, D.; et al. The family Parvoviridae. *Arch. Virol.* **2014**, *159*, 1239–1247. [CrossRef] [PubMed]
3. Reed, A.P.; Jones, E.V.; Miller, T.J. Nucleotide sequence and genome organization of canine parvovirus. *J. Virol.* **1988**, *62*, 266–276. [CrossRef] [PubMed]
4. Hueffer, K.; Parker, J.S.; Weichert, W.S.; Geisel, R.E.; Sgro, J.Y.; Parrish, C.R. The natural host range shift and subsequent evolution of canine parvovirus resulted from virus-specific binding to the canine transferrin receptor. *J. Virol.* **2003**, *77*, 1718–1726. [CrossRef] [PubMed]

5. Parker, J.S.; Parrish, C.R. Cellular uptake and infection by canine parvovirus involves rapid dynamin-regulated clathrin-mediated endocytosis, followed by slower intracellular trafficking. *J. Virol.* **2000**, *74*, 1919–1930. [CrossRef] [PubMed]
6. Miranda, C.; Thompson, G. Canine parvovirus: The worldwide occurrence of antigenic variants. *J. Gen. Virol.* **2016**, *97*, 2043–2057. [CrossRef]
7. Chang, S.F.; Sgro, J.Y.; Parrish, C.R. Multiple amino acids in the capsid structure of canine parvovirus coordinately determine the canine host range and specific antigenic and hemagglutination properties. *J. Virol.* **1992**, *66*, 6858–6867. [CrossRef]
8. Truyen, U.; Evermann, J.F.; Vieler, E.; Parrish, C.R. Evolution of canine parvovirus involved loss and gain of feline host range. *Virology* **1996**, *215*, 186–189. [CrossRef]
9. Parrish, C.R. Host range relationships and the evolution of canine parvovirus. *Vet. Microbiol.* **1999**, *69*, 29–40. [CrossRef]
10. Parrish, C.R.; Aquadro, C.F.; Strassheim, M.L.; Evermann, J.F.; Sgro, J.Y.; Mohammed, H.O. Rapid antigenic-type replacement and DNA sequence evolution of canine parvovirus. *J. Virol.* **1991**, *65*, 6544–6552. [CrossRef]
11. Tsao, J.; Chapman, M.S.; Agbandje, M.; Keller, W.; Smith, K.; Wu, H.; Luo, M.; Smith, T.J.; Rossmann, M.G.; Compans, R.W.; et al. The three-dimensional structure of canine parvovirus and its functional implications. *Science* **1991**, *251*, 1456–1464. [CrossRef] [PubMed]
12. Buonavoglia, C.; Martella, V.; Pratelli, A.; Tempesta, M.; Cavalli, A.; Buonavoglia, D.; Bozzo, G.; Elia, G.; Decaro, N.; Carmichael, L. Evidence for evolution of canine parvovirus type 2 in Italy. *J. Gen. Virol.* **2001**, *82 Pt 12*, 3021–3025. [CrossRef]
13. Decaro, N.; Buonavoglia, C. Canine parvovirus—A review of epidemiological and diagnostic aspects, with emphasis on type 2c. *Vet. Microbiol.* **2012**, *155*, 1–12. [CrossRef] [PubMed]
14. Duque-Garcia, Y.; Echeverri-Zuluaga, M.; Trejos-Suarez, J.; Ruiz-Saenz, J. Prevalence and molecular epidemiology of Canine parvovirus 2 in diarrheic dogs in Colombia, South America: A possible new CPV-2a is emerging? *Vet. Microbiol.* **2017**, *201*, 56–61. [CrossRef]
15. Zhou, P.; Zeng, W.; Zhang, X.; Li, S. The genetic evolution of canine parvovirus—A new perspective. *PLoS ONE* **2017**, *12*, e0175035. [CrossRef] [PubMed]
16. Grecco, S.; Iraola, G.; Decaro, N.; Alfieri, A.; Alfieri, A.; Gallo Calderon, M.; da Silva, A.P.; Name, D.; Aldaz, J.; Calleros, L.; et al. Inter- and intracontinental migrations and local differentiation have shaped the contemporary epidemiological landscape of canine parvovirus in South America. *Virus Evol.* **2018**, *4*, vey011. [CrossRef]
17. Faz, M.; Martinez, J.S.; Gomez, L.B.; Quijano-Hernandez, I.; Fajardo, R.; Del Angel-Caraza, J. Origin and genetic diversity of canine parvovirus 2c circulating in Mexico. *Arch. Virol.* **2019**, *164*, 371–379. [CrossRef]
18. Ikeda, Y.; Mochizuki, M.; Naito, R.; Nakamura, K.; Miyazawa, T.; Mikami, T.; Takahashi, E. Predominance of canine parvovirus (CPV) in unvaccinated cat populations and emergence of new antigenic types of CPVs in cats. *Virology* **2000**, *278*, 13–19. [CrossRef]
19. Xiao, C.; Rossmann, M.G. Interpretation of electron density with stereographic roadmap projections. *J. Struct. Biol.* **2007**, *158*, 182–187. [CrossRef]
20. Mylonakis, M.E.; Kalli, I.; Rallis, T.S. Canine parvoviral enteritis: An update on the clinical diagnosis, treatment, and prevention. *Vet. Med. (Auckland)* **2016**, *7*, 91–100. [CrossRef]
21. Desario, C.; Decaro, N.; Campolo, M.; Cavalli, A.; Cirone, F.; Elia, G.; Martella, V.; Lorusso, E.; Camero, M.; Buonavoglia, C. Canine parvovirus infection: Which diagnostic test for virus? *J. Virol. Methods* **2005**, *126*, 179–185. [CrossRef] [PubMed]
22. Decaro, N.; Desario, C.; Elia, G.; Martella, V.; Mari, V.; Lavazza, A.; Nardi, M.; Buonavoglia, C. Evidence for immunisation failure in vaccinated adult dogs infected with canine parvovirus type 2c. *New Microbiol.* **2008**, *31*, 125–130. [PubMed]
23. Nakamura, M.; Tohya, Y.; Miyazawa, T.; Mochizuki, M.; Phung, H.T.; Nguyen, N.H.; Huynh, L.M.; Nguyen, L.T.; Nguyen, P.N.; Nguyen, P.V.; et al. A novel antigenic variant of Canine parvovirus from a Vietnamese dog. *Arch. Virol.* **2004**, *149*, 2261–2269. [CrossRef] [PubMed]
24. Perez, R.; Bianchi, P.; Calleros, L.; Francia, L.; Hernandez, M.; Maya, L.; Panzera, Y.; Sosa, K.; Zoller, S. Recent spreading of a divergent canine parvovirus type 2a (CPV-2a) strain in a CPV-2c homogenous population. *Vet. Microbiol.* **2012**, *155*, 214–219. [CrossRef] [PubMed]

25. Palermo, L.M.; Hafenstein, S.L.; Parrish, C.R. Purified feline and canine transferrin receptors reveal complex interactions with the capsids of canine and feline parvoviruses that correspond to their host ranges. *J. Virol.* **2006**, *80*, 8482–8492. [CrossRef] [PubMed]
26. Palermo, L.M.; Hueffer, K.; Parrish, C.R. Residues in the apical domain of the feline and canine transferrin receptors control host-specific binding and cell infection of canine and feline parvoviruses. *J. Virol.* **2003**, *77*, 8915–8923. [CrossRef] [PubMed]
27. Jeoung, S.Y.; Ahn, S.J.; Kim, D. Genetic analysis of VP2 gene of canine parvovirus isolates in Korea. *J. Vet. Med. Sci.* **2008**, *70*, 719–722. [CrossRef]
28. Zhang, R.; Yang, S.; Zhang, W.; Zhang, T.; Xie, Z.; Feng, H.; Wang, S.; Xia, X. Phylogenetic analysis of the VP2 gene of canine parvoviruses circulating in China. *Virus Genes* **2010**, *40*, 397–402. [CrossRef]
29. Hong, C.; Decaro, N.; Desario, C.; Tanner, P.; Pardo, M.C.; Sanchez, S.; Buonavoglia, C.; Saliki, J.T. Occurrence of canine parvovirus type 2c in the United States. *J. Vet. Diagn. Investig.* **2007**, *19*, 535–539. [CrossRef]
30. Decaro, N.; Desario, C.; Addie, D.D.; Martella, V.; Vieira, M.J.; Elia, G.; Zicola, A.; Davis, C.; Thompson, G.; Thiry, E.; et al. The study molecular epidemiology of canine parvovirus, Europe. *Emerg. Infect. Dis.* **2007**, *13*, 1222–1224. [CrossRef]
31. Wang, H.; Jin, H.; Li, Q.; Zhao, G.; Cheng, N.; Feng, N.; Zheng, X.; Wang, J.; Zhao, Y.; Li, L.; et al. Isolation and sequence analysis of the complete NS1 and VP2 genes of canine parvovirus from domestic dogs in 2013 and 2014 in China. *Arch. Virol.* **2016**, *161*, 385–393. [CrossRef] [PubMed]
32. Mira, F.; Purpari, G.; Lorusso, E.; Di Bella, S.; Gucciardi, F.; Desario, C.; Macaluso, G.; Decaro, N.; Guercio, A. Introduction of Asian canine parvovirus in Europe through dog importation. *Transbound. Emerg. Dis.* **2018**, *65*, 16–21. [CrossRef] [PubMed]
33. Mira, F.; Purpari, G.; Di Bella, S.; Colaianni, M.L.; Schiro, G.; Chiaramonte, G.; Gucciardi, F.; Pisano, P.; Lastra, A.; Decaro, N.; et al. Spreading of canine parvovirus type 2c mutants of Asian origin in southern Italy. *Transbound. Emerg. Dis.* **2019**, *66*, 2297–2304. [CrossRef] [PubMed]
34. Voorhees, I.E.H.; Lee, H.; Allison, A.B.; Lopez-Astacio, R.; Goodman, L.B.; Oyesola, O.O.; Omobowale, O.; Fagbohun, O.; Dubovi, E.J.; Hafenstein, S.L.; et al. Limited intra-host diversity and background evolution accompany 40 years of canine parvovirus host adaptation and spread. *J. Virol.* **2019**, *94*. [CrossRef] [PubMed]
35. Lee, H.; Callaway, H.M.; Cifuente, J.O.; Bator, C.M.; Parrish, C.R.; Hafenstein, S.L. Transferrin receptor binds virus capsid with dynamic motion. *Proc. Natl. Acad. Sci. USA* **2019**, *116*, 20462–20471. [CrossRef] [PubMed]
36. De la Torre, D.; Mafla, E.; Puga, B.; Erazo, L.; Astolfi-Ferreira, C.; Ferreira, A.P. Molecular characterization of canine parvovirus variants (CPV-2a, CPV-2b, and CPV-2c) based on the VP2 gene in affected domestic dogs in Ecuador. *Vet. World* **2018**, *11*, 480–487. [CrossRef]
37. Mukhopadhyay, H.K.; Matta, S.L.; Amsaveni, S.; Antony, P.X.; Thanislass, J.; Pillai, R.M. Phylogenetic analysis of canine parvovirus partial VP2 gene in India. *Virus Genes* **2014**, *48*, 89–95. [CrossRef]
38. Mukhopadhyay, H.K.; Nookala, M.; Thangamani, N.R.; Sivaprakasam, A.; Antony, P.X.; Thanislass, J.; Srinivas, M.V.; Pillai, R.M. Molecular characterisation of parvoviruses from domestic cats reveals emergence of newer variants in India. *J. Feline Med. Surg.* **2017**, *19*, 846–852. [CrossRef]
39. Magee, D.; Taylor, J.E.; Scotch, M. The Effects of Sampling Location and Predictor Point Estimate Certainty on Posterior Support in Bayesian Phylogeographic Generalized Linear Models. *Sci. Rep.* **2018**, *8*, 5905. [CrossRef]

MDPI

Article

Molecular Investigation of Canine Parvovirus-2 (CPV-2) Outbreak in Nevis Island: Analysis of the Nearly Complete Genomes of CPV-2 Strains from the Caribbean Region

Kerry Gainor [1,†], April Bowen [2,†], Pompei Bolfa [1], Andrea Peda [3], Yashpal S. Malik [4] and Souvik Ghosh [1,*]

1 Department of Biomedical Sciences, Ross University School of Veterinary Medicine, Basseterre P.O. Box 334, Saint Kitts and Nevis; KerryGainor@students.rossu.edu (K.G.); pbolfa@rossvet.edu.kn (P.B.)
2 Nevis Animal Speak, Cades Bay Nevis, Basserrete, Saint Kitts and Nevis; aprilbowenlvt@outlook.com
3 Department of Clinical Sciences, Ross University School of Veterinary Medicine, Basseterre P.O. Box 334, Saint Kitts and Nevis; apeda@rossvet.edu.kn
4 College of Animal Biotechnology, Guru Angad Dev Veterinary and Animal Science University, Ludhiana, Punjab 141001, India; malikyps@gmail.com
* Correspondence: souvikrota@gmail.com or sghosh@rossu.edu; Tel.: +18-(69)-4654161 (ext. 401-1202)
† These authors contributed equally to this work.

Abstract: To date, there is a dearth of information on canine parvovirus-2 (CPV-2) from the Caribbean region. During August–October 2020, the veterinary clinic on the Caribbean island of Nevis reported 64 household dogs with CPV-2-like clinical signs (hemorrhagic/non-hemorrhagic diarrhea and vomiting), of which 27 animals died. Rectal swabs/fecal samples were obtained from 43 dogs. A total of 39 of the 43 dogs tested positive for CPV-2 antigen and/or DNA, while 4 samples, negative for CPV-2 antigen, were not available for PCR. Among the 21 untested dogs, 15 had CPV-2 positive littermates. Analysis of the complete VP2 sequences of 32 strains identified new CPV-2a (CPV-2a with Ser297Ala in VP2) as the predominant CPV-2 on Nevis Island. Two nonsynonymous mutations, one rare (Asp373Asn) and the other uncommon (Ala262Thr), were observed in a few VP2 sequences. It was intriguing that new CPV-2a was associated with an outbreak of gastroenteritis on Nevis while found at low frequencies in sporadic cases of diarrhea on the neighboring island of St. Kitts. The nearly complete CPV-2 genomes (4 CPV-2 strains from St. Kitts and Nevis (SKN)) were reported for the first time from the Caribbean region. Eleven substitutions were found among the SKN genomes, which included nine synonymous substitutions, five of which have been rarely reported, and the two nonsynonymous substitutions. Phylogenetically, the SKN CPV-2 sequences formed a distinct cluster, with CPV-2b/USA/1998 strains constituting the nearest cluster. Our findings suggested that new CPV-2a is endemic in the region, with the potential to cause severe outbreaks, warranting further studies across the Caribbean Islands. Analysis of the SKN CPV-2 genomes corroborated the hypothesis that recurrent parallel evolution and reversion might play important roles in the evolution of CPV-2.

Keywords: canine parvovirus; Caribbean region; new CPV-2a; outbreak; endemic; nearly complete genomes; virus evolution

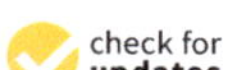

Citation: Gainor, K.; Bowen, A.; Bolfa, P.; Peda, A.; Malik, Y.S.; Ghosh, S. Molecular Investigation of Canine Parvovirus-2 (CPV-2) Outbreak in Nevis Island: Analysis of the Nearly Complete Genomes of CPV-2 Strains from the Caribbean Region. *Viruses* **2021**, *13*, 1083. https://doi.org/10.3390/v13061083

Academic Editor: Giorgio Gallinella

Received: 10 May 2021
Accepted: 3 June 2021
Published: 6 June 2021

1. Introduction

Canine parvovirus-2 (CPV-2), members of the genus *Protoparvovirus* within the family *Parvoviridae*, are highly contagious enteric pathogens of household dogs, often causing fatal hemorrhagic gastroenteritis in puppies [1–4]. Morphologically, CPV-2 are small, nonenveloped viruses containing a single-stranded, negative-sense DNA genome (~5200 bp in size) [2,5]. The CPV-2 genome possesses at least two major open reading frames (ORFs), designated as ORF1 and ORF2. The ORF1 encodes two nonstructural (NS1 and NS2) proteins, while ORF2 codes for two structural (VP1 and VP2) proteins, translated through alternative splicing of the same viral mRNAs [2,5].

The CPV-2 nonstructural proteins have been shown to be crucial for viral replication, DNA packaging, cytotoxicity, and pathogenesis [6–8]. On the other hand, the structural proteins form the viral capsid, of which VP2 is the major component [9,10]. The VP2 protein plays important roles in determining host range, tissue tropisms, and virus-host interactions, and is highly antigenic, forming the basis of the currently licensed CPV-2 vaccines [2,3,11,12]. To date, the majority of the studies on CPV-2 are based on the VP2 encoding gene [2,13,14], while limited information is available on the genetic variations in the nonstructural genes [15,16].

Based on amino acid (aa) differences in the VP2 protein, CPV-2 strains have been classified into four major antigenic variants: CPV-2, CPV-2a, CPV-2b, and CPV-2c [2,13,14]. The earliest CPV-2 strains that emerged in dogs during the late 1970s are referred to as the CPV-2 variant [2,16]. By the end of 1980, CPV-2 was rapidly replaced by a new antigenic variant, designated as the CPV-2a variant [17,18]. The other CPV-2 variants, CPV-2b and CPV-2c, were first reported in 1984 and 2000, respectively [19,20].

The antigenic differences between CPV-2a, CPV-2b, and CPV-2c have been attributed to a single aa mismatch (Asn in CPV-2a, Asp in CPV-2b, and Glu in CPV-2c) at residue 426 of the VP2 protein [2]. However, by phylogenetic analysis of complete/nearly full-length CPV-2 sequences, the CPV-2a, CPV-2b, and CPV-2c variants lacked clear monophyletic segregation, and were grouped together into a single 'CPV-2a clade', which was distinct from the cluster of the earliest CPV-2 strains, referred to as the CPV-2 clade [16]. Other nonsynonymous mutations have also been observed in the CPV-2 variants [2,13,14,16,21]. Notable among these aa changes is the presence of Ser297Ala in VP2 of several CPV-2a and CPV-2b strains, sometimes referred to as new CPV-2a and new CPV-2b, respectively [13]. Currently, the CPV-2a, CPV-2b, and CPV-2c variants and their mutants, such as new CPV-2a and new CPV-2b, are circulating worldwide, with different relative frequencies between countries and between sampling periods [2,13,14,16,21–23].

To date, there is only a single report on the detection and molecular characterization of CPV-2 from the Caribbean region [24]. During February 2015–August 2016, new CPV-2a was detected (25/104 dogs tested CPV-2 positive, 20 samples were sequenced) in sporadic cases of diarrhea in household dogs on St. Kitts Island [24]. In the present study, we report a molecular investigation of a new CPV-2a associated severe outbreak of canine gastroenteritis in the Caribbean island of Nevis that resulted in the death of 27 animals.

Since there are no reports on complete CPV-2 genomes from the Caribbean region, the nearly full-length genomes of 3 CPV-2 strains (representing each of the new CPV-2a mutants with a nonsynonymous substitution in the VP2 gene) detected during the outbreak on Nevis and that of a previously reported CPV-2 strain from St. Kitts were analyzed in the present study.

2. Materials and Methods

2.1. Ethics Statement

The present study was submitted to the Institutional Animal Care and Use Committee (IACUC) of the Ross University School of Veterinary Medicine (RUSVM), St. Kitts Island. Ethical review and approval were waived for this study by the RUSVM IACUC as the research study was based on leftover samples that were collected for diagnostic purposes at the veterinary clinic on Nevis Island (RUSVM IACUC sample/tissue notification letter number TSU 1.23.21).

2.2. Sampling

Nevis is a small Caribbean island (total area of 93 km^2, human population ~12,000) located in the lesser Antilles, and together with the neighboring island of St. Kitts, constitutes the twin Federation of St. Kitts and Nevis (https://www.paho.org/, accessed on 19 April 2021) (Figure 1A). Although there are no official records on the canine population in Nevis Island, many of the houses keep dogs as pets, with the island mix breed (a cross between a breed native to the Caribbean Islands and another canine breed) representing the majority of the domestic dogs.

(A)

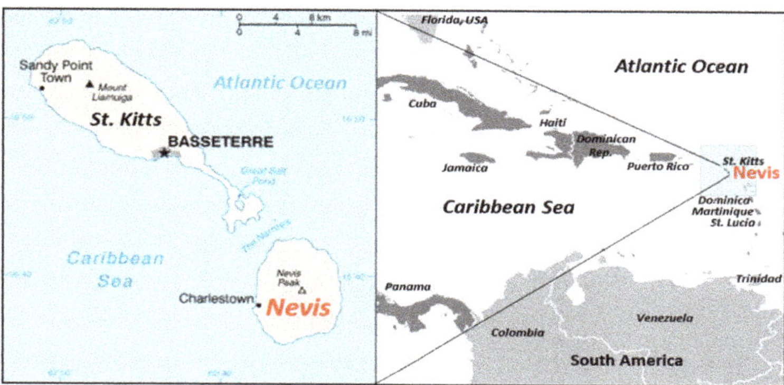

(B)

Figure 1. (**A**) Geographical location of Nevis Island. (**B**) Map of Nevis Island showing the locations of the 64 dogs (highlighted with pins) with canine parvovirus-2-like (CPV-2-like) clinical signs. Red pin, dogs that tested positive for CPV-2 by the SNAP® Parvo Test (IDEXX, Westbrook, ME, USA) and/or PCR; Yellow pin, untested dogs with CPV-2-like clinical signs that had a CPV-2 positive littermate (by the SNAP® Parvo Test and/or PCR); Blue, untested dogs with CPV-2-like clinical signs. The pins were inserted into the map using the Map Maker software (Maps.co, Mountain View, CA, USA). The maps for Figures 1A and 1B were obtained from https://www.cia.gov/library/publications/the-world-factbook (accessed on 1 April 2021) and https://www.google.com/maps (accessed on 30 March 2021), respectively.

From 1 August 2020 to 17 October 2020, 64 household dogs with CPV-2-like clinical signs were presented at the veterinary clinic on Nevis Island. A total of 44 rectal swabs/fecal samples were obtained from 43 of the dogs. One dog was sampled twice (5 September 2020 and 5 November 2020).

2.3. Screening

The samples were screened for the presence of CPV-2 antigen using the SNAP® Parvo Test (IDEXX, Westbrook, ME, USA) following the manufacturer's instructions.

2.4. Amplification of CPV-2 Genome

Viral DNA was extracted using the QIAamp Fast DNA Stool Mini Kit (Qiagen Sciences, Germantown, MD, USA) according to the manufacturer's instructions. Primers used for amplification of the CPV-2 genome and/or obtaining CPV-2 genome sequences were designed in the present study (Supplementary Material S1). The complete VP2 ORF of CPV-2 was amplified by two overlapping PCRs, while three additional overlapping PCRs were employed to amplify the nearly full-length CPV-2 genomes (spanning all the coding regions, corresponding to nucleotide (nt) 272-nt 4585 of reference CPV-2 strain CPV-N (GenBank accession number M19296)) (Supplementary Material S1). PCRs were performed using the Platinum™ Taq DNA Polymerase (Invitrogen™, Thermo Fisher Scientific Corporation, Waltham, MA, USA) following the instructions provided by the manufacturer. Sterile water was used as a negative control during the PCR reactions.

2.5. Nucleotide Sequencing

The PCR amplicons were purified using the Wizard® SV Gel and PCR Clean-Up kit (Promega, Madison, WI, USA) according to the manufacturer's instructions and sequenced in both directions using forward and reverse primers (Supplementary Material S1). Nucleotide sequences were obtained using the ABI Prism Big Dye Terminator Cycle Sequencing Ready Reaction Kit (Applied Biosystems, Foster City, CA, USA) on an ABI 3730XL Genetic Analyzer (Applied Biosystems, Foster City, CA, USA).

2.6. Sequence Analysis

Putative ORFs encoding the CPV-2 NS1 and VP2 proteins were identified using the ORF finder (https://www.ncbi.nlm.nih.gov/orffinder/, accessed on 5 April 2021), while those coding for NS2 and VP1 proteins were determined by alignment of the obtained CPV-2 nt sequences with published CPV-2 coding sequences. The standard BLASTN and BLASTP program (Basic Local Alignment Search Tool, www.ncbi.nlm.nih.gov/blast, accessed on 2 April 2021) was used to perform a homology search for related cognate nt and deduced aa sequences, respectively. Multiple alignments of nt and deduced aa sequences were carried out using the CLUSTALW program (version ddbj, http://clustalw.ddbj.nig.ac.jp/, accessed on 2 April 2021) with default parameters. The nearly complete CPV-2 genome sequences were examined for recombination events using the RDP4 program with default parameters, as described previously [16,25]. Briefly, a CPV-2 sequence was identified as a recombinant if it was supported by two or more detection methods (3Seq, BOOTSCAN, CHIMERA, GENECONV, MAXCHI, RDP, and SISCAN) with a highest acceptable p-value of $p < 0.01$ with Bonferroni's correction.

A data set (excluding recombinant sequences) of 210 nearly complete CPV-2 sequences from domestic and wild canids and 6 FPV sequences were created for phylogenetic analysis, based on those reported in previous studies [16,21]. Phylogenetic analysis was performed by the maximum likelihood (ML) method using the MEGA7 software [26], with gamma-distributed rate variation among sites and 1000 bootstrap replicates. Phylogenetic trees were constructed using both the Hasegawa-Kishino-Yano (HKY) and general time-reversible (GTR) substitution models.

2.7. GenBank Accession Numbers

The GenBank accession numbers for the CPV-2 genome sequences determined in this study are MW595661-MW595693 (complete VP2 ORF sequences) and MW616469-MW616472 (nearly full-length CPV-2 genome sequences).

3. Results and Discussion

3.1. Molecular Investigation of CPV-2 Outbreak in Nevis Island

From 1 August 2020 to 17 October 2020, the veterinary clinic on Nevis Island reported 64 household dogs with anorexia, gastroenteritis (with or without blood in feces), lethargy, and vomiting (CPV-2-like clinical signs). A total of 27 (42.2%) of the dogs died. Based on case histories, the outbreak of gastroenteritis appeared to have started at Charlestown, the capital of Nevis, and eventually spread to other parts of the island, with most cases mapped to the more densely populated western coastal region of the island (Figure 1B).

A total of 43 dogs were tested for CPV-2 antigen using the SNAP® Parvo Test (IDEXX, Westbrook, ME, USA), of which 32 samples were available for PCR. A total of 39 (90.6%) of the 43 dogs tested CPV-2 positive by SNAP® Parvo Test (24/43 dogs, 55.8%) and/or PCR (32/32 dogs, 100%) (Table 1), while 4 samples that were negative for CPV-2 antigen could not be tested by PCR. Fifteen of the PCR positive dogs tested negative by the SNAP® Parvo Test (Table 1), which might be attributed to the lower sensitivities of the CPV-2 antigen tests compared to PCR/qPCR assays, even in dogs with CPV-2-like clinical signs, as reported in previous studies [2,27]. We did not observe any differences in clinical severity between dogs that tested negative and positive to the SNAP® Parvo Test. Although samples could not be obtained from 21 dogs with CPV-2-like clinical signs, 15 of the animals had littermates that were positive for CPV-2 antigen and/or DNA. Most of the sick dogs were <6 months of age (89%, 57/64 dogs with CPV-2-like clinical signs) and were either not vaccinated or received incomplete immunization (did not receive all doses of the vaccine) against CPV-2 (95.3%, 61/64 dogs with CPV-2-like clinical signs), corroborating previous observations on the increased risk of CPV-2 infection in unvaccinated puppies [2,4]. Since not all the dogs were tested for CPV-2, and none of the obtained samples were screened for other enteric pathogens, we could not establish whether CPV-2 was the sole etiological agent in the outbreak of gastroenteritis on Nevis Island.

In order to determine the CPV-2 variant/s circulating during the outbreak in Nevis, complete VP2 ORF sequences were obtained from the 32 PCR positive samples. Based on the presence of 297Ala and 426Asn in the putative VP2 proteins [13], all the CPV-2 strains from Nevis were classified as new CPV-2a (Table 2; Supplementary Material S2). In a previous study (2015–2016), new CPV-2a was only identified in sporadic cases of diarrhea on the neighboring island of St. Kitts [24]. These observations suggested that new CPV-2a is endemic and might be predominant in this part of the world. In South America, new CPV-2a, or other mutants of CPV-2a (VP2 Ser297Asn, Phe267Tyr, Tyr324Ile, and Thr440Ala) have been found to coexist with CPV-2b and/or CPV-2c variants at various frequencies, emerging as the major strain in a few studies [13,21,28–33]. In studies from North (Canada and USA) and Central (Mexico) Americas, CPV-2b, or CPV-2c was most prevalent, while a single report from Alaska detected both new CPV-2a and new CPV-2b in an outbreak of canine gastroenteritis [13,34–37]. Taken together, the molecular epidemiology of CPV-2 in the Caribbean region appears to differ from those reported in nearby North, Central, and South American countries, although further studies on the other islands are required to validate this observation.

Table 1. Details of the dogs that tested positive for canine parvovirus-2 (CPV-2) by the SNAP® Parvo Test (IDEXX, Westbrook, ME, USA) and/or PCR on Nevis Island. The CPV-2 strains from Nevis are denoted with the prefix CN (Canine Nevis).

Sample/Strain	Date of Sampling	Age	Sex	Breed	Vaccination against CPV-2 [1]	SNAP® Parvo Test [2]	PCR [3]	GenBank Accession Number (VP2 Gene, Complete ORF)	CPV-2 Variant
CN1	1 August 2020	3 months	Female	Shih Tzu	None	Positive	Not performed	Not sequenced	Not determined
CN2	11 August 2020	4 months	Male	Bulldog	Incomplete [4]	Positive	Not performed	Not sequenced	Not determined
CN3	14 August 2020	2 months	Male	Pit bull	None	Positive	Not performed	Not sequenced	Not determined
CN4	15 August 2020	3 months 15 days	Male	Pit bull	None	Positive	Not performed	Not sequenced	Not determined
CN5	15 August 2020	4 months	Female	Pit bull	None	Positive	Not performed	Not sequenced	Not determined
CN6	21 August 2020	3 months	Female	Pit bull	Not available	Positive	Not performed	Not sequenced	Not determined
CN7	22 August 2020	2 months	Male	Island mix [5]	None	Positive	Positive	MW595661	New CPV-2a [6]
CN8	24 August 2020	2 months 15 days	Male	Pit bull	None	Negative	Positive	MW595662	New CPV-2a
CN9	25 August 2020	2 months	Female	Pit bull	None	Positive	Not performed	Not sequenced	Not determined
CN10	26 August 2020	4 months	Male	Island mix	None	Positive	Positive	MW595663	New CPV-2a
CN11	28 August 2020	4 months 15 days	Male	Island mix	Incomplete [4]	Positive	Positive	MW595664	New CPV-2a
CN12	28 August 2020	7 months	Male	Pit bull	None	Positive	Positive	MW595665	New CPV-2a
CN13	28 August 2020	2 year	Male	Rottweiler mix	None	Negative	Positive	MW595666	New CPV-2a
CN14	28 August 2020	7 months	Female	Island mix	Not available	Positive	Positive	MW595667	New CPV-2a
CN15	29 August 2020	2 months 15 days	Female	Island mix	None	Negative	Positive	MW595668	New CPV-2a
CN16	1 September 2020	10 months	Male	Island mix	Complete	Positive	Positive	MW595669	New CPV-2a
CN17	1 September 2020	2 months	Male	Bulldog x Mastiff	None	Positive	Positive	MW595670	New CPV-2a
CN18	1 September 2020	4 months	Male	Island mix	None	Negative	Positive	MW595671	New CPV-2a
CN19	1 September 2020	4 months	Female	Pit bull	Not available	Negative	Positive	MW595672	New CPV-2a
CN20	1 September 2020	4 months	Male	Pit bull	Not available	Positive	Positive	MW595673	New CPV-2a
CN21	2 September 2020	1 year	Female	Island mix	None	Negative	Positive	MW595674	New CPV-2a
CN22	3 September 2020	6 months	Male	Island mix	Complete	Positive	Positive	MW595675	New CPV-2a

Table 1. Cont.

Sample/ Strain	Date of Sampling	Age	Sex	Breed	Vaccination against CPV-2 [1]	SNAP® Parvo Test [2]	PCR [3]	GenBank Accession Number (VP2 Gene, Complete ORF)	CPV-2 Variant
CN23	3 September 2020	2 months	Female	Pit bull	None	Negative	Positive	MW595676	New CPV-2a
CN24	3 September 2020	4 months	Female	Not available	None	Negative	Positive	MW595677	New CPV-2a
CN25	4 September 2020	2 months	Female	Island mix	None	Negative	Positive	MW595678	New CPV-2a
CN26	4 September 2020	2 months	Female	Island mix	None	Negative	Positive	MW595679	New CPV-2a
CN27 [7]	5 September 2020	1 month 15 days	Male	Great Dane	None	Negative	Positive	MW595680	New CPV-2a
CN28	8 September 2020	6 months	Male	Pit bull	None	Positive	Positive	MW595681	New CPV-2a
CN29	11 September 2020	2 months	Male	Pit bull	None	Positive	Positive	MW595682	New CPV-2a
CN30	13 September 2020	2 months	Male	Pit bull	None	Positive	Positive	MW595683	New CPV-2a
CN31	14 September 2020	7 months	Female	Pit bull mix	Complete	Positive	Positive	MW595684	New CPV-2a
CN32	25 September 2020	2 months	Female	Island mix	None	Negative	Positive	MW595685	New CPV-2a
CN33	29 September 2020	2 months 15 days	Female	Island mix	None	Positive	Positive	MW595686	New CPV-2a
CN34	7 October 2020	7 months	Male	Island mix	None	Positive	Positive	MW595687	New CPV-2a
CN35	8 October 2020	4 months	Female	Island mix	Incomplete [4]	Positive	Positive	MW595688	New CPV-2a
CN36	8 October 2020	4 months	Male	Island mix	None	Negative	Positive	MW595689	New CPV-2a
CN37	9 October 2020	4 months	Male	Island mix	None	Negative	Positive	MW595690	New CPV-2a
CN38	9 October 2020	2 months	Male	Island mix	Incomplete [4]	Negative	Positive	MW595691	New CPV-2a
CN39	17 October 2020	1 month 15 days	Male	Island mix	None	Positive	Positive	MW595692	New CPV-2a
CN40 [7]	5 November 2020	3 months 15 days	Male	Great Dane	None	Negative	Positive	MW595693	New CPV-2a

[1] Indicates vaccination status at the time of sampling; [2] The SNAP® Parvo Test (IDEXX, Westbrook, ME, USA) detects CPV-2 antigen in feces; [3] Based on amplification of the CPV-2 VP2 gene (Supplementary Material S1); [4] Did not receive all doses of the vaccine, as recommended by the American Animal Hospital Association (https://www.aaha.org/globalassets/02-guidelines/canine-vaccination, accessed on 2 March 2021); [5] Indicates a cross between a local canine breed and another breed; [6] CPV-2a with Ser297Ala in VP2; [7] Samples CN27 (collected during hemorrhagic gastroenteritis) and CN40 (collected from an asymptomatic animal) were from the same dog.

Table 2. Comparison of key amino acid (aa) residues of putative VP2 proteins of canine parvovirus-2 (CPV-2) strains detected on Nevis Island with those of other CPV-2 strains. Strain RVC50 represents the new CPV-2a (CPV-2a with Ser297Ala in VP2) strains detected during a previous study (February 2015–August 2016) in the neighboring island of St. Kitts [24]. The CPV-2 strains from Nevis are shown with italic font, while strain RVC50 from St. Kitts is underlined. Identical aa residues are shown with the same color. Amino acid mismatches between the CPV-2 strains from Nevis and strain RVC50 are shown in red font. Positions of aa residues correspond to those of strain CPV-b/USA/1978. Alignment of the deduced VP2 aa and ORF sequences of the CPV-2 strains from Nevis is shown in Supplementary Materials S2 and S3, respectively.

Amino Acid Position / **Strain/Place/Year**	87	101	262	267	297	300	305	321	324	373	375	426	440	555	570	Variant
CPV-b/USA/1978	Met	Ile	Ala	Phe	Ser	Ala	Asp	Asn	Tyr	Asp	Asn	Asn	Thr	Val	Lys	CPV-2
CPV-15/USA/1984	Leu	Thr	Ala	Phe	Ser	Gly	Tyr	Asn	Tyr	Asp	Asp	Asn	Thr	Ile	Lys	CPV-2a
CPV-39/USA/1984	Leu	Thr	Ala	Phe	Ser	Gly	Tyr	Asn	Tyr	Asp	Asp	Asp	Thr	Val	Lys	CPV-2b
219/08-13/ITA/2008	Leu	Thr	Ala	Phe	Ala	Gly	Tyr	Asn	Tyr	Asp	Asp	Glu	Thr	Val	Lys	CPV-2c
CPV-435/USA/2003	Leu	Thr	Ala	Phe	Ala	Gly	Tyr	Asn	Tyr	Asp	Asp	Asn	Thr	Val	Lys	New CPV-2a
<u>RVC50/St. Kitts/2016</u>	Leu	Thr	Ala	Phe	Ala	Gly	Tyr	Asn	Tyr	Asp	Asp	Asn	Thr	Val	Lys	New CPV-2a
GX304/CHN/2017	Leu	Thr	Thr	Tyr	Ala	Gly	Tyr	Asn	Ile	Asp	Asp	Asn	Ala	Val	Lys	New CPV-2a
Beaumaris/AUS/2017	Leu	Thr	Ala	Phe	Ala	Gly	Tyr	Asn	Ile	Asn	Asp	Asn	Thr	Val	Lys	New CPV-2a
CPV-436/USA/2003	Leu	Thr	Ala	Phe	Ala	Gly	Tyr	Asn	Tyr	Asp	Asp	Asp	Thr	Val	Lys	New CPV-2b
CN7/Nevis/2020	Leu	Thr	Ala	Phe	Ala	Gly	Tyr	Asn	Tyr	Asp	Asp	Asn	Thr	Val	Lys	New CPV-2a
CN8/Nevis/2020	Leu	Thr	Ala	Phe	Ala	Gly	Tyr	Asn	Tyr	Asp	Asp	Asn	Thr	Val	Lys	New CPV-2a
CN10/Nevis/2020	Leu	Thr	Ala	Phe	Ala	Gly	Tyr	Asn	Tyr	Asp	Asp	Asn	Thr	Val	Lys	New CPV-2a
CN11/Nevis/2020	Leu	Thr	Ala	Phe	Ala	Gly	Tyr	Asn	Tyr	Asp	Asp	Asn	Thr	Val	Lys	New CPV-2a
CN12/Nevis/2020	Leu	Thr	Ala	Phe	Ala	Gly	Tyr	Asn	Tyr	Asp	Asp	Asn	Thr	Val	Lys	New CPV-2a
CN13/Nevis/2020	Leu	Thr	Ala	Phe	Ala	Gly	Tyr	Asn	Tyr	Asp	Asp	Asn	Thr	Val	Lys	New CPV-2a
CN14/Nevis/2020	Leu	Thr	Ala	Phe	Ala	Gly	Tyr	Asn	Tyr	Asn	Asp	Asn	Thr	Val	Lys	New CPV-2a
CN15/Nevis/2020	Leu	Thr	Ala	Phe	Ala	Gly	Tyr	Asn	Tyr	Asp	Asp	Asn	Thr	Val	Lys	New CPV-2a
CN16/Nevis/2020	Leu	Thr	Ala	Phe	Ala	Gly	Tyr	Asn	Tyr	Asp	Asp	Asn	Thr	Val	Lys	New CPV-2a
CN17/Nevis/2020	Leu	Thr	Ala	Phe	Ala	Gly	Tyr	Asn	Tyr	Asp	Asp	Asn	Thr	Val	Lys	New CPV-2a
CN18/Nevis/2020	Leu	Thr	Ala	Phe	Ala	Gly	Tyr	Asn	Tyr	Asp	Asp	Asn	Thr	Val	Lys	New CPV-2a
CN19/Nevis/2020	Leu	Thr	Thr	Phe	Ala	Gly	Tyr	Asn	Tyr	Asp	Asp	Asn	Thr	Val	Lys	New CPV-2a
CN20/Nevis/2020	Leu	Thr	Thr	Phe	Ala	Gly	Tyr	Asn	Tyr	Asp	Asp	Asn	Thr	Val	Lys	New CPV-2a
CN21/Nevis/2020	Leu	Thr	Ala	Phe	Ala	Gly	Tyr	Asn	Tyr	Asp	Asp	Asn	Thr	Val	Lys	New CPV-2a
CN22/Nevis/2020	Leu	Thr	Ala	Phe	Ala	Gly	Tyr	Asn	Tyr	Asp	Asp	Asn	Thr	Val	Lys	New CPV-2a
CN23/Nevis/2020	Leu	Thr	Ala	Phe	Ala	Gly	Tyr	Asn	Tyr	Asn	Asp	Asn	Thr	Val	Lys	New CPV-2a

Table 2. *Cont.*

Amino Acid Position / Strain/Place/Year	87	101	262	267	297	300	305	321	324	373	375	426	440	555	570	Variant
CN24/Nevis/2020	Leu	Thr	Thr	Phe	Ala	Gly	Tyr	Asn	Tyr	Asp	Asp	Asn	Thr	Val	Lys	New CPV-2a
CN25/Nevis/2020	Leu	Thr	Ala	Phe	Ala	Gly	Tyr	Asn	Tyr	Asn	Asp	Asn	Thr	Val	Lys	New CPV-2a
CN26/Nevis/2020	Leu	Thr	Ala	Phe	Ala	Gly	Tyr	Asn	Tyr	Asp	Asp	Asn	Thr	Val	Lys	New CPV-2a
CN27/Nevis/2020 [1]	Leu	Thr	Thr	Phe	Ala	Gly	Tyr	Asn	Tyr	Asp	Asp	Asn	Thr	Val	Lys	New CPV-2a
CN28/Nevis/2020	Leu	Thr	Ala	Phe	Ala	Gly	Tyr	Asn	Tyr	Asp	Asp	Asn	Thr	Val	Lys	New CPV-2a
CN29/Nevis/2020	Leu	Thr	Ala	Phe	Ala	Gly	Tyr	Asn	Tyr	Asp	Asp	Asn	Thr	Val	Lys	New CPV-2a
CN30/Nevis/2020	Leu	Thr	Ala	Phe	Ala	Gly	Tyr	Asn	Tyr	Asp	Asp	Asn	Thr	Val	Lys	New CPV-2a
CN31/Nevis/2020	Leu	Thr	Ala	Phe	Ala	Gly	Tyr	Asn	Tyr	Asp	Asp	Asn	Thr	Val	Lys	New CPV-2a
CN32/Nevis/2020	Leu	Thr	Ala	Phe	Ala	Gly	Tyr	Asn	Tyr	Asp	Asp	Asn	Thr	Val	Lys	New CPV-2a
CN33/Nevis/2020	Leu	Thr	Ala	Phe	Ala	Gly	Tyr	Asn	Tyr	Asp	Asp	Asn	Thr	Val	Lys	New CPV-2a
CN34/Nevis/2020	Leu	Thr	Ala	Phe	Ala	Gly	Tyr	Asn	Tyr	Asp	Asp	Asn	Thr	Val	Lys	New CPV-2a
CN35/Nevis/2020	Leu	Thr	Ala	Phe	Ala	Gly	Tyr	Asn	Tyr	Asp	Asp	Asn	Thr	Val	Lys	New CPV-2a
CN36/Nevis/2020	Leu	Thr	Ala	Phe	Ala	Gly	Tyr	Asn	Tyr	Asp	Asp	Asn	Thr	Val	Lys	New CPV-2a
CN37/Nevis/2020	Leu	Thr	Ala	Phe	Ala	Gly	Tyr	Asn	Tyr	Asp	Asp	Asn	Thr	Val	Lys	New CPV-2a
CN38/Nevis/2020	Leu	Thr	Ala	Phe	Ala	Gly	Tyr	Asn	Tyr	Asp	Asp	Asn	Thr	Val	Lys	New CPV-2a
CN39/Nevis/2020	Leu	Thr	Ala	Phe	Ala	Gly	Tyr	Asn	Tyr	Asp	Asp	Asn	Thr	Val	Lys	New CPV-2a
CN40/Nevis/2020 [1]	Leu	Thr	Ala	Phe	Ala	Gly	Tyr	Asn	Tyr	Asp	Asp	Asn	Thr	Val	Lys	New CPV-2a
VANGUARD/vaccine	Arg	Ile	Ala	Phe	Ser	Ala	Asp	Asn	Tyr	Asp	Glu	Asn	Thr	Val	Lys	CPV-2
Duramune/vaccine	Leu	Thr	Ala	Phe	Ala	Gly	Tyr	Lys	Tyr	Asp	Asp	Asp	Thr	Val	Glu	New CPV-2b

[1] Strain CN27 and strain CN40 were detected in the same dog on 5 September 2020 and 5 November 2020, respectively.

The new CPV-2a strains from Nevis shared absolute deduced VP2 aa identities between themselves and those of new CPV-2a strains reported previously from St. Kitts [24], except for Ala262Thr in 4 CPV-2 strains and Asp373Asn in 3 other CPV-2 strains (Table 2; Supplementary Material S2). Although the significance of the aa at residue 262 of VP2 is not yet known, VP2 Ala262Thr has been described as a novel mutation [38], reported in new CPV-2a and new CPV-2b strains from Western Australia [38], two new CPV-2a strains from India (GenBank accession numbers DQ182624 and KU866399), and a single new CPV-2a strain from China (MH177301). The other nonsynonymous mutation, VP2 Asp373Asn, has been rarely reported in CPV-2 sequences, found in two new CPV-2a strains (one each from Australia (MN259033) and Thailand (GQ379047)), a single CPV-2 strain from a cat in Taiwan (KY010491), and two feline panleukopenia virus strains (FPV) (MH559110 and MK570710). The aa residue at 373 of VP2 is located within the VP2 flexible loop (a surface loop between VP2 residues 359 and 375), which is a pH-sensitive structure that governs binding to divalent ions in FPV and CPV-2 [39,40]. Structural studies on FPV at pH 7.5 have shown that the ion density, adjacent to the flexible loop, is coordinated by Asp 373 and Asp 375, and carbonyl oxygen atoms of Arg 361 and Gly 362 [39]. However, the implication/s of VP2 Asp373Asn remains to be determined. In addition to the two aa mismatches, five synonymous mutations were observed among the VP2 sequences of the new CPV-2a strains from Nevis (Supplementary Material S3).

Following natural CPV-2 infection or immunization with the commercially available modified live virus (MLV) vaccines, dogs have been shown to shed viral DNA for as long as 50 days post-infection [41]. In the present study, one of the dogs was sampled twice (5 September 2020 and 5 November 2020), during hemorrhagic gastroenteritis and after two months, when it was apparently healthy and presented at the clinic for vaccination (Table 1). Both the samples (CN27 and CN40) tested positive for CPV-2 by PCR. Surprisingly, the new CPV-2a strain with VP2 Ala262Thr was identified in the first sample, while the new CPV-2a with VP2 262Ala, detected in the majority of the samples from Nevis and during 2015-2016 in St. Kitts, was detected in the second sample (Table 2). Vaccinated dogs with protective antibody titers have been shown to shed low amounts of CPV-2 field strains in the feces [41,42]. Although the viral DNA was not quantified by qPCR, we observed weak amplification following PCR of the second sample, indicating a low viral load. Therefore, it might be possible that the dog developed immunity following initial infection with the new CPV-2a VP2 Ala262Thr, and eventually was asymptomatically infected with the new CPV-2a VP2 262Ala strain. Alternatively, reversion of VP2 Ala262Thr to the more prevalent new CPV-2a VP2 262Ala might be possible, as reverse mutations have been described in the CPV-2 genome in previous studies [13,16,38,43].

Most of the animals in the present study were not vaccinated or received incomplete vaccination. On the other hand, 3 CPV-2 positive dogs that were completely immunized (vaccine NOBIVAC® CANINE 1-DAPPv, Merck Animal Health, Elkhorn, NE, USA) against the virus suffered from severe clinical disease (Table 1). The vaccine NOBIVAC® CANINE 1-DAPPv contains a CPV-2b variant, while the dogs were infected with new CPV-2a. Although the efficacy of the commercially available CPV-2 MLV vaccines against the different CPV-2 antigenic variants has been debated, they have been shown to be effective in significantly reducing the clinical severity of CPV-2 disease caused by the non-vaccine field variants [3]. Since the dogs were aged $\geq$ 6 months, it is unlikely that maternal antibodies interfered with the efficacy of the vaccine. Other factors, such as vaccine-related errors (issues with vaccine storage, transport and/or administration) and/or host-related factors (impaired immune status, non-responders, and/or malnutrition), might have contributed to vaccine failure [3].

In a previous study on St. Kitts, new CPV-2a was sporadically detected in 20 diarrheic dogs (25/104 dogs were CPV-2 positive; 5 samples could not be sequenced) during a period of 1 year and 7 months [24]. During and around the duration of the outbreak in Nevis, the veterinary clinic on St. Kitts reported only five sporadic cases of CPV-2 infection (three and two dogs tested CPV-2 positive by the SNAP® Parvo Test and PCR, respectively), of which two dogs died (Supplementary Material S4). Analysis of the complete CPV-2 VP2 sequences from the two dead dogs (strain CK81, GenBank accession number MW616470, and strain CK84, MW616472) revealed 100% deduced aa identities with those of the new CPV-2a reported in our study (Supplementary Material S2). Since the canine breeds, environmental conditions, vaccination trends, husbandry practices, and veterinary care are similar between the two islands, we found it intriguing that new CPV-2a was associated with an outbreak of gastroenteritis on Nevis, while found at low frequencies in sporadic cases of diarrhea on the neighboring island of St. Kitts.

There is limited movement of animals between St. Kitts and Nevis, as the twin islands are connected by ferry service. It might be possible that a new CPV-2a was recently introduced into Nevis from St. Kitts and that the canine population on Nevis Island was naive to infection with the virus, resulting in an outbreak situation. However, sporadic cases of CPV-2 have been previously reported in Nevis (based on old case records at the veterinary clinic in Nevis), although none of the samples were molecularly characterized to identify the CPV-2 variant. Therefore, we could not determine whether a new CPV-

2a was circulating in Nevis before the outbreak. Furthermore, based on the available information, we could not ascertain if the two nonsynonymous mutations (Ala262Thr and Asp373Asn) in the VP2 genes of a few new CPV-2a strains appeared during the outbreak or were already circulating at low frequencies in the island canine population. Nevis has a sizable population of stray dogs, which could have facilitated the spread of the virus across the island.

3.2. Analysis of the Nearly Complete Genomes of CPV-2 Strains from St. Kitts and Nevis Islands

Considering the lack of information on CPV-2 genomes from the Caribbean region, we decided to determine the nearly full-length CPV-2 genome sequences (4269 nt, possessing the entire NS and VP coding regions) of 2 strains representing the new CPV-2a circulating in St. Kitts and Nevis (strain CN10 from the outbreak in Nevis, and strain RVC50 from our previous study in St. Kitts [24]), and one of each of the new CPV-2a with a nonsynonymous mutation in the VP2 gene (strain CN20 with VP2 Ala262Thr and strain CN14 with VP2 Asp373Asn).

The nearly complete genomes of CPV-2 strains from St. Kitts and Nevis (henceforth, collectively referred to as SKN strains) shared nt sequence identities of 99.77–99.93% between themselves (Supplementary Material S5). Absolute deduced NS1 and NS2 aa identities were observed between the SKN strains, while the putative VP1/VP2 proteins of strains CN14 and CN20 differed in an aa residue with those of the other SKN strains (Tables 2 and 3; Supplementary Material S6). By BLASTN analysis, the SKN strains shared $\geq$ 99% nt sequence identities with several CPV-2a, CPV-2b, and CPV-2c variants, corroborating previous observations that the complete/nearly complete genomes of all CPV-2 variants are ~99% identical in nt sequence [16]. Although the SKN strains were assigned to new CPV-2a, they shared higher nt sequence identities with the nearly complete genomes of CPV-2c (99.46–99.63% with GenBank accession number KX434458) and CPV-2b (99.39–99.58% with EU659121) strains than those of CPV-2a variants (identities of $\leq$99.32–99.48% with EU659118).

A total of 11 substitutions (5 and 6 within the NS and VP coding region, respectively) were found among the nearly complete SKN CPV-2 genome sequences that included the two nonsynonymous substitutions in the VP coding region (Table 3, Supplementary Materials S5 and S6). The evolution of CPV-2 has been characterized by only a limited number of substitutions that became fixed or widespread during the last 40 years since the emergence of the virus [16]. A previous study identified 38 mutations (15 synonymous and 23 nonsynonymous mutations) that differed in an nt from the earliest CPV-2 strains (the CPV-2 antigenic variant) and were present in >10% of other CPV-2 strains (CPV-2a, CPV-2b, and CPV-2c variants) [16]. In the present study, 35 substitutions (24 synonymous and 11 nonsynonymous substitutions) were observed between the SKN CPV-2 sequences and that of strain CPV12 (representing the earliest CPV-2 strains from the late 1970s, GenBank accession number MN451655), of which 5 synonymous and 1 nonsynonymous substitution have been rarely reported in other CPV-2 strains (<10 CPV-2 sequences) (Table 3, Supplementary Materials S5 and S6). All the four SKN strains retained the five nonsynonymous substitutions in the VP coding region that was characteristic of the global sweep from CPV-2 to CPV-2a during the late 1980s (Table 3, Supplementary Material S5) [16,18].

Table 3. Nucleotide (nt) mismatches (highlighted with yellow) between the nearly complete genome sequences (4269 nt, spanning all the coding regions) of canine parvovirus-2 (CPV-2) strain CPV12 (representing the earliest CPV-2 strains from the late 1970s) and CPV-2 strains detected on St. Kitts (strain RVC50) and Nevis (strains CN10, CN14, and CN20) islands. Nonsynonymous mutations that differed in an nt from the CPV12 sequence are shown in red. Alignment of the nearly full-length nt and complete deduced amino acid sequences of the CPV-2 strains from St. Kitts and Nevis with those of strain CPV12 are shown in Supplementary Materials S5 and S6, respectively.

Nt Position [1]	CPV-2 Strain (Year/Place of Detection)					Nt Change → Translational Effect	Coding Region
	CPV-12 (1978/USA)	RVC50 (2016/St. Kitts)	CN10 (2020/Nevis)	CN14 (2020/Nevis)	CN20 (2020/Nevis)		
516	G	A	A	A	A	G516A → Synonymous	Within the NS coding region
562	T	C	C	C	C	T562C → Synonymous	
726	G	G	A	G	A	G726A → Synonymous	
753	A	A	G	A	G	A753G [2] → Synonymous	
1104	T	T	C	T	C	T1104C [2] → Synonymous	
1164	A	A	G	A	G	A1164G → Synonymous	
1209	T	C	C	C	C	T1209C → Synonymous	
1305	T	C	C	C	C	T1305C [2] → Synonymous	
1623	A	G	G	A	G	A1623G → Synonymous	
1752	A	G	G	G	G	A1752G → NS2 Thr94Ala	
1923	G	A	A	A	A	G1923A → NS2 Asp151Asn	
1926	A	G	G	G	G	A1926G → NS2 Met152Val	
1975	T	C	C	C	C	T1975C → Synonymous	
2086	A	G	G	G	G	A2086G → VP1 Intron	
2154	G	A	A	A	A	G2154A → Synonymous	Within the VP coding region
2436	A	A	G	A	G	A2436G [2] → Synonymous	
2574	T	A	A	A	A	T2754A → Synonymous	
2773	A	T	T	T	T	A2773T → VP2 Met87Leu [3]	
2816	T	C	C	C	C	T2816C → VP2 Ile101Thr [3]	
2923	G	A	A	A	A	G2923A → Synonymous	
2940	A	G	G	G	G	A2940G → Synonymous	
3006	C	T	T	T	T	C3006T → Synonymous	
3039	T	T	G	T	T	T3039G [2] → Synonymous	
3117	A	A	G	A	A	A3117G → Synonymous	
3297	G	G	G	G	A	G3297A → VP2 Ala262Thr	
3403	T	G	G	G	G	T3403G → VP2 Ser297Ala	
3413	C	G	G	G	G	C3413G → VP2 Ala300Gly [3]	
3427	G	T	T	T	T	G3427T → VP2 Asp305Tyr [3]	
3582	T	C	C	C	C	T3582C → Synonymous	
3631	G	G	G	A	G	G3631A [2] → VP2 Asp373Asn	
3637	A	G	G	G	G	A3637G → VP2 Asn375Asp [3]	
3702	G	G	G	A	G	G3702A → Synonymous	
3894	A	G	G	G	G	A3894G → Synonymous	
4017	A	G	G	G	G	A4017G → Synonymous	
4030	T	C	C	C	C	T4030C → Synonymous	

[1] Nucleotide positions are those of the nearly complete genome sequence of strain CPV12 (GenBank accession number MN451655); [2] Present in <10 published CPV-2 sequences, as revealed by BLASTN analysis (https://blast.ncbi.nlm.nih.gov/, accessed on 9 April 2021) and multiple alignments of the dataset of 210 nearly complete CPV-2 sequences; [3] Nucleotide substitutions (and corresponding amino acid changes) that emerged during the global sweep from CPV-2 to CPV-2a [16].

Since recombinant sequences might influence the outcomes of phylogenetic analysis [16,21], the nearly complete SKN CPV-2 genome sequences were screened for recombination events, as described previously [16,25]. However, no recombination breakpoints were detected in any of the SKN CPV-2 sequences. To rule out biases in clustering patterns, phylogenetic trees were created using both the HKY and GTR substitution models, as the

HKY model was determined as the best-fit model using the 'find best model' function of MEGA7, while the GTR model was employed in a recent, important study on the evolution of CPV-2 [16]. Similar clustering patterns were observed with both the models (Figure 2; Supplementary Materials S7 and S8). Findings from previous studies, such as (i) phylogenetic classification of CPV-2 strains into two major clades: CPV-2 (comprising the earliest CPV-2 strains from the late 1970s) and CPV-2a (consisting of CPV-2a, CPV-2b, and CPV-2c variants) [16], (ii) a lack of phylogenetic resolution within the 'CPV-2a clade', characterized by low bootstrap values for most clusters [15,16], (iii) a lack of monophyletic segregation based on CPV-2 variants (CPV-2a, CPV-2b, and CPV-2c) [15,16], and (iv) presence of Asian and Western (CPV-2 sequences from the Americas and Europe) clades [21,44] were retained in the phylogenetic analysis (Figure 2; Supplementary Materials S7 and S8). Phylogenetically, the new CPV-2a SKN strains were placed with the Western CPV-2 strains and retained the signature aa residues that have been previously described as evolutionarily significant and characteristic to the Western clades (Figure 2; Table 4; Supplementary Materials S7 and S8) [21,44]. Within the Western clade, the SKN strains formed a distinct cluster. The nearest cluster to the SKN strains was that of CPV-2b strains detected in the USA during 1998, followed by the major Western CPV-2c cluster (consisting of primarily South American and Italian strains, designated as Western clade-III in a previous study [44]). However, the clade consisting of the SKN strains and the CPV-2b/USA/1998 strains was supported by a low bootstrap value (bootstrap value of 45 and 47 by the GTR and HKY models, respectively). The clustering patterns were corroborated by nt sequence identities and the geographical proximity of St. Kitts and Nevis to the USA and the South American countries. Except for VP2 Ala262Thr (strain CN20) and VP2 Asp373Asn (strain CN14), only a single nonsynonymous mutation (VP2 residue 426) that constitutes the basis of differentiation of CPV-2 into the CPV-2a, CPV-2b, and CPV-2c antigenic variants was observed between the new CPV-2a SKN strains and CPV-2 strains belonging to the CPV-2b/USA/1998 cluster, or the major Western CPV-2c cluster.

Table 4. Evolutionarily relevant amino acid residues in canine parvovirus-2 (CPV-2) protein sequences, as described in previous studies [21,44]. The two nonsynonymous substitutions (VP2 Ala262Thr and VP2 Asp373Asn) observed among the CPV-2 strains from St. Kitts and Nevis are also shown.

Amino Acid Residue	NS1					NS2	VP2				
	60	544	545	572	630	152	262	267	324	373	426
Clade [1]											
CPV-2 origin	Ile	Tyr	Glu	Glu	Leu	Met	Ala	Phe	Tyr	Asp	Asn
Asian	Ile/Val	Phe/Tyr	Glu/Val	Lys	Leu/Pro	Val	Ala	Phe/Tyr	Ile/Tyr	Asp	Asn/Asp/Glu
Western	Ile	Phe/Tyr	Glu	Glu	Leu	Met/Val	Ala	Phe	Leu/Tyr	Asp	Asn/Asp/Glu
Strain/Place/Year											
CN10/Nevis/2020	Ile	Tyr	Glu	Glu	Leu	Val	Ala	Phe	Tyr	Asp	Asn
CN14/Nevis/2020	Ile	Tyr	Glu	Glu	Leu	Val	Ala	Phe	Tyr	Asn [3]	Asn
CN20/Nevis/2020	Ile	Tyr	Glu	Glu	Leu	Val	Thr [2]	Phe	Tyr	Asp	Asn
RVC50/St. Kitts/2016	Ile	Tyr	Glu	Glu	Leu	Val	Ala	Phe	Tyr	Asp	Asn

[1] As described in previous studies [21,44]. Western: CPV-2 strains from Europe and the Americas; [2] Reported in new CPV-2a and new CPV-2b strains from Western Australia [38], 2 new CPV-2a strains from India (GenBank accession numbers DQ182624 and KU866399), and a single new CPV-2a strain from China (MH177301); [3] Found in 2 new CPV-2a strains (one each from Australia (MN259033) and Thailand (GQ379047)), a single CPV-2 strain from a cat in Taiwan (KY010491), and 2 feline panleukopenia virus strains (MH559110 and MK570710).

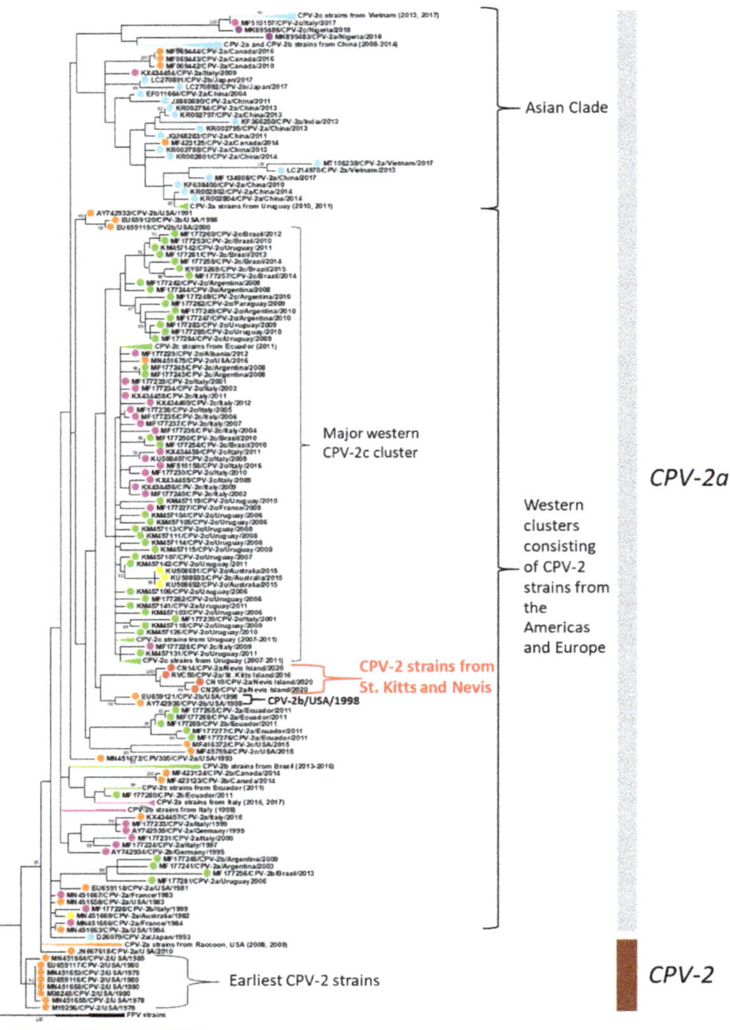

Figure 2. Phylogenetic analysis of the nearly complete genomes of canine parvovirus-2 (CPV-2) strains from St. Kitts and Nevis with those of other CPV-2 strains. The tree was created using the Hasegawa-Kishino-Yano model with gamma-distributed rate variation among sites and 1000 bootstrap replicates. The name of the strain/CPV-2 antigenic variant/place/year of detection is shown for the CPV-2 strains from St. Kitts and Nevis, while the GenBank accession number/CPV-2 antigenic variant/place/year of detection have been mentioned for the other CPV-2 strains. The two major phylogenetic clades (CPV-2 and CPV-2a) are demarcated with a brown and a light gray bar, respectively. The CPV-2 clade consists of the earliest CPV-2 strains from the late 1970s (the CPV-2 antigenic variants), while the 'CPV-2a clade' is composed of the CPV-2a, CPV-2b, and CPV-2c antigenic variants [16]. Feline panleukopenia virus (FPV) strains were included in the analysis. Sky blue, purple, yellow, pink, orange, green, and red circles indicate that the CPV-2 strain was detected in Asia, Africa, Australia, Europe, North America, South America, and St. Kitts and Nevis, respectively. Scale bar, 0.01 substitutions per nucleotide. Bootstrap values of <70 are not shown. Some of the clusters have been compressed to accommodate the entire tree. The complete phylogenetic tree is shown in Supplementary Material S7. Phylogenetic analysis using the General Time Reversible model is shown in Supplementary Material S8.

Corroborating previous observations [15,16,21,44], phylogenetically, the clustering of the nearly complete SKN CPV-2 genomes with those of other CPV-2 strains did not correspond to the clustering patterns based on VP2 coding sequences (Figure 2; Supplementary Materials S7–S9). The complete VP2 nt sequences of the SKN strains clustered near that of CPV-2a strain CPV-435/USA/2003 (GenBank accession number AY742953), although the clade was supported by a low bootstrap value of 45 (Supplementary Material S9). Since the complete/nearly complete genome sequence of CPV-435/USA/2003 was not available in the GenBank database, we could not include the strain in the analysis of nearly complete CPV-2 genomes.

Taken together, these findings supported previous observations that recurrent parallel evolution and reversion might play important roles in the evolution of CPV-2 and that the recently circulating CPV-2 strains are minor variants of a common 'pan genome' template that emerged after the global sweep from CPV-2 to CPV-2a [15,16].

It is likely that CPV-2 was imported into St. Kitts and Nevis Islands from another country. Since more humans (and pet dogs) travel to St. Kitts from the USA than from any other country, it might be possible the virus was introduced into the twin islands from the USA. On the other hand, St. Kitts and Nevis have a large population of the small Indian mongoose (*Urva auropunctata*) that thrive in wild, rural, and urban habitats [45]. A previous study had reported high rates of detection of parvovirus DNA (58%, n = 99) and antibodies (90%, n = 20) in the Egyptian mongoose (*Herpestes ichneumon*) [46]. Considering the role of wild canids as a potential source of CPV-2 to pet dogs [2,13,17,46,47], it would be interesting to investigate whether the new CPV-2a and/or new CPV-2a mutants (VP2 Ala262Thr and VP2 Asp373Asn) circulating in dogs on St. Kitts and Nevis were derived from the local mongoose population.

3.3. Conclusions

Although the currently licensed CPV-2 vaccines have been extensively used in veterinary practice and shown to confer protection against the different CPV-2 antigenic variants, CPV-2 continues to remain one of the leading causes of mortality and morbidity in domestic dogs [2–4]. To date, there is a dearth of data on CPV-2 from the Caribbean region. Our findings suggested that new CPV-2a might be endemic in this part of the world, with the potential to cause severe outbreaks facilitated by low vaccination rates among the canine population in the region. Recently, the neighboring islands of St. Eustatius and St. Lucia experienced large outbreaks of CPV-2 [48,49]. However, no information was available on the nature of CPV-2 variant/s circulating during the outbreaks. These observations underscore the importance of continuous molecular epidemiological studies on CPV-2 in all the Caribbean Islands alongside creating public awareness on the disease and vaccination.

The nearly complete CPV-2 genomes were reported for the first time from the Caribbean region, providing important insights into the overall genetic makeup of the CPV-2 strains circulating in St. Kitts and Nevis, especially the identification of six substitutions (five synonymous and one nonsynonymous substitution) that have been rarely reported in other CPV-2 sequences. Overall, analysis of the SKN CPV-2 genomes corroborated the hypothesis that recurrent parallel evolution and reversion might play important roles in the evolution of CPV [15,16]. Once again, it was revealed that the analysis of CPV-2 VP2 aa sequences do not reflect the true evolutionary patterns of CPV-2 strains.

Supplementary Materials: The following are available online at https://www.mdpi.com/article/10.3390/v13061083/s1, Supplementary Material S1: Primers used in the present study, Supplementary Material S2: Multiple alignments of the putative VP2 proteins of canine parvovirus-2 (CPV-2) strains detected in Nevis (designated with prefix CN) with those reported from St. Kitts in 2016 (strain RVC50) and in 2020 (strains CK81 and CK84), Supplementary Material S3: Multiple alignments of the VP2 coding sequences of canine parvovirus-2 (CPV-2) strains detected in Nevis (designated with prefix CN) during 2020 with those reported from the neighboring island of St. Kitts in 2016 (strain RVC50) and in 2020 (strains CK81 and CK84), Supplementary Material S4: Gross (A) and microscopic (B) lesions observed in the small intestine of a dog with hemorrhagic gastroenteritis that eventually died on the island of St. Kitts. Small intestinal scrapings collected during necropsy tested positive for canine parvovirus-2 (CPV-2) by PCR. Analysis of the complete deduced VP2 amino acid sequence identified the CPV-2 strain as new CPV-2a, Supplementary Material S5: Nucleotide (nt) mismatches between the nearly complete genome sequences (4269 nt, spanning all the coding regions) of canine parvovirus-2 (CPV-2) strain CPV12 (representing the earliest CPV-2 strains from the late 1970s) and CPV-2 strains detected on St. Kitts in 2016 (strain RVC50) and Nevis in 2020 (strains CN10, CN14, and CN20), Supplementary Material S6: Multiple alignments of the putative NS1 (A), NS2 (B), VP1 (C) and VP2 (D) proteins of canine parvovirus-2 (CPV-2) strain CPV12 (GenBank accession number MN451655, representing the earliest CPV-2 strains from the late 1970s) and CPV-2 strains detected on St. Kitts in 2016 (strain RVC50) and Nevis in 2020 (strains CN10, CN14, and CN20), Supplementary Material S7: Phylogenetic analysis of the nearly complete genomes of canine parvovirus-2 (CPV-2) strains from St. Kitts and Nevis with those of other CPV-2 strains. The tree was created using the Hasegawa-Kishino-Yano model with gamma-distributed rate variation among sites and 1000 bootstrap replicates, Supplementary Material S8: Phylogenetic analysis of the nearly complete genomes of canine parvovirus-2 (CPV-2) strains from St. Kitts and Nevis with those of other CPV-2 strains. The tree was created using the General Time Reversible model with gamma-distributed rate variation among sites and 1000 bootstrap replicates, Supplementary Material S9: Phylogenetic analysis of the complete VP2 coding sequences of canine parvovirus-2 (CPV-2) strains from St. Kitts and Nevis with those of other CPV-2 strains.

Author Contributions: Conceptualization, K.G. and S.G.; collected samples, A.B. and P.B.; collected epidemiological data, A.B., A.P., K.G., P.B. and S.G.; secured funding, S.G.; contributed reagents, S.G. screened samples, A.B., K.G. and S.G.; performed laboratory work, K.G. and S.G.; performed data analysis, K.G. and S.G.; wrote the manuscript, K.G. and S.G.; edited and finalized the manuscript A.B., A.P., P.B., K.G., S.G. and Y.S.M. All authors have read and agreed to the published version of the manuscript.

Funding: The present study was funded by intramural grant # Viruses 41001-21 entitled 'Detection and molecular characterization of viruses in pigs and wildlife in the Caribbean and Central America from the One Health Center for Zoonoses and Tropical Veterinary Medicine, Ross University School of Veterinary Medicine, St. Kitts and Nevis.

Institutional Review Board Statement: The present study was submitted to the Institutional Animal Care and Use Committee (IACUC) of the Ross University School of Veterinary Medicine (RUSVM) St. Kitts Island. Ethical review and approval were waived for this study by the RUSVM IACUC as the research study was based on leftover samples that were collected for diagnostic purposes at the veterinary clinic on Nevis Island (RUSVM IACUC sample/tissue notification letter number TSU 1.23.21).

Acknowledgments: We would like to thank the staff members of Nevis Animal Speak, Nevis Island for helping with the sampling and the division of pathology, Ross University School of Veterinary Medicine, St. Kitts Island, for performing the necropsies on the two dead dogs.

Conflicts of Interest: The authors declare no conflict of interest.

References

1. Cotmore, S.F.; Agbandje-McKenna, M.; Canuti, M.; Chiorini, J.A.; Eis-Hubinger, A.-M.; Hughes, J.; Mietzsch, M.; Modha, S. Ogliastro, M.; Pénzes, J.J.; et al. ICTV Virus Taxonomy Profile: Parvoviridae. *J. Gen. Virol.* **2019**, *100*, 367–368. [CrossRef]
2. Decaro, N.; Buonavoglia, C. Canine parvovirus—A review of epidemiological and diagnostic aspects, with emphasis on type 2. *Vet. Microbiol.* **2012**, *155*, 1–12. [CrossRef]
3. Decaro, N.; Buonavoglia, C.; Barrs, V. Canine parvovirus vaccination and immunisation failures: Are we far from disease eradication? *Vet. Microbiol.* **2020**, *247*, 108760. [CrossRef]
4. Mazzaferro, E.M. Update on Canine Parvoviral Enteritis. *Vet. Clin. N. Am. Small Anim. Pr.* **2020**, *50*, 1307–1325. [CrossRef]
5. Reed, A.P.; Jones, E.V.; Miller, T.J. Nucleotide sequence and genome organization of canine parvovirus. *J. Virol.* **1988**, *62*, 266–27 [CrossRef]
6. Niskanen, E.A.; Kalliolinna, O.; Ihalainen, T.O.; Häkkinen, M.; Vihinen-Ranta, M. Mutations in DNA Binding and Transactivatio Domains Affect the Dynamics of Parvovirus NS1 Protein. *J. Virol.* **2013**, *87*, 11762–11774. [CrossRef] [PubMed]
7. Niskanen, E.A.; Ihalainen, T.O.; Kalliolinna, O.; Häkkinen, M.M.; Vihinen-Ranta, M. Effect of ATP Binding and Hydrolysis on Dynamics of Canine Parvovirus NS1. *J. Virol.* **2010**, *84*, 5391–5403. [CrossRef] [PubMed]

8. Wang, D.; Yuan, W.; Davis, I.; Parrish, C.R. Nonstructural Protein-2 and the Replication of Canine Parvovirus. *Virology* **1998**, *240*, 273–281. [CrossRef] [PubMed]
9. Mietzsch, M.; Pénzes, J.J.; Mc Kenna, A. Twenty-Five Years of Structural Parvovirology. *Viruses* **2019**, *11*, 362. [CrossRef] [PubMed]
10. Tsao, J.; Chapman, M.S.; Agbandje, M.; Keller, W.; Smith, K.; Wu, H.; Luo, M.; Smith, T.J.; Rossmann, M.G.; Compans, R.W.; et al. The three-dimensional structure of canine parvovirus and its functional implications. *Science* **1991**, *251*, 1456–1464. [CrossRef] [PubMed]
11. Hueffer, K.; Parker, J.; Weichert, W.S.; Geisel, R.E.; Sgro, J.-Y.; Parrish, C.R. The Natural Host Range Shift and Subsequent Evolution of Canine Parvovirus Resulted from Virus-Specific Binding to the Canine Transferrin Receptor. *J. Virol.* **2003**, *77*, 1718–1726. [CrossRef]
12. Nelson, C.D.; Palermo, L.M.; Hafenstein, S.L.; Parrish, C.R. Different mechanisms of antibody-mediated neutralization of parvoviruses revealed using the Fab fragments of monoclonal antibodies. *Virology* **2007**, *361*, 283–293. [CrossRef] [PubMed]
13. Miranda, C.; Thompson, G. Canine parvovirus: The worldwide occurrence of antigenic variants. *J. Gen. Virol.* **2016**, *97*, 2043–2057. [CrossRef] [PubMed]
14. Zhou, P.; Zeng, W.; Zhang, X.; Li, S. The genetic evolution of canine parvovirus—A new perspective. *PLoS ONE* **2017**, *12*, e0175035. [CrossRef]
15. Mira, F.; Canuti, M.; Purpari, G.; Cannella, V.; Di Bella, S.; Occhiogrosso, L.; Schirò, G.; Chiaramonte, G.; Barreca, S.; Pisano, P.; et al. Molecular Characterization and Evolutionary Analyses of Carnivore Protoparvovirus 1 NS1 Gene. *Viruses* **2019**, *11*, 308. [CrossRef]
16. Voorhees, I.E.H.; Lee, H.; Allison, A.B.; Lopez-Astacio, R.; Goodman, L.; Oyesola, O.O.; Omobowale, O.; Fagbohun, O.; Dubovi, E.J.; Hafenstein, S.L.; et al. Limited Intrahost Diversity and Background Evolution Accompany 40 Years of Canine Parvovirus Host Adaptation and Spread. *J. Virol.* **2019**, *94*, 01162-19. [CrossRef] [PubMed]
17. Parrish, C.R. Host range relationships and the evolution of canine parvovirus. *Vet. Microbiol.* **1999**, *69*, 29–40. [CrossRef]
18. Truyen, U.; Evermann, J.F.; Vieler, E.; Parrish, C.R. Evolution of Canine Parvovirus Involved Loss and Gain of Feline Host Range. *Virology* **1996**, *215*, 186–189. [CrossRef]
19. Parrish, C.R.; Aquadro, C.F.; Strassheim, M.L.; Evermann, J.F.; Sgro, J.Y.; Mohammed, H.O. Rapid antigenic-type replacement and DNA sequence evolution of canine parvovirus. *J. Virol.* **1991**, *65*, 6544–6552. [CrossRef]
20. Buonavoglia, C.; Martella, V.; Pratelli, A.; Tempesta, M.; Cavalli, A.; Buonavoglia, D.; Bozzo, G.; Elia, G.; Decaro, N.; Carmichael, L. Evidence for evolution of canine parvovirus type 2 in Italy. *J. Gen. Virol.* **2001**, *82*, 3021–3025. [CrossRef]
21. Grecco, S.; Iraola, G.; DeCaro, N.; Alfieri, A.; Alfieri, A.; Calderón, M.G.; Da Silva, A.P.; Name, D.; Aldaz, J.; Calleros, L.; et al. Inter- and intracontinental migrations and local differentiation have shaped the contemporary epidemiological landscape of canine parvovirus in South America. *Virus Evol.* **2018**, *4*, vey011. [CrossRef]
22. Touihri, L.; Bouzid, I.; Daoud, R.; Desario, C.; EL Goulli, A.F.; DeCaro, N.; Ghorbel, A.; Buonavoglia, C.; Bahloul, C. Molecular characterization of canine parvovirus-2 variants circulating in Tunisia. *Virus Genes* **2008**, *38*, 249–258. [CrossRef]
23. Martella, V.; DeCaro, N.; Elia, G.; Buonavoglia, C. Surveillance Activity for Canine Parvovirus in Italy. *J. Vet. Med. Ser. B* **2005**, *52*, 312–315. [CrossRef] [PubMed]
24. Navarro, R.; Nair, R.; Peda, A.; Aung, M.S.; Ashwinie, G.S.; Gallagher, C.A.; Malik, Y.S.; Kobayashi, N.; Ghosh, S. Molecular characterization of canine parvovirus and canine enteric coronavirus in diarrheic dogs on the island of St. Kitts: First report from the Caribbean region. *Virus Res.* **2017**, *240*, 154–160. [CrossRef]
25. Martin, D.P.; Murrell, B.; Golden, M.; Khoosal, A.; Muhire, B. RDP4: Detection and analysis of recombination patterns in virus genomes. *Virus Evol.* **2015**, *1*, vev003. [CrossRef] [PubMed]
26. Kumar, S.; Stecher, G.; Tamura, K. MEGA7: Molecular Evolutionary Genetics Analysis Version 7.0 for Bigger Datasets. *Mol. Biol. Evol.* **2016**, *33*, 1870–1874. [CrossRef]
27. Yip, H.Y.E.; Peaston, A.; Woolford, L.; Khuu, S.J.; Wallace, G.; Kumar, R.S.; Patel, K.; Azari, A.A.; Akbarzadeh, M.; Sharifian, M.; et al. Diagnostic Challenges in Canine Parvovirus 2c in Vaccine Failure Cases. *Viruses* **2020**, *12*, 980. [CrossRef]
28. De Oliveira, P.S.B.; Cargnelutti, J.F.; Masuda, E.K.; Weiblen, R.; Flores, E.F. New variants of canine parvovirus in dogs in southern Brazil. *Arch. Virol.* **2019**, *164*, 1361–1369. [CrossRef] [PubMed]
29. Castillo, C.; Neira, V.; Aniñir, P.; Grecco, S.; Pérez, R.; Panzera, Y.; Zegpi, N.-A.; Sandoval, A.; Sandoval, D.; Cofre, S.; et al. First Molecular Identification of Canine Parvovirus Type 2 (CPV2) in Chile Reveals High Occurrence of CPV2c Antigenic Variant. *Front. Vet. Sci.* **2020**, *7*, 194. [CrossRef]
30. De La Torre, D.; Mafla, E.; Puga, B.; Erazo, L.; Astolfi-Ferreira, C.; Ferreira, A.P. Molecular characterization of canine parvovirus variants (CPV-2a, CPV-2b, and CPV-2c) based on the VP2 gene in affected domestic dogs in Ecuador. *Vet. World* **2018**, *11*, 480–487. [CrossRef]
31. Maya, L.; Calleros, L.; Francia, L.; Hernández, M.; Iraola, G.; Panzera, Y.; Sosa, K.; Pérez, R. Phylodynamics analysis of canine parvovirus in Uruguay: Evidence of two successive invasions by different variants. *Arch. Virol.* **2013**, *158*, 1133–1141. [CrossRef] [PubMed]
32. Giraldo-Ramirez, S.; Rendon-Marin, S.; Ruiz-Saenz, J. Phylogenetic, Evolutionary and Structural Analysis of Canine Parvovirus (CPV-2) Antigenic Variants Circulating in Colombia. *Viruses* **2020**, *12*, 500. [CrossRef]
33. Quispe, R.Q.; Espinoza, L.L.; Beltrán, R.R.; Alcántara, R.H.R.; Hernández, L.M. Canine parvovirus types 2a and 2c detection from dogs with suspected parvoviral enteritis in Peru. *Virus Dis.* **2018**, *29*, 109–112. [CrossRef] [PubMed]

34. Kapil, S.; Cooper, E.; Lamm, C.; Murray, B.; Rezabek, G.; Johnston, L.; Campbell, G.; Johnson, B. Canine Parvovirus Types 2c and 2b Circulating in North American Dogs in 2006 and 2007. *J. Clin. Microbiol.* **2007**, *45*, 4044–4047. [CrossRef]
35. Parker, J.; Murphy, M.; Hueffer, K.; Chen, J. Investigation of a Canine Parvovirus Outbreak using Next Generation Sequencing. *Sci. Rep.* **2017**, *7*, 9633. [CrossRef] [PubMed]
36. Gagnon, C.A.; Allard, V.; Cloutier, G. Canine parvovirus type 2b is the most prevalent genomic variant strain found in parvovirus antigen positive diarrheic dog feces samples across Canada. *Can. Vet. J.* **2016**, *57*, 29–31.
37. Faz, M.; Martínez, J.S.; Gómez, L.B.; Quijano-Hernández, I.; Fajardo, R.; Del Ángel-Caraza, J. Origin and genetic diversity of canine parvovirus 2c circulating in Mexico. *Arch. Virol.* **2018**, *164*, 371–379. [CrossRef] [PubMed]
38. Kwan, E.; Carrai, M.; Lanave, G.; Hill, J.; Parry, K.; Kelman, M.; Meers, J.; Decaro, N.; Beatty, J.A.; Martella, V.; et al. Analysis of canine parvoviruses circulating in Australia reveals predominance of variant 2b and identifies feline parvovirus-like mutations in the capsid proteins. *Transbound. Emerg. Dis.* **2021**, *68*, 656–666. [CrossRef] [PubMed]
39. Simpson, A.A.; Chandrasekar, V.; Hébert, B.; Sullivan, G.M.; Rossmann, M.G.; Parrish, C.R. Host range and variability of calcium binding by surface loops in the capsids of canine and feline parvoviruses. *J. Mol. Biol.* **2000**, *300*, 597–610. [CrossRef]
40. Organtini, L.J.; Allison, A.B.; Lukk, T.; Parrish, C.R.; Hafenstein, S. Global Displacement of Canine Parvovirus by a Host-Adapted Variant: Structural Comparison between Pandemic Viruses with Distinct Host Ranges. *J. Virol.* **2014**, *89*, 1909–1912. [CrossRef] [PubMed]
41. Decaro, N.; Buonavoglia, C. Canine parvovirus post-vaccination shedding: Interference with diagnostic assays and correlation with host immune status. *Vet. J.* **2017**, *221*, 23–24. [CrossRef] [PubMed]
42. Freisl, M.; Speck, S.; Truyen, U.; Reese, S.; Proksch, A.-L.; Hartmann, K. Faecal shedding of canine parvovirus after modified-live vaccination in healthy adult dogs. *Vet. J.* **2017**, *219*, 15–21. [CrossRef] [PubMed]
43. Allison, A.B.; Kohler, D.J.; Ortega, A.; Hoover, E.A.; Grove, D.M.; Holmes, E.C.; Parrish, C.R. Host-Specific Parvovirus Evolution in Nature Is Recapitulated by In Vitro Adaptation to Different Carnivore Species. *PLoS Pathog.* **2014**, *10*, e1004475. [CrossRef] [PubMed]
44. Manh, T.N.; Piewbang, C.; Rungsipipat, A.; Techangamsuwan, S. Molecular and phylogenetic analysis of Vietnamese canine parvovirus 2C originated from dogs reveals a new Asia-IV clade. *Transbound. Emerg. Dis.* **2021**, *68*, 1445–1453. [CrossRef]
45. Kleymann, A.; Becker, A.A.M.J.; Malik, Y.S.; Kobayashi, N.; Ghosh, S. Detection and Molecular Characterization of Picobirnaviruses (PBVs) in the Mongoose: Identification of a Novel PBV Using an Alternative Genetic Code. *Viruses* **2020**, *12*, 99. [CrossRef] [PubMed]
46. Duarte, M.D.; Henriques, A.M.; Barros, S.C.; Fagulha, T.; Mendonça, P.; Carvalho, P.; Monteiro, M.; Fevereiro, M.; Basto, M.P.; Rosalino, L.M.; et al. Snapshot of Viral Infections in Wild Carnivores Reveals Ubiquity of Parvovirus and Susceptibility of Egyptian Mongoose to Feline Panleukopenia Virus. *PLoS ONE* **2013**, *8*, e59399. [CrossRef]
47. Steinel, A.; Parrish, C.R.; Bloom, M.E.; Truyen, U. Parvovirus Infections in Wild Carnivores. *J. Wildl. Dis.* **2001**, *37*, 594–607. [CrossRef]
48. Suspected Outbreak of Virus among Statia Dogs. Available online: https://www.thedailyherald.sx/islands/suspected-outbreak-of-virus-among-statia-dogs (accessed on 25 April 2021).
49. St Lucia Seeing Alarming Increase in Deadly Canine Parvovirus. Available online: https://www.loopslu.com/content/st-lucia-seeing-alarming-increase-deadly-canine-parvovirus (accessed on 25 April 2021).

Review

Battling Neurodegenerative Diseases with Adeno-Associated Virus-Based Approaches

Olja Mijanović [1], Ana Branković [2], Anton V. Borovjagin [3], Denis V. Butnaru [4], Evgeny A. Bezrukov [5], Roman B. Sukhanov [5], Anastasia Shpichka [4], Peter Timashev [4,6,7,8] and Ilya Ulasov [1,*]

1 Group of Experimental Biotherapy and Diagnostics, Institute for Regenerative Medicine, Sechenov First Moscow State Medical University, 119991 Moscow, Russia; olja.mijanovic@gmail.com
2 Department of Forensics, University of Criminal Investigation and Police Studies, 11000 Belgrade, Serbia; sonny_brankovic@yahoo.com
3 Department of Biomedical Engineering, University of Alabama at Birmingham, Birmingham, AL 35294, USA; aborovjagin@gmail.com
4 Institute for Regenerative Medicine, Sechenov First Moscow State Medical University, 119991 Moscow, Russia; butnaru_dv@mail.ru (D.V.B.); ana-shpichka@yandex.ru (A.S.); timashev.peter@gmail.com (P.T.)
5 Institute for Uronephrology and Reproductive Health, Sechenov First Moscow State Medical University, 119991 Moscow, Russia; eabezrukov@rambler.ru (E.A.B.); rb_suhanov@mail.ru (R.B.S.)
6 Institute of Photonic Technologies, Research Center "Crystallography and Photonics", Russian Academy of Sciences, Troitsk, 142190 Moscow, Russia
7 Department of Polymers and Composites, N.N. Semenov Institute of Chemical Physics, 119991 Moscow, Russia
8 Chemistry Department, Lomonosov Moscow State University, 119991 Moscow, Russia
* Correspondence: ulasov75@yahoo.com

Received: 25 February 2020; Accepted: 15 April 2020; Published: 18 April 2020

Abstract: Neurodegenerative diseases (NDDs) are most commonly found in adults and remain essentially incurable. Gene therapy using AAV vectors is a rapidly-growing field of experimental medicine that holds promise for the treatment of NDDs. To date, effective delivery of a therapeutic gene into target cells via AAV has been a major obstacle in the field. Ideally, transgenes should be delivered into the target cells specifically and efficiently, while promiscuous or off-target gene delivery should be minimized to avoid toxicity. In the pursuit of an ideal vehicle for NDD gene therapy, a broad variety of vector systems have been explored. Here we specifically outline the advantages of adeno-associated virus (AAV)-based vector systems for NDD therapy application. In contrast to many reviews on NDDs that can be found in the literature, this review is rather focused on AAV vector selection and their testing in experimental and preclinical NDD models. Preclinical and in vitro data reveal the strong potential of AAV for NDD-related diagnostics and therapeutic strategies.

Keywords: AAV; neuro-degenerative disease; gene therapy

---

1. Background

Neurodegenerative diseases (NDDs) are the most common primary pathologies of the central nervous system (CNS) in adults. They affect the majority of the elderly population and arise either as a consequence of an accumulation of previously acquired genetic modification in the human genome or de novo—i.e., without preceding neurodegenerative pathologies—as a result of an accumulation of neurotoxic proteins in the cytoplasm of neurons (de novo disease). The lack of effective therapies to control NDD made the progression of the disease inevitable [1]. Systemic effects from NDD development are usually associated with infiltration of the brain with peripheral immune cells [2], which contributes to neuroinflammation [3] and neurodegeneration [4]. The cause of systemic and molecular abnormalities developing throughout the NDD progression remains to be investigated as any developments in the NDD therapy field require a detailed understanding of the molecular mechanism of neurodegeneration process as well as how to intervene with the disease process.

A significant determinant of NDD progression is an accumulation of toxic proteins on a subcellular level [5,6]. Intracellular accumulation of specific abnormally-aggregated proteins in the form of inclusion bodies is associated with different pathologies of most common NDDs: while Alzheimer's destroys memory, Parkinson's and Huntington's diseases affect movement. Previous research has suggested that all three diseases are caused by the death of neurons and other cells in the brain, which is associated with abnormally-folding and aggregating amyloid proteins, different in each disease. For instance, a misfolded amyloid-β and mutated tau, α-synuclein, huntingtin, and ALS-linked trans-activation response (TAR) DNA-binding protein 43 (*TARDBP*) along with FUS are implicated in molecular mechanisms of Alzheimer's, Parkinson's, Huntington's, and amyotrophic lateral sclerosis (ALS), respectively [7]. Those all form insoluble clumps inside brain cells that ultimately rupture the endocytic vesicles. It was suggested that the latter mechanism is conserved and, therefore, common treatment strategies should be possible that would involve boosting a brain cell's ability to degrade protein clumps and damaged vesicles [8]. This observation offers a possibility for therapeutic intervention, as it suggests the involvement of only a few signaling pathways that recently emerged as potential targets for neuronal dysregulation. Among them is autophagy signaling, whose dysfunction permits the accumulation of misfolded proteins in various structural parts of the neural cells [9,10]. All the dysregulation results in excessive inhibition of adenosine triphosphate (ATP)-dependent chaperones and proteolytic machinery [11] that represses neurogenesis [12], stimulates damage via oxidative stress [13], and promotes neuronal cell death. Recent studies suggest that inactivated autophagy-related proteins in damaged neurons [14] represent a suitable target for experimental NDD therapy [15]. Although there is extensive literature on the accumulation of toxic proteins inside neuronal cells as a result of autophagy repression and its pharmacological correction [16], relatively few studies have addressed the possibility of targeting neuronal cells with aberrant signaling [17–21] by using alternate approaches, such as gene therapy. Below, we summarized the benefits of gene therapy application for NDD treatment.

Neurodegenerative diseases are typically caused by numerous genetic abnormalities, whose effect is enhanced by environmental factors [22] and epigenetic events [23,24]. Basic research has identified numerous pathways that contribute to NDD pathogenesis. Based on this rapidly growing knowledge, multiple drugs for NDD therapy have been approved to-date [25–29]. Interestingly, Chinese herbal medicine may also represent a promising alternative for NDD treatment [30]. In the past three decades, the scientific community has made significant efforts in utilizing newly developed molecular technologies to selectively deliver a therapeutic payload into target cells via a nonconventional method, known as "gene therapy" [31,32].

Viral vector-based gene therapy offers a new therapeutic opportunity against an irreversible lethal impact of many inheritable or genetic diseases. For instance, in the case of Spinal Muscular Atrophy type 1 (SMA1), a single dose of intravenously delivered AAV9 vector carrying a transgene encoding the missing SMN protein resulted in longer survival of patients [33]. Another example is the study by Massaro et al., who used a mouse model of an untreatable acute form of infantile neuronopathic Gaucher disease to inject an adeno-associated virus (AAV)-based vector encoding neuronal glucocerebrosidase that abolished neurodegeneration and ameliorated neuroinflammation [34]. In a later study, gene therapy with an AAV vector improved motor skills and increased lifespan of animals in a mouse model of inherited lysosomal storage disorders, known as the neuronal ceroid-lipofuscinoses (NCLs) or more commonly referred to as Batten disease, a form of NCLs with a deficiency in the membrane-bound protein CLN6 (CLN6 disease) [35]. Although it is too early for evaluating the clinical efficacy of the developed AAV gene therapy vectors in human patients, some AAV-based gene therapies demonstrate a remarkable therapeutic effect in preclinical studies. For example, an AAV-based gene therapy aimed at correcting a rare metabolic aromatic L-amino acid decarboxylase deficiency (AADC) was granted a biologics license application (BLA)-ready status by the FDA in 2018 [36].

2. Viral Vector-Based Gene Therapy: Efficiency and Selectivity of Delivery

2.1. Choosing an Ideal Vector

Gene therapy for NDDs represents a rapidly developing novel therapeutic approach with a potential capability of restoring normal neuronal functions [37] by intracellular delivery of exogenous genetic material, carrying a disease correcting/therapeutic information in the form of recombinant deoxyribonucleic acid (DNA) or ribonucleic acid (RNA) molecules engineered into a delivery vehicle, known as vector. Once inside a recipient (host) cell, the transferred genetic material utilizes cellular machinery to biosynthesize (translate the encoded information into) therapeutic proteins aimed at replacing their defective counterparts, produced in neurons by mutated endogenous gene, inactivating the mutated gene's product or inserting a foreign transgene with therapeutic properties to battle the disease. Despite the variability in the existing delivery vectors, routes of delivery, and delivery techniques, the intracellular mechanism of the delivered gene's expression remains essentially the same. The transduced or transfected host cells then begin synthesizing protein(s) aimed at repairing a congenital disorder or treat an acquired disease [38]. Depending on the nature of target cells, gene delivery can be classified as either somatic or germline. While the goal of somatic gene delivery is to directly express the transgene in target cells and correct the disease at the individual cell level [39], germline gene transfer is aimed at integrating a therapeutic or correction gene (fully functional wild type copy) into the germline cells, thereby creating inheritable changes, affecting every cell of the next generation. Although more advanced techniques of gene delivery are constantly being developed, the most common approach to correct NDDs involves a direct transfection or transduction (when using virus-based delivery vectors) of isolated patient-derived cells or modification of target cells in situ—i.e., directly in patient's tissue. The former strategy of gene delivery, known as ex vivo gene therapy, involves isolation and in vitro culturing of patient-derived cells for exogenous genetic modification or gene correction and their subsequent injection back into the patient [40]. In contrast, in vivo gene therapy involves the delivery of genetic modifications to target cells directly by systemic [41]—i.e., intraperitoneal (ip) [42,43] or intravenous (iv) [44–46] vector administration. However, large doses of systemically-delivered AAV vectors elicit immune responses affecting transgene expression, treatment longevity and even patient safety. To circumvent those drawbacks, alternate routes of AAV administration to the central nervous system (CNS) have been explored that allowed administration of lower doses of the gene therapy vector directly by into the brain parenchyma or by injection into the cerebrospinal fluid via the intracerebroventricular or intrathecal (cisternal or lumbar) routes. This strategy results in higher transgene expression along with the decreased immune responses. However, the remaining critical concern of the direct CNS-delivery is AAV-associated neuroinflammation, manifested in preclinical studies by the dorsal root ganglion (DRG) and spinal cord pathology with mononuclear cell infiltration [47].

Since the precision of neural cell targeting as well as the expression magnitude of a therapeutic payload are critical characteristics for the efficient repairing of genetic or metabolic abnormalities, the use of viral vector-based gene therapy, well-suited for modulation of those characteristics, offers a significant advantage over the utilization of non-viral vectors.

Due to an essential condition for an effective gene therapy intervention is an efficient and specific delivery of a therapeutic transgene to target cells, an ideal vector should be inherently suitable for or possibly tailored (genetically engineered) to specific target cells of interest. In addition, it should not trigger a robust immune response in the host and be capable of providing a sustained transgene expression for the duration of disease treatment [48]. By these criteria, some virus-based vectors became preferred gene delivery vehicles in contemporary gene therapy applications. Since the natural mechanism of viral cell entry involves attachment to target cell's surface receptor(s), most virus-based gene therapy vectors can be engineered by retargeting the viral particle (capsid) surface to primary receptors abundant on or even specific to target cells by using genetic and other molecular technologies developed in the past two decades. Viral gene therapy vectors specifically designed for NDD applications should be able to efficiently cross the blood–brain barrier (BBB), if delivered by iv infusion [44,49], and support an efficient transgene expression, regardless of aberrant cell signaling caused by the accumulation of toxic proteins or restoration of an autophagic state in neuronal cells [50]. For instance, vectors based on genomes of adeno-associated virus (AAV) [44,49,51–56], lentivirus [57], type 1 herpes simplex virus [58], and Semliki Forest virus (SFV) [59] have been broadly used for neuron targeting. Extensive use of the above viruses in neuroscience is dictated by their natural ability to efficiently and selectively transduce and provide high transgene expression in neurons. Moreover, those viral vectors minimally perturb stress pathways in the infected neural cells [60–62], which makes them attractive tools specifically for the NDD gene therapy applications. When the efficiency and the safety of AAV and lentiviral vectors were compared, AAV demonstrated a significantly higher transgene expression levels and minimal interaction with the innate immune effectors. In contrast, lentiviral transduction resulted in modest transgene expression efficiency and provoked a rapid self-limiting proinflammatory response [63].

In contrast to the majority of viral vectors utilized for gene therapy interventions, AAV-based vectors exhibit the highest potential and success rate in the treatment of several monogenic diseases, as demonstrated by clinical trials [64,65].

2.2. Advantages of AAV-Based Vectors

What makes AAV an exceptional candidate vector for NDD gene therapy applications is its low pathogenicity in humans: none of the AAV subtypes appear to cause a life-threatening systemic response in humans. This is an important feature since patient safety is the most critical condition for the successful development of gene therapy interventions of NDD and other human diseases. Natural tropism of most AAV serotypes, particularly that of AAV2, towards various human tissue types that include the retina and neural tissues [66], liver [67,68], vascular tissue [69,70], lung [71], and skeletal muscles [72] is the other important criterion making AAV a preferred gene therapy vector for NDD [73,74]. The broad tropism of AAV is owed to the diversity of its natural cell-binding targets, such as primary receptors heparan sulfate proteoglycan (HSPG) [75] and/or *N*-glycans with terminal galactose [48], and several known co-receptors, such as fibroblast growth factor (FGF) receptor 1 [76], platelet-derived growth factor (PDGF) receptor [77], hepatocyte growth factor receptor [78], $\alpha_V\beta_1$ and $\alpha_V\beta_5$ integrins [79,80], O-linked sialic acid [81] and laminin receptor (LRP/LR), previously found in liver tissue [82], and recently also discovered in neurons [83]. In fact, cell surface receptors have been definitively identified only for some AAV serotypes: HSPG for AAV-3, O-linked sialic acid for AAV4, PDGF for AAV5, and LRP/LR for AAV serotypes 2, 3, 8, and 9 [49]. To date, there have been 11 identified AAV serotypes (AAV1-AAV11). Due to the fact that the transduction of human neural cells with AAV2 serotype is limited, and humans are seropositive with regard to AAV2, the other AAV serotypes have been explored as alternative gene therapy vectors. This allowed achieving more efficient transduction of selective subsets of brain cells in animal NDD models [84].

The versatility of AAV tropism can be attributed to variability in the amino acid (aa) sequence of the AAV structural (capsid) proteins VP1, VP2, and VP3, whose mixture (in the ratio of 1:1:10, respectively) makes icosahedral symmetry capsid of 60 monomers that are involved in cell surface attachment of the viral particles. The variability in the amino-acid (aa) sequence is confined to the 9 surface-exposed (VR-I to VR-IX) domains of the AAV structural/capsid proteins. AAV tropism alteration for tailoring of the vectors to specific gene therapy applications can be achieved by virus pseudotyping—i.e., creating artificial hybrid AAV particles by combining genome from one strain/serotype with capsid proteins from a different strain/serotype [85,86]. AAV capsid engineering is a two-decade-old approach that embodies two distinct strategies: directed evolution and rational design. The former involves shuffling of capsid genes from available serotypes [87], random peptide insertion into known sites of AAV capsid [88,89] or phage display [90] and ultimately involves a selection process that requires multiple generations of screening to identify functional capsids. Alternatively, a rational design strategy utilizes a structural knowledge to refine capsid structure and/or replace native cell-binding motifs with foreign motifs of desired affinity [90–93]). Recently, a more advanced method of AAV targeting, known as barcoded rational AAV vector evolution (BRAVE), was proposed that encompasses all of the benefits of the rational design approach while maintaining the broad screening diversity permitted by directed evolution, but with only a single generation screening. This new strategy is based on the viral library production approach, where each viral particle displays a protein-derived peptide on the surface, which is linked to a unique barcode in the packaged genome [94].

Naturally, AAV does not produce replication factors and, therefore, is unable to replicate (propagate) without a helper virus (adenovirus, herpes simplex virus (HSV) or vaccinia virus), which provides activating proteins to AAV to enable its replication. Furthermore, in the presence of some helper viruses, such as adenovirus, AAV can even exhibit a lysogenic behavior, which can rarely occur without a helper. Replication deficiency of AAV adds to the biological safety of AAV as yet another benefit for its application as a gene therapy vector.

3. AAV in NDD Experimental Therapy

3.1. *In Vitro vs. In Vivo Models*

With regard to NDD gene therapy, an experimental tissue model for the blood-brain barrier (BBB) is of the highest relevance and importance, and the most developed disease models include those for Alzheimer's, Parkinson's, and Huntington's diseases. These and other NDD models are discussed in detail elsewhere [95]. Numerous NDD disease platforms have been created to date by using various approaches [96–99]. This includes several in vitro disease models developed in the last three years, which were successfully used in a few NDD gene therapy studies. One of the crucial criteria in the efficacy assessment of viral vector-based gene therapy of NDD is the ability of delivery vectors to efficiently cross the BBB, as iv administration remains the primary route of vector delivery. In contrast to in vivo (animal) models, in vitro models allow us to easily understand the mechanism of viral trafficking. In this regard, Merkel et al. showed that primary human brain microvascular endothelial cells, as a model of the human BBB, can be effectively utilized as a tool to characterize trans-endothelial movement and transduction kinetics of various AAV serotypes in vitro and even their effect on BBB integrity. Specifically, by using this in vitro BBB model, the authors found that AAV9 penetrates brain microvascular endothelial cell barriers more effectively than AAV2, although it exhibits relatively lower transduction efficiency [100].

A proper design of NDD in vivo models should allow assessing the robustness of therapies in the settings that mimic clinical endpoints. Although information obtained from studying in vitro models is usually capable of providing a sufficient understanding of disease pathogenesis, it fails to characterize the aspect of disease progression. Therefore, the use of in vivo disease models allows investigators to observe and analyze symptomatic manifestation of NDDs alongside with the therapeutic outcomes of their experimental treatment. This also aids in getting a comprehensive insight into the complex interconnection between etiological and symptomatic changes during the progression of NDDs, such as neuronal ceroid lipofuscinosis (NCL or Batten disease) [35], spinal muscular atrophy [101], or Niemann-Pick disease [102], whose development is typically associated with genetic perturbations in multiple genes. Therefore, in contrast to in vitro models, a mouse model of NDD would allow recapitulating onset and progression of the counterpart human disease of a given degree of severity. For instance, the studies involving AAV9-based vectors [103–106] aimed at restoring expression of target genes with genetic abnormalities require testing of the experimental strategies in a preclinical setting and under different routes of vector delivery to affected cells. Hudry et al. [49] were able to evaluate the efficacy of AAV-mediated intracellular delivery of reporter transgene in a CNS disorder.

It was found that, depending on the viral capsid composition and the route of vector administration, AAV9 is capable of providing a high level of transduction in mixed astrocyte and neuron culture, although vector biodistribution upon intracerebroventricular injection can vary. As far as the route of injection is concerned, a direct injection of experimental therapeutics into cerebrospinal fluid allows their fast and efficient delivery and spread throughout CNS and spinal cord. Therefore, intrathecal injection is nowadays commonly used in various experimental strategies as an effective route of viral agent delivery and achieving therapeutic effects [107–110].

3.2. Selection of Optimal Gene Therapy Strategy

The choice of a suitable gene therapy strategy for NDD is commonly dictated by the known alterations in signaling pathways associated with a particular NDD. Multiple scientific reports suggest that under normal physiological conditions a microtubule-associated protein tau (MAPT) is sensitive to autophagy-associated ubiquitin-binding protein p62/SQSTM1, which promotes degradation of the misfolded, microtubule-dissociated tau protein. However, in tauopathy, an insoluble form of mutant MAPT becomes resistant to recognition by and binding to p62/SQSTM1 and as a result, fails to degrade inside autophagosomes [6]. This study revealed that the pathology caused by the intracellular formation of neurofibrillary tangles (NFTs) by the hyperphosphorylated form of mutant MAPT, can be suppressed by p62/SQSTM1 overexpression resulting from delivery of AAV 2/9-SQSTM1 expression vector. Earlier, Janda et al. demonstrated that several NDDs are associated with an onset of oxidative stress that inhibits LC3 lipidation and autophagic flux in the affected neurons [111], suggesting multiple reasons for the reduction of basal autophagy and the resulting promotion of NDD. Since the discovery of the relevance of defective autophagy signaling to NDDs, a few studies utilized AAV vectors to overcome the autophagy signaling defects. It was reported that defects in autophagy signaling can be overcome by activation of lysosomal transport via intracellular expression of hydrolase glucocerebrosidase [112] or reduction of tau protein phosphorylation via arginase 1 [113] that eliminates the cargo from vulnerable neurons. In the aggregate, the above-referenced studies strongly suggest that rescuing impairments in the autophagy pathway inside diseased cells is a promising therapeutic strategy for some NDDs. Moreover, aberrant signaling that is attributed to the NDD progression can also be corrected via AAV-based vectors. Recently, it was shown that the expression of triggering receptor expressed on myeloid cells 2 (TREM2) is associated with high risk for AD development. Therefore, AAV-mediated expression of the soluble form of TREM2 (sTREM2), a proteolytic product of TREM2, reduces amyloid plaque load and rescues functional deficits of spatial memory [114]. Overall, the delivery of a therapeutic payload via a modified AAV contributes to the restoration of the disease-affected pathway (Figure 1).

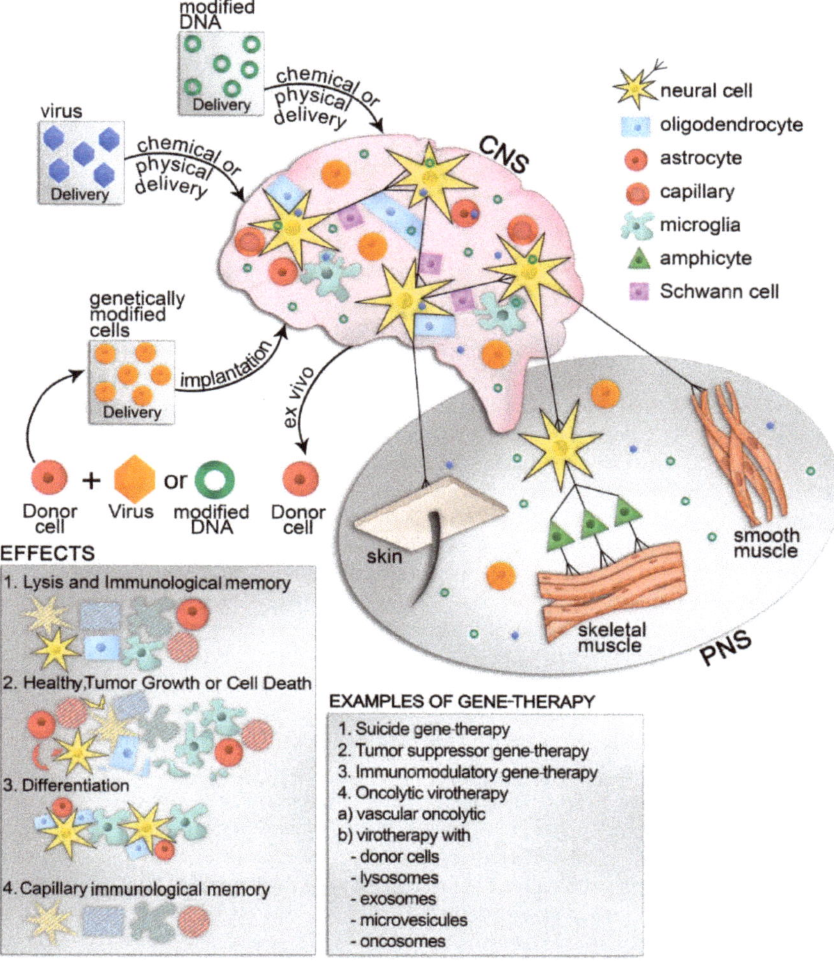

Figure 1. Potential applications of gene targeting for therapy of neurodegenerative diseases.

3.3. *AAV-Based Gene Therapy for Parkinson's Disease*

The movement disorder, known as Parkinson's Disease (PD), the second most common neurodegenerative disorder, occurs largely due to the loss of dopaminergic neurons of the substantia nigra, resulting in striatal depletion of the neurotransmitter dopamine. AAV-based gene therapy vectors could be effectively used for gene therapy of PD through increasing dopamine levels in target cells [115]. Another strategy proposed for gene therapy of PD is based on targeting of neural parenchyma by AAV-based α-synuclein expression vector (AAV-PHP.B-GBA1) delivered via an iv administration, as opposed to commonly used injection in the mouse forebrain. This allowed vector to permeate and diffuse throughout the neural parenchyma, resulting in targeting of both the central and the peripheral nervous systems in a global pattern and recovering animal behavior by reducing synucleinopathy [44].

Gutekunst et al. demonstrated that the delivery of C3-ADP ribosyltransferase (C3) selectively and irreversibly inhibits activation of RhoA GTPase (a key intracellular regulator of axon regrowth in the mammalian CNS) in neurons or neuronal progenitors and effectively prevents inhibition of axonal regrowth leading to axon regeneration [116]. Furthermore, the use of AAV-mediated C3 delivery allowed the authors to determine a critical C3 concentration promoting neuron outgrowth on chondroitin sulfate substrate. Another example of AAV vector system implementation for gene therapy of PD is the delivery of rat ATP6V0C (a pore-forming stalk c-subunit of the V0 sector of the vacuolar proton ATPase contributing to release of serotonin, acetylcholine, and dopamine by stromal cells) into the substantia nigra of the diseased mice, in which high potassium stimulation increased overflow of the endogenous dopamine (DA) [117,118].

3.4. AAV Vectors and Gene Targeting

In addition to gene delivery for therapeutic protein expression/overexpression purposes in neural cells, AAV vectors have also been used for gene targeting (knockdown and knockout) applications, which include shRNA and miRNA expression strategies. It becomes increasingly popular to knockdown the expression of cellular factors or enzymes contributing to NDD pathobiology to restore normal physiological conditions in disease-affected neural cells. For instance, an AAV8-based vector has been utilized to modulate the expression of cellular proteins implemented in the pathology of the disease. Lu et al. [119] demonstrated that an elevated level of SM synthase-1 promotes the lysosomal degradation of BACE1. However, due to the dominant-negative nature of some genetic mutations, direct sequestration of aberrant (mutant) proteins with normal function-inhibiting properties or even genetic knock-out of mutant genes (excision/destruction of mutant genes) or their complete replacement with corrected/wild type counterparts becomes a necessary and the most rational strategy.

Mutations in genes that control normal CNS development and metabolism give rise to several genetic (inheritable) NDDs. For instance, Krabbe disease is caused by a genetic abnormality that is associated with mutations in the galactosylceramidase (GALC) gene. The GALC enzyme is contained inside lysosomes, where it hydrolyzes specific galactolipids, including galactosylceramide and psychosine (galactosylsphingosine). While the breakdown of those lipids is part of the normal myelin turnover, accumulation of psychosine metabolite in the absence of galactosylceramidase is toxic for neurons and needs to be prevented by a quick restoration of the galactosylceramidase function. Gene therapy for Krabbe disease can thus be accomplished through the GALC gene correction/replacement strategy in the mouse model [46] as well as in the tissue of human patients.

Mutations in superoxide dismutase 1 gene (SOD1) are known to be one of the causes for amyotrophic lateral sclerosis (ALS), a fatal neurodegenerative disease caused by progressive loss of upper and lower motor neurons in human CNS. Although responsible for only about 20% of familial (inherited) ALS cases, *SOD1* mutations contribute to the loss of motor neurons during ALS progression via a toxic gain of function mechanism that involves NF-κB-dependent mechanism, resulting in intracellular aggregation of the aberrant protein [120]. A strategy using an AAV9 delivery vector, engineered to express shRNA targeting SOD1 mRNA, was employed by Frakes et al. to reduce (knock-down) SOD1 expression in both astrocytes and motor neurons. The AAV9 vector was able to efficiently penetrate the BBB and increase the survival of the ALS mice [121]. The same vector was used in the AAV9-mediated knockdown of neuritin, which resulted in the reduction of synaptic transmission in the medial prefrontal cortex (mPFC) pyramidal neurons in mice [122]. Selective suppression of mutant huntingtin aggregation and neuronal dysfunction in a rat model of Huntington's disease (HD) by applying AAV5-miHTT-451, induced functional improvements in the HD pathological process without causing an activation of microglia or astrocytes immune response [123].

Despite the remarkable therapeutic efficacy of RNAi-based gene silencing in target cells, most of the shRNA-mediated NDD gene therapy applications only partially reduce the *SOD1* gene expression, since neither a complete knockout of the target mRNA [124–126], nor a sustained gene knockdown can be technically achieved [124,127]. Therefore, the novel gene targeting technologies, based on the recent discovery of transcription activator-like effector nucleases (TALENs) and clustered regularly interspaced short palindromic repeat (CRISPR)-Cas9 nuclease complex (CRISPR-Cas9 nuclease) as components of bacterial natural "immunity" against viruses, completely revolutionized the genome editing [128] field by dramatically improving gene targeting efficiency and specificity, thereby allowing a highly accurate cleavage of any DNA sequence at any given point [129]. The precision of DNA cleavage mediated by CRISP-Cas9 nucleases has been validated in recent studies aimed at correcting genetic mutations on a single nucleotide level [130]. Wen et al. reported that the deletion of the human insulin growth factor-1 (IGF-1) gene, implicated in hyperpolarization of mitochondrial electrical transmembrane potential, resulted in the accumulation of anti-apoptotic protein factors B-cell lymphoma-extra large (BCL-XL) and B-cell lymphoma 2 (BCL-2). Moreover, the deletion of the *IGF1* gene via an intratracheal injection of an AAV-delivered CRISPR-Cas9 expression system restored mitophagy and reduced protective effect on mitochondria, suggesting a new strategy for treating ALS [131]. A strong therapeutic effect was reported by Raikwar et al. [131] and Gaj et al. [132], who used the CRISPR-Cas9 gene targeting (Figure 1) approach to achieve a reduced expression of the glial maturation factor in glial cells that contributed to the formation of Alzheimer's disease plaques and corrected an autosomal mutation in the *SOD1* gene, responsible for ALS progression, respectively. Most recently, Ekman et al. pointed out that correction of mutant exon 1 of the huntingtin gene (HTT) results in a ~50% decrease of neuronal inclusions and improved motor deficits [133].

Recently, it was suggested that downregulation of some miRNAs may be involved in modulation of apoptotic response and autophagy defects during the NDD progression [134,135]. Miyazaki et al. found that bulbar muscular atrophy (SBMA) is caused by the expansion of the polyglutamine (poly Q) tract in the androgen receptor (AR-poly Q) protein sequence [136]. It was predicted that several miRNAs, such as miR-196a [136], miR-298 or miR-328 [137], might regulate the activity of NDD-related proteins, such as Elav-like family member 2 (CELF2) that enhances the stability of AR mRNA, or beta-amyloid precursor protein-converting enzyme (BACE), respectively. AAV-based delivery and expression of miR-196a promotes decay of the AR mRNA by silencing its stabilizing factor CUGBP, an Elav-like family member 2 (CELF2), and thereby ameliorates the SBMA phenotypes in the mouse model. These studies suggest that AAV vector is a highly efficient tool for the delivery and the expression of recombinant DNA or regulatory, mRNA-silencing miRNAs in monogenic inherited disease applications. Furthermore, in the case of polygenetic NDD diseases, utilization of AAV for targeting a mutant allele to silence the expression of a mutant protein in patients could become a potent therapeutic approach in the future. One of the remaining challenges of such gene silencing strategy that should be addressed in AAV vector design is how to avoid targeting of the wild type or normal (non-mutated) gene copy in the diseased cells.

4. Modulation of Interaction Between the CNS and the Environment

Inflammation is an essential hallmark of various neurodegenerative diseases. An increasing number of clinical studies demonstrate that upon activation microglia and astrocytes produce proinflammatory chemokines, such as tumor necrosis factor alpha (TNFα) and interleukin type 1 (IL1), around amyloid plaques. In most of those cases, administration of interleukin type 2 (IL2) and interleukin type 4 (IL4) has been shown to stimulate an anti-inflammatory response. On the other hand, multiple lines of evidence suggest that the delivery of growth factors via viral vectors can provide significant therapeutic effects. For instance, the nerve growth factor (NGF) was expressed in the brain cells of patients with AD by means of AAV2-based vector delivery in a clinical trial conducted by Tuszynski et al. (Trial Registration: NCT00087789 and NCT00017940). It was found that neurons of the degenerating brain retain the ability to respond to the growth factor administration by axonal sprouting, cell hypertrophy, and activation of functional markers for up to 10 years [138]. Other studies used either lentiviral vector- or AAV-based delivery of a brain-derived neurotrophic factor (BDNF), progranulin (PGRN), or cerebral dopamine neurotrophic factor (CDNF) in mouse models of AD or PD, respectively [71–73]. Delivery of BDNF in transgenic APP mouse model (mutant mice carrying two amyloid precursor protein (APP) mutations associated with early-onset familial Alzheimer's disease) resulted in significant amelioration of cell loss by BDNF and marked improvements in the hippocampal-dependent behavior (contextual fear conditioning), compared with control-treated APP mice [72]. Elevation of PGRN expression in nigrostriatal neurons upon AAV-mediated PGRN delivery protected nigrostriatal neurons from MPTP toxicity, reduced inflammation and apoptosis, as well as completely preserved locomotor function in the mouse PD model [71]. Striatal delivery of AAV2.CDNF allowed recovery of 6-OHDA-induced behavior deficits and resulted in a significant restoration of tyrosine hydroxylase immunoreactive (TH-ir) neurons in the substantia nigra as well as functional recovery of dopaminergic neurons.

An impressive therapeutic outcome was observed in yet another innovative cell/gene therapy combinational approach utilizing the human umbilical cord blood cells (hUCBCs) transduced with adenoviral (Ad) vectors encoding human vascular endothelial growth factor (VEGF), glial cell-line derived neurotrophic factor (GDNF) [139] and/or neural cell adhesion molecule 1 (NCAM) genes [74–76]. After hUCBCs were transplanted into transgenic amyotrophic lateral sclerosis (model) mice a significant improvement in animals' behavioral performance (open-field and grip-strength tests), as well as an increased life-span were observed. Expression of NCAM-VEGF or NCAM-GDNF was observed in ALS mice 10 weeks after delivering genetically modified hUCBCs, whereas the cell vehicles were detectable for 5 months following the transplantation [140].

Although some NDDs are caused by dysregulation of neuron's synaptic function, AVV-based gene therapy could still offer a therapeutic solution. Specifically, an AAV-mediated delivery of the retinoschisin transgene resulted in the restored structure and function of the photoreceptor cell synapse in the mouse disease models. This restoration of operational synapse in the animal model led to a human clinical trial for gene therapy of X-linked retinoschisis [141]. A recent study established the formation of synaptic connections between NeuroD1-converted neurons with already present neurons. These results indicate possibilities of NeuroD1 AAV-based gene therapy for functional brain repair after ischemic injury through in vivo astrocyte-to-neuron conversion [142].

5. Conclusions and Future Directions

Outcomes of clinical trials for NDDs can be compromised by limited data from the prior animal (preclinical) studies, by suboptimal expression vector's cassette design or by the immunogenicity of either delivery vector itself, or its expression content (a therapeutic transgene). The Clinicaltrials.gov database contains information about 6128 clinical trials worldwide that were aimed at treating various neurodegenerative diseases, including 103 trials using gene therapy approaches 29 of which utilized natural mechanisms of viral replication [143]. At this point, two clinical trials (NCT02418598 and NCT00643890) have been terminated by the organizers (both due to financial reasons) and one (NCT03381729) was suspended by the FDA. The latter study involved 27 patients selected out of 51 with spinal muscular atrophy 1 (SMA1) disorder to receive an injection of AAV9-based vector (6×10^{13}, 1.2×10^{14} and 2.4×10^{14} vp per patient) designed to express the SMN protein under the control of hybrid CMV/β-actin promoter (ZOLGENSMA). Currently, the future of ZOLGENSMA use via intrathecal route remains unclear, but the available clinical safety information suggests using caution, when infusing ZOLGENSMA intravenously, and inclusion pre-treatment with corticosteroids, checking blood biochemistry and liver enzymes (https://www.zolgensma.com/how-zolgensma-works). Compared to non-viral vectors (e.g., polyplexes [144]), AAV vectors appear to be highly suitable for NDD gene therapy applications and, therefore, became highly popular for NDD gene therapy studies in the recent years. However, the problem with AAV crossing BBB along with the vector's immunogenicity [47] in human patients remains the main obstacle even for those vectors that are capable of effectively preventing NDD progression.

Table 1. Worldwide trials using AAV vectors against neurodegenerative disease.

Vectors	Transgene	Disease	Phase	Patients Enrolled	Outcome	Study name	Clinicaltrail.gov	Status
AAV	Beta-nerve growth factor (NGF)	Alzheimer's disease	1	10	A Phase I, Dose-Escalating Study to Assess the Safety and Tolerability of CERE-110 [Adeno-Associated Virus (AAV)-Based Vector-Mediated Delivery of Beta-Nerve Growth Factor (NGF)] in Subjects with Mild to Moderate Alzheimer's Disease	CERE-110 in Subjects With Mild to Moderate Alzheimer's Disease	**NCT00087789** CERE-110 2.0 × 10^{10} vg, CERE-110 1.0 × 10^{11} vg, CERE-110 2.0 × 10^{11} vg	Completed
AAV	Beta-nerve growth factor (NGF)	Alzheimer's disease	2	49	A Double-Blind, Placebo-Controlled (Sham Surgery), Randomized, Multicenter Study Evaluating CERE-110 Gene Delivery in Subjects with Mild to Moderate Alzheimer's Disease	Randomized, Controlled Study Evaluating CERE-110 in Subjects With Mild to Moderate Alzheimer's Disease	**NCT00876863** CERE-110 2.0 × 10^{11} vg	Completed
AAV	Neurotrophic (growth) factor (Neurturin)	Parkinson's disease	1/2	60 est/57 fact	A Phase 1/2 Trial Assessing the Safety and Efficacy of Bilateral Intraputaminal and Intranigral Administration of CERE-120 (Adeno-Associated Virus Serotype 2 [AAV2]-Neurturin [NTN]) in Subjects with Idiopathic Parkinson's Disease	Safety and Efficacy of CERE-120 in Subjects With Parkinson's Disease	**NCT00985517** CERE-120 2.4 × 10^{12} vg	Completed
AAV	Glutamic acid decarboxylase (GAD)	Parkinson's disease	1	12	Phase I Study of Subthalamic GAD Gene Transfer in Medically Refractory Parkinson's Disease	Safety Study of Subthalamic Nucleus Gene Therapy for Parkinson's Disease	**NCT00195143**	Completed
AAV	Glutamic acid decarboxylase (GAD)	Parkinson's disease	2	44 (est)	Phase 2 Safety and Efficacy Study Evaluating Glutamic Acid Decarboxylase Gene Transfer to Subthalamic Nuclei in Subjects with Advanced Parkinson's Disease	Study of AAV-GAD Gene Transfer Into the Subthalamic Nucleus for Parkinson's Disease	**NCT00643890** One-time bilateral administration of rAAV-GAD at 1 × 10^{12} vector genomes in 35 uL	Terminated (Financial reasons)
AAV	Glutamic acid decarboxylase GAD	Parkinson's disease		40 est/0 fact	N/A	Long Term Follow-Up Study for rAAV-GAD Treated Subjects	**NCT01301573**	Terminated (Financial reasons)

Table 1. *Cont.*

Vectors	Transgene	Disease	Phase	Patients Enrolled	Outcome	Study name	Clinicaltrail.gov	Status
AAV	Aromatic L-amino acid decarboxylase (hAADC-2)	Parkinson's disease	1	10	A Phase1 Open-Label Safety Study of Intrastriatal Infusion of Adeno-Associated Virus Encoding Human Aromatic L-Amino Acid Decarboxylase (AAV-hAADC-2) in Subjects with Parkinson's Disease [AAV-hAADC-2-003]	A Study of AAV-hAADC-2 in Subjects With Parkinson's Disease	**NCT00229736** 9×10^{10} vector genomes (vg) of AAV-hAADC-2 in a single dose of 200 μL bilaterally infused over 4 striatal targets 3×10^{11} vector genomes (vg) of AAV-hAADC-2 in a single dose of 200 μL bilaterally infused over 4 striatal targets	Completed
AAV	Aromatic L-amino acid decarboxylase (hAADC-2)	Parkinson's disease	1/2	6 est/2 fact	A Phase I /II Study of Intra-Putaminal Infusion of Adeno-Associated Virus Encoding Human Aromatic L-Amino Acid Decarboxylase in Subjects with Parkinson's Disease	AADC Gene Therapy for Parkinson's Disease	**NCT02418598** AAV-hAADC-2 is administered via bilateral intra-putaminal infusion. The number of vector genomes (vg) cohort 1: 3×10^{11} vg/subject cohort 2: 9×10^{11} vg/subject.	Terminated (Another clinical study for regulatory approval is planned)
AAV1	Neurotrophin factor 3 (NTF3)	Charcot–Marie–Tooth disease	1/2a	9 est/0 fact	Phase I/II a Trial Evaluating scAAV1.tMCK.NTF3 for Treatment of Charcot–Marie–Tooth Neuropathy Type 1A (CMT1A)	Phase I/II a Trial of scAAV1.tMCK.NTF3 for Treatment of CMT1A	**NCT03520751** $N = 3$: intramuscular injection of (scAAV1.tMCK.NTF3) distributed bilaterally between both limbs at low dose (2×10^{12} vg/kg). $N = 6$: intramuscular injection of (scAAV1.tMCK.NTF3) distributed bilaterally between both limbs at low dose (6×10^{12} vg/kg)	Not yet recruiting
AAV2	Neurotrophic (growth) factor (Neurturin)	Parkinson's disease	2	58 est/51 fact	Multicenter, Randomized, Double-Blind, Sham Surgery-Controlled Study of CERE-120 (Adeno-Associated Virus Serotype 2 [AAV2]-Neurturin [NTN]) to Assess the Efficacy and Safety of Bilateral Intraputaminal (IPu) Delivery in Subjects with Idiopathic Parkinson's Disease	Double-Blind, Multicenter, Sham Surgery Controlled Study of CERE-120 in Subjects With Idiopathic Parkinson's Disease	**NCT00400634**, CERE-120, bilaterally: 5.4×10^{11} vg	Completed

Table 1. *Cont.*

Vectors	Transgene	Disease	Phase	Patients Enrolled	Outcome	Study name	Clinicaltrail.gov	Status
AAV2	Neurotrophic (growth) factor (Neurturin)	Parkinson's disease	1	12 est	A Phase I, Open-Label Study of CERE-120 (Adeno-Associated Virus Serotype 2 [AAV2]-Neurturin [NTN] to Assess the Safety and Tolerability of Intrastriatal Delivery to Subjects with Idiopathic Parkinson's Disease	Safety of CERE-120 (AAV2-NTN) in Subjects With Idiopathic Parkinson's Disease	**NCT00252850**	Completed
AAV2	Human aromatic L-amino acid decarboxylase (AADC) gene	Parkinson's disease	1	15 est/10 fact	An Open-label Safety and Efficacy Study of VY-AADC01 Administered by MRI-Guided Convective Infusion into the Putamen of Subjects with Parkinson's Disease with Fluctuating Responses to Levodopa	Safety Study of AADC Gene Therapy (VY-AADC01) for Parkinson's Disease (AADC)	**NCT01973543** VY-AADC01; Single dose, neurosurgically-infused, bilaterally into the striatum: 7.5×10^{11} vg, 1.5×10^{12} vg, 4.7×10^{12} vg	Active, not recruiting
AAV2	Human aromatic L-amino acid decarboxylase (AADC) gene	Parkinson's disease	2	42 est	A Randomized, Placebo Surgery Controlled, Double-Blinded, Multi-center, Phase 2 Clinical Trial, Evaluating the Efficacy and Safety of VY-AADC02 in Advanced Parkinson's Disease with Motor Fluctuations	VY-AADC02 for Parkinson's Disease With Motor Fluctuations	**NCT03562494**, VY-AADC02 infusion, 2.5×10^{12}	Recruiting
AAV2	Human aromatic L-amino acid decarboxylase (AADC) gene	Parkinson's disease		50 est	An Observational, Long-Term Extension Study for Participants of Prior VY-AADC01 or VY-AADC02 Clinical Studies	Observational, Long-term, Extension Study for Participants of Prior VY-AADC01 or VY-AADC02 Studies	**NCT03733496**, Participants who received VY-AADC01 or VY-AADC02	Enrolling by invitation
AAV2	Human CLN2	Late infantile neuronal ceroid lipofuscinosis (LINCL)	1	11 est/10 fact	Administration of a Replication Deficient Adeno-Associated Virus Gene Transfer Vector Expressing the Human CLN2 cDNA to the Brain of Children with Late Infantile Neuronal Ceroid Lipofuscinosis	Safety Study of a Gene Transfer Vector for Children With Late Infantile Neuronal Ceroid Lipofuscinosis	**NCT00151216** AAV2CUhCLN2, $N = 5, 3 \times 10^{12}$ $N = 6, 3 \times 10^{12}$	Active, not recruiting
AAV2	Glial cell line-derived neurotrophic factor (GDNF)	Parkinson's disease	1	28 est/25 fact	A Phase 1 Open-Label Dose Escalation Safety Study of Convection Enhanced Delivery (CED) of Adeno-Associated Virus Encoding Glial Cell Line-Derived Neurotrophic Factor (AAV2-GDNF) in Subjects with Advanced Parkinson's Disease	AAV2-GDNF for Advanced Parkinson s Disease	**NCT01621581** 9×10^{10} vg, 3×10^{11} vg, 9×10^{11} vg 3×10^{12} vg	Active, not recruiting

Table 1. *Cont.*

Vectors	Transgene	Disease	Phase	Patients Enrolled	Outcome	Study name	Clinicaltrail.gov	Status
AAV2	Human ND4	Leber's congenital amaurosis	3	90 est	Efficacy and Safety of Bilateral Intravitreal Injection of GS010: A Randomized, Double-Masked, Placebo-Controlled Trial in Subjects Affected with G11778A ND4 Leber's Hereditary Optic Neuropathy for Up to One Year	Safety and Efficacy Study of Gene Therapy for The Treatment of Leber's Hereditary Optic Neuropathy	**NCT03293524**, GS010, IVT eye, 9×10^{10} vg	Active, not recruiting
AAV2	Human ND4	Leber's congenital amaurosis	2 + 3	159 est/48 fact	Safety and Efficacy Study of Gene Therapy for The Treatment of Leber's Hereditary Optic Neuropathy		**NCT03153293**, Single IVT injection, 1×10^{10} vg/0.05 mL	Active, not recruiting
AAV2	Human ND4	Leber's hereditary optic neuropathy	3	36 est	A Randomized, Double-Masked, Sham-Controlled Clinical Trial to Evaluate the Efficacy of a Single Intravitreal Injection of GS010 in Subjects Affected for 6 Months or Less by LHON Due to the G11778A Mutation in the Mitochondrial ND4 Gene	Efficacy Study of GS010 for the Treatment of Vision Loss up to 6 Months From Onset in LHON Due to the ND4 Mutation (RESCUE)	**NCT02652767**, GS010	Active, not recruiting
AAV2	Human ND4	Leber's hereditary optic neuropathy	3	37 est/36 fact	Randomized, Double-Masked, Sham-Controlled Clinical Trial to Evaluate the Efficacy of a Single Intravitreal Injection of GS010 in Subjects Affected for More Than 6 Months and to 12 Months by LHON Due to the G11778A Mutation in the ND4 Gene	Efficacy Study of GS010 for Treatment of Vision Loss From 7 Months to 1 Year From Onset in LHON Due to the ND4 Mutation (REVERSE) (REVERSE)	**NCT02652780**, rAAV2/2-ND4 intravitreal, 9×10^{10} vg	Completed
AAV2	Human ND4	Leber's hereditary optic neuropathy		74 est	Long-Term Follow-Up of ND4 LHON Subjects Treated with GS010 Ocular Gene Therapy in the RESCUE or REVERSE Phase III Clinical Trials	RESCUE and REVERSE Long-term Follow-up (RESCUE/REVERSE)	**NCT03406104**, GS010, intravitreal injection	Recruiting
AAV2	Human ND4	Leber's hereditary optic neuropathy			EAP Single Patient: Safety of Bilateral Intravitreal Injection of GS010 in a Single Subject Affected with G11778A ND4 Leber's Hereditary Optic Neuropathy	EAP_GS010_single Patient	**NCT03672968**	Available
AAV2	Human ND4	Leber's hereditary optic neuropathy	1	30 est/27 fact	An Open-Label Dose Escalation Study of an Adeno-Associated Virus Vector (scAAV2-P1ND4v2) for Gene Therapy of Leber's Hereditary Optic Neuropathy (LHON) Caused by the G11778A Mutation in Mitochondrial DNA	Safety Study of an Adeno-associated Virus Vector for Gene Therapy of Leber's Hereditary Optic Neuropathy (LHON)	**NCT02161380** scAAV2-P1ND4v2 intravitreal: 1.18×10^{9} vg 5.81×10^{9} vg 2.4×10^{10} vg 1.0×10^{11} vg	Recruiting

Table 1. *Cont.*

Vectors	Transgene	Disease	Phase	Patients Enrolled	Outcome	Study name	Clinicaltrail.gov	Status
AAV9	CLN6	CLN6, Batten disease	1/2a	13 est/6 fact	Phase I/II a Gene Transfer Clinical Trial for Variant Late Infantile Neuronal Ceroid Lipofuscinosis, Delivering the CLN6 Gene by Self-Complementary AAV9	Gene Therapy for Children With Variant Late Infantile Neuronal Ceroid Lipofuscinosis 6 (vLINCL6) Disease	**NCT02725580** AT-GTX-501 (scAAV9.CB.CLN6)	Active, not recruiting
AAV9	CLN3	CLN3, neuronal ceroid-lipofuscinosis	1/2a	7 est	Phase I/II a Gene Transfer Clinical Trial for Juvenile Neuronal Ceroid Lipofuscinosis, Delivering the CLN3 Gene by Self-Complementary AAV9	Gene Therapy for Children With CLN3 Batten Disease	**NCT03770572** AT-GTX-502 High dose Low dose	Recruiting
AAV9	Human survival motor neuron (SMN)	Spinal muscular atrophy type 1 (SMA1)	1	15 est/9 fact	Phase I Gene Transfer Clinical Trial for Spinal Muscular Atrophy Type 1 Delivering AVXS-101	Gene Transfer Clinical Trial for Spinal Muscular Atrophy Type 1	**NCT02122952** iv $N = 3, 6.7 \times 10^{13}$ vg/kg $N = 3, 2.0 \times 10^{14}$ vg/kg	Completed
AAV9	Human survival motor neuron (SMN)	SMA	1	51 est/27 fact	Phase I, Open-Label, Dose Comparison Study of AVXS-101 for Sitting but Non-Ambulatory Patients with Spinal Muscular Atrophy	Study of Intrathecal Administration of Onasemnogene Abeparvovec-xioi for Spinal Muscular Atrophy (STRONG)	**NCT03381729** Intrathecal Onasemnogene abeparvovec-xioi 6.0×10^{13} vg 1.2×10^{14} vg 2.4×10^{14} vg	Suspended
AAV9	Human survival motor neuron (SMN)	Spinal muscular atrophy type 1 (SMA1)	3	22est/15 fact	Phase 3, Open-Label, Single-Arm, Single-Dose Gene Replacement Therapy Clinical Trial for Patients with Spinal Muscular Atrophy Type 1 With One or Two SMN2 Copies Delivering AVXS-101 by Intravenous Infusion	Gene Replacement Therapy Clinical Trial for Patients With Spinal Muscular Atrophy Type 1 (STR1VE)	**NCT03306277** iv Onasemnogene abeparvovec-xioi	Completed
AAV9	Human survival motor neuron (SMN)	Spinal muscular atrophy type 1 (SMA1)	3	44 est/30 fact	A Global Study of a Single, One-Time Dose of AVXS-101 Delivered to Infants with Genetically Diagnosed and Pre-Symptomatic Spinal Muscular Atrophy with Multiple Copies of SMN2	Pre-Symptomatic Study of Intravenous Onasemnogene Abeparvovec-xioi in Spinal Muscular Atrophy (SMA) for Patients With Multiple Copies of SMN2 (SPR1NT)	**NCT03505099** iv, Onasemnogene abeparvovec-xioi 1.1×10^{14} vg/kg	Active, not recruiting

Table 1. *Cont.*

Vectors	Transgene	Disease	Phase	Patients Enrolled	Outcome	Study name	Clinicaltrail.gov	Status
AAV9	Human survival motor neuron (SMN)	Spinal muscular atrophy (SMA) type 1	3	33est/30 fact	European, Phase 3, Open-Label, Single-Arm, Single-Dose Gene Replacement Therapy Clinical Trial for Patients with Spinal Muscular Atrophy Type 1 with One or Two SMN2 Copies Delivering AVXS-101 by Intravenous Infusion	Single-Dose Gene Replacement Therapy Clinical Trial for Patients With Spinal Muscular Atrophy Type 1 (STRIVE-EU)	**NCT03461289** iv Onasemnogene Abeparvovec-xioi	Active, not recruiting
AAV10	Human SGSH and SUMF1 cDNAs	Sanfilippo type A syndrome	1 + 2	4 est	An Open-label, Single Arm, Monocentric, Phase I/II Clinical Study of Intracerebral Administration of Adeno-associated Viral Vector Serotype 10 Carrying the Human SGSH and SUMF1 cDNAs for the Treatment of Sanfilippo Type A Syndrome	Intracerebral Gene Therapy for Sanfilippo Type A Syndrome	**NCT01474343** Intracerebral SAF-301	Completed
AAV10	Human SGSH and SUMF1 cDNAs	Sanfilippo type A syndrome	1 + 2	4 est/	Long-Term Follow-Up of Patient with Sanfilippo Type A Syndrome Who Have Previously Been Treated in the P1-SAF-301 Clinical Study Evaluating the Tolerability and Safety of the Intracerebral Administration of SAF-301	Long-term Follow-up of Sanfilippo Type A Patients Treated by Intracerebral SAF-301 Gene Therapy	**NCT02053064** Intracerebral SAF-301 Long term effect	Completed
AAVrh10	Human apolipoprotein E2 (APOE2)	Alzheimer's disease	1	15 est/0 fact	Maximum Tolerated Dose of Intracisternal delivery of AAVrh.10hAPOE2 (no results)	Gene Therapy for APOE4 Homozygote of Alzheimer's Disease	**NCT03634007** AAVrh.10hAPOE2: 8.0×10^{10} gc/kg, 2.5×10^{11} gc/kg 8.0×10^{11} gc/kg	Recruiting
AAVrh10	Human CLN2	Late-infantile neuronal ceroid Lipofuscinosis	1	25 est/16fact	Direct CNS Administration of a Replication Deficient Adeno-associated Virus Gene Transfer Vector Serotype rh.10 Expressing the Human CLN2 cDNA to Children with Late Infantile Neuronal Ceroid Lipofuscinosis (LINCL)	Safety Study of a Gene Transfer Vector (Rh.10) for Children With Late Infantile Neuronal Ceroid Lipofuscinosis (LINCL)	**NCT01161576** AAVrh10.CUhCLN2, $N = 6, 9 \times 10^{11}$ $N = 10, 2.85 \times 10^{11}$	Active, not recruiting
AAVrh10	Human CLN2	Late infantile neuronal ceroid lipofuscinosis	1 + 2	16 est/8 fact	Improved Results on Weill Cornell LINCL Scale and Mullen Scale (no results posted)	AAVRh.10 Administered to Children With Late Infantile Neuronal Ceroid Lipofuscinosis	**NCT01414985** AAVrh10.CUhCLN2, $N = 2, 9 \times 10^{11}$ $N = 6, 2.85 \times 10^{11}$	Active, not recruiting

In this review, we outlined the major basic and clinical research directions in the field of NDD gene therapy using viral vectors, and particularly AAV, for delivery of therapeutic payload aimed at restoration of autophagy and metabolic defects in neurons. Encouragingly, many of the reported studies demonstrate strong and long-lasting therapeutic effects (Table 1). The summarized basic research data appears to be in good agreement with the results of preclinical studies, suggesting strong efficacies of neuronal cell targeting with AAV. In addition, findings from several basic studies suggest existence of a crosstalk between the brain microenvironment and neuronal cell signaling. A combination of AAV-based gene therapy with other therapeutic strategies targeting autophagy signaling components could result in even more favorable outcomes. Such co-therapeutics include traditional and herbal medicines [145]. Therefore, dysfunction in neuronal signaling is closely interconnected with the brain environment, and a combinational intervention using these two disease components could become a promising therapeutic strategy, which may provide a clue for an effective solution for the ultimate NDD cure.

Author Contributions: Writing: O.M., A.B., A.S., A.V.B., P.S., and I.U.; Figures: O.M. and I.U.; Editing: A.B., O.M., D.V.B., E.A.B., R.B.S., A.V.B., P.T., A.S., and I.U. All authors have read and agreed to the published version of the manuscript.

Funding: This work was supported into frameworks of the state task for ICP RAS 0082-2019-0012 (State registration number AAAA-A20-120021190043-7) and by the Russian Academic Excellence Project "5–100".

Conflicts of Interest: Authors declare no conflict of interest.

Abbreviations

AAV	Adeno-associated virus
Ad	Adenovirus
BBB	Blood–brain barrier
CNS	Central neural system
DNA	Deoxyribonucleic acid
GDNF	Glial cell-line derived neurotrophic factor
FGF	Fibroblast growth factor
HSPG	Heparan sulfate proteoglycan
HSV	Herpes simplex virus
NDD	Neuro-degenerative disease
PDGFR	Platelet-derived growth factor receptor
RNA	Ribonucleic acid
SFV	Semliki forest virus
SMN	Survival motor neuron protein
VEGF	Vascular endothelial growth factor

References

1. Bordi, M.; Berg, M.J.; Mohan, P.S.; Peterhoff, C.M.; Alldred, M.J.; Che, S.; Ginsberg, S.D.; Nixon, R.A. Autophagy flux in CA1 neurons of Alzheimer hippocampus: Increased induction overburdens failing lysosomes to propel neuritic dystrophy. *Autophagy* **2016**, *12*, 2467–2483. [CrossRef]
2. Ingelsson, M.; Fukumoto, H.; Newell, K.L.; Growdon, J.H.; Hedley-Whyte, E.T.; Frosch, M.P.; Albert, M.S.; Hyman, B.T.; Irizarry, M.C. Early Abeta accumulation and progressive synaptic loss, gliosis, and tangle formation in AD brain. *Neurology* **2004**, *62*, 925–931. [CrossRef]
3. Monahan, J.A.; Warren, M.; Carvey, P.M. Neuroinflammation and peripheral immune infiltration in Parkinson's disease: An autoimmune hypothesis. *Cell Transplant.* **2008**, *17*, 363–372. [CrossRef]
4. Brochard, V.; Combadière, B.; Prigent, A.; Laouar, Y.; Perrin, A.; Beray-Berthat, V.; Bonduelle, O.; Alvarez-Fischer, D.; Callebert, J.; Launay, J.-M.; et al. Infiltration of CD4+ lymphocytes into the brain contributes to neurodegeneration in a mouse model of Parkinson disease. *J. Clin. Investig.* **2008**, *119*, 182–192. [CrossRef]

5. Rousseaux, M.W.; De Haro, M.; Lasagna-Reeves, C.A.; De Maio, A.; Park, J.; Jafar-Nejad, P.; Al-Ramahi, I.; Sharma, A.; See, L.; Lu, N.; et al. TRIM28 regulates the nuclear accumulation and toxicity of both alpha-synuclein and tau. *eLife* **2016**, *5*, 18748. [CrossRef]
6. Duggan, M.; Torkzaban, B.; Ahooyi, T.M.; Khalili, K.; Gordon, J. Age-related neurodegenerative diseases. *J. Cell. Physiol.* **2019**, *235*, 3131–3141. [CrossRef]
7. Min, B.; Chung, K.C. New insight into transglutaminase 2 and link to neurodegenerative diseases. *BMB Rep.* **2018**, *51*, 5–13. [CrossRef]
8. Flavin, W.P.; Bousset, L.; Green, Z.C.; Chu, Y.; Skarpathiotis, S.; Chaney, M.J.; Kordower, J.H.; Melki, R.; Campbell, E.M. Endocytic vesicle rupture is a conserved mechanism of cellular invasion by amyloid proteins. *Acta Neuropathol.* **2017**, *134*, 629–653. [CrossRef]
9. Gorantla, N.V.; Chinnathambi, S. Tau Protein Squired by Molecular Chaperones During Alzheimer's Disease. *J. Mol. Neurosci.* **2018**, *66*, 356–368. [CrossRef]
10. Cristofani, R.; Crippa, V.; Rusmini, P.; Cicardi, M.E.; Meroni, M.; Licata, N.V.; Sala, G.; Giorgetti, E.; Grunseich, C.; Galbiati, M.; et al. Inhibition of retrograde transport modulates misfolded protein accumulation and clearance in motoneuron diseases. *Autophagy* **2017**, *13*, 1280–1303. [CrossRef]
11. Soares, T.; Reis, S.; Pinho, B.; Duchen, M.R.; Oliveira, J.M. Targeting the proteostasis network in Huntington's disease. *Ageing Res. Rev.* **2019**, *49*, 92–103. [CrossRef]
12. Lim, J.; Bang, Y.; Choi, H.J. Abnormal hippocampal neurogenesis in Parkinson's disease: Relevance to a new therapeutic target for depression with Parkinson's disease. *Arch. Pharmacal Res.* **2018**, *41*, 943–954. [CrossRef]
13. Shah, S.Z.A.; Zhao, D.; Hussain, T.; Sabir, N.; Mangi, M.H.; Yang, L. p62-Keap1-NRF2-ARE Pathway: A Contentious Player for Selective Targeting of Autophagy, Oxidative Stress and Mitochondrial Dysfunction in Prion Diseases. *Front. Mol. Neurosci.* **2018**, *11*, 310. [CrossRef]
14. Ariosa, A.R.; Klionsky, D.J. Autophagy core machinery: Overcoming spatial barriers in neurons. *J. Mol. Med.* **2016**, *94*, 1217–1227. [CrossRef]
15. Kumar, A.; Dhawan, A.; Kadam, A.; Shinde, A. Autophagy and Mitochondria: Targets in Neurodegenerative Disorders. *CNS Neurol. Disord. Drug Targets* **2018**, *17*, 696–705. [CrossRef]
16. Tramutola, A.; Lanzillotta, C.; Di Domenico, F. Targeting mTOR to reduce Alzheimer-related cognitive decline: From current hits to future therapies. *Expert Rev. Neurother.* **2016**, *17*, 33–45. [CrossRef]
17. Lee, C.-C.; Chang, C.-P.; Lin, C.-J.; Lai, H.-L.; Kao, Y.-H.; Cheng, S.-J.; Chen, H.-M.; Liao, Y.-P.; Faivre, E.; Buée, L.; et al. Adenosine Augmentation Evoked by an ENT1 Inhibitor Improves Memory Impairment and Neuronal Plasticity in the APP/PS1 Mouse Model of Alzheimer's Disease. *Mol. Neurobiol.* **2018**, *55*, 8936–8952. [CrossRef]
18. Tramutola, A.; Lanzillotta, C.; Barone, E.; Arena, A.; Zuliani, I.; Mosca, L.; Blarzino, C.; Butterfield, D.A.; Perluigi, M.; Di Domenico, F. Intranasal rapamycin ameliorates Alzheimer-like cognitive decline in a mouse model of Down syndrome. *Transl. Neurodegener.* **2018**, *7*, 28. [CrossRef]
19. Musial, T.; Molina-Campos, E.; Bean, L.A.; Ybarra, N.; Borenstein, R.; Russo, M.; Buss, E.; Justus, D.; Neuman, K.M.; Ayala, G.D.; et al. Store depletion-induced h-channel plasticity rescues a channelopathy linked to Alzheimer's disease. *Neurobiol. Learn. Mem.* **2018**, *154*, 141–157. [CrossRef]
20. Guo, N.; Xie, W.; Xiong, P.; Li, H.; Wang, S.; Chen, G.; Gao, Y.; Zhou, J.; Zhang, Y.; Bu, G.; et al. Cyclin-dependent kinase 5-mediated phosphorylation of chloride intracellular channel 4 promotes oxidative stress-induced neuronal death. *Cell Death Dis.* **2018**, *9*, 951. [CrossRef]
21. Picconi, B.; De Leonibus, E.; Calabresi, P. Synaptic plasticity and levodopa-induced dyskinesia: Electrophysiological and structural abnormalities. *J. Neural Transm.* **2018**, *125*, 1263–1271. [CrossRef]
22. Cacabelos, R. Parkinson's Disease: From Pathogenesis to Pharmacogenomics. *Int. J. Mol. Sci.* **2017**, *18*, 551. [CrossRef]
23. Burchfield, J.S.; Li, Q.; Wang, H.Y.; Wang, R.-F. JMJD3 as an epigenetic regulator in development and disease. *Int. J. Biochem. Cell Boil.* **2015**, *67*, 148–157. [CrossRef]
24. Li, P.; Marshall, L.; Oh, G.; Jakubowski, J.L.; Groot, D.; He, Y.; Wang, T.; Petronis, A.; Labrie, V. Epigenetic dysregulation of enhancers in neurons is associated with Alzheimer's disease pathology and cognitive symptoms. *Nat. Commun.* **2019**, *10*, 2246. [CrossRef]
25. Vucic, S.; Ryder, J.; Mekhael, L.; Rd, H.; Mathers, S.; Needham, M.; Dw, S.; Mc, K. TEALS study group Phase 2 randomized placebo controlled double blind study to assess the efficacy and safety of tecfidera in patients with amyotrophic lateral sclerosis (TEALS Study). *Medicine* **2020**, *99*, e18904. [CrossRef]

26. Petramfar, P.; Hajari, F.; Yousefi, G.; Azadi, S.; Hamedi, A. Efficacy of oral administration of licorice as an adjunct therapy on improving the symptoms of patients with Parkinson's disease, A randomized double blinded clinical trial. *J. Ethnopharmacol.* **2020**, *247*, 112226. [CrossRef]
27. Andrade, S.; Ramalho, M.J.; Loureiro, J.A.; Pereira, M.D.C. Natural Compounds for Alzheimer's Disease Therapy: A Systematic Review of Preclinical and Clinical Studies. *Int. J. Mol. Sci.* **2019**, *20*, 2313. [CrossRef]
28. Gasperi, V.; Sibilano, M.; Savini, I.; Catani, M.V. Niacin in the Central Nervous System: An Update of Biological Aspects and Clinical Applications. *Int. J. Mol. Sci.* **2019**, *20*, 974. [CrossRef] [PubMed]
29. Farlow, M.R.; Thompson, R.E.; Wei, L.-J.; Tuchman, A.J.; Grenier, E.; Crockford, D.; Wilke, S.; Benison, J.; Alkon, D.L. A Randomized, Double-Blind, Placebo-Controlled, Phase II Study Assessing Safety, Tolerability, and Efficacy of Bryostatin in the Treatment of Moderately Severe to Severe Alzheimer's Disease. *J. Alzheimer's Dis.* **2019**, *67*, 555–570. [CrossRef]
30. Bao, X.-X.; Ma, H.-H.; Ding, H.; Li, W.-W.; Zhu, M. Preliminary optimization of a Chinese herbal medicine formula based on the neuroprotective effects in a rat model of rotenone-induced Parkinson's disease. *J. Integr. Med.* **2018**, *16*, 290–296. [CrossRef]
31. Johnstone, D.M.; Moro, C.; Stone, J.; Benabid, A.-L.; Mitrofanis, J. Turning On Lights to Stop Neurodegeneration: The Potential of Near Infrared Light Therapy in Alzheimer's and Parkinson's Disease. *Front. Mol. Neurosci.* **2016**, *9*, 75001. [CrossRef] [PubMed]
32. Keeler, A.; Elmallah, M.; Flotte, T.R. Gene Therapy 2017: Progress and Future Directions. *Clin. Transl. Sci.* **2017**, *10*, 242–248. [CrossRef] [PubMed]
33. Mendell, J.R.; Al-Zaidy, S.; Shell, R.; Arnold, W.D.; Rodino-Klapac, L.R.; Prior, T.W.; Lowes, L.P.; Alfano, L.; Berry, K.; Church, K.; et al. Single-Dose Gene-Replacement Therapy for Spinal Muscular Atrophy. *N. Engl. J. Med.* **2017**, *377*, 1713–1722. [CrossRef]
34. Massaro, G.; Mattar, C.N.Z.; Wong, A.M.S.; Sirka, E.; Buckley, S.M.K.; Herbert, B.R.; Karlsson, S.; Perocheau, D.P.; Burke, D.; Heales, S.; et al. Fetal gene therapy for neurodegenerative disease of infants. *Nat. Med.* **2018**, *24*, 1317–1323. [CrossRef]
35. Holthaus, S.-M.K.; Herranz-Martin, S.; Massaro, G.; Aristorena, M.; Hoke, J.; Hughes, M.P.; Maswood, R.; Semenyuk, O.; Basche, M.; Shah, A.Z.; et al. Neonatal brain-directed gene therapy rescues a mouse model of neurodegenerative CLN6 Batten disease. *Hum. Mol. Genet.* **2019**, *28*, 3867–3879. [CrossRef]
36. Das, S.; Huang, S.; Lo, A.W. Acceleration of rare disease therapeutic development: A case study of AGIL-AADC. *Drug Discov. Today* **2019**, *24*, 678–684. [CrossRef]
37. Gessler, D.J.; Gao, G. Gene Therapy for the Treatment of Neurological Disorders: Metabolic Disorders. *Breast Cancer* **2016**, *1382*, 429–465.
38. Conlon, T.J.; Mavilio, F. The Pharmacology of Gene and Cell Therapy. *Mol. Ther. Methods Clin. Dev.* **2018**, *8*, 181–182. [CrossRef]
39. Cwetsch, A.W.; Pinto, B.; Savardi, A.; Cancedda, L. In vivo methods for acute modulation of gene expression in the central nervous system. *Prog. Neurobiol.* **2018**, *168*, 69–85. [CrossRef]
40. Mavilio, F.; Ferrari, G. Genetic modification of somatic stem cells. The progress, problems and prospects of a new therapeutic technology. *EMBO Rep.* **2008**, *9* (Suppl. 1), S64–S69. [CrossRef]
41. Vagner, T.; Dvorzhak, A.; Wójtowicz, A.M.; Harms, C.; Grantyn, R. Systemic application of AAV vectors targeting GFAP-expressing astrocytes in Z -Q175-KI Huntington's disease mice. *Mol. Cell. Neurosci.* **2016**, *77*, 76–86. [CrossRef] [PubMed]
42. Shahaduzzaman, M.; Nash, K.; Hudson, C.; Sharif, M.; Grimmig, B.; Lin, X.; Bai, G.; Liu, H.; Ugen, K.E.; Cao, C.; et al. Anti-Human α-Synuclein N-Terminal Peptide Antibody Protects against Dopaminergic Cell Death and Ameliorates Behavioral Deficits in an AAV-α-Synuclein Rat Model of Parkinson's Disease. *PLoS ONE* **2015**, *10*, e0116841. [CrossRef] [PubMed]
43. Chapdelaine, P.; Gerard, C.; Sanchez, N.; Cherif, K.; Rousseau, J.; Ouellet, D.L.; Jauvin, D.; Tremblay, J.P. Development of an AAV9 coding for a 3XFLAG-TALEfrat#8-VP64 able to increase in vivo the human frataxin in YG8R mice. *Gene Ther.* **2016**, *23*, 606–614. [PubMed]
44. Morabito, G.; Giannelli, S.G.; Ordazzo, G.; Bido, S.; Castoldi, V.; Indrigo, M.; Cabassi, T.; Cattaneo, S.; Luoni, M.; Cancellieri, C.; et al. AAV-PHP.B-Mediated Global-Scale Expression in the Mouse Nervous System Enables GBA1 Gene Therapy for Wide Protection from Synucleinopathy. *Mol. Ther.* **2017**, *25*, 2727–2742. [CrossRef] [PubMed]

45. Foust, K.D.; Nurre, E.; Montgomery, C.L.; Hernandez, A.; Chan, C.M.; Kaspar, B.K. Intravascular AAV9 preferentially targets neonatal neurons and adult astrocytes. *Nat. Biotechnol.* **2008**, *27*, 59–65. [CrossRef] [PubMed]
46. Pan, X.; Sands, S.A.; Yue, Y.; Zhang, K.; Levine, S.M.; Duan, N. An Engineered Galactosylceramidase Construct Improves AAV Gene Therapy for Krabbe Disease in Twitcher Mice. *Hum. Gene Ther.* **2019**, *30*, 1039–1051. [CrossRef]
47. Perez, B.A.; Shutterly, A.; Chan, Y.; Byrne, B.; Corti, M. Management of Neuroinflammatory Responses to AAV-Mediated Gene Therapies for Neurodegenerative Diseases. *Brain Sci.* **2020**, *10*, 119. [CrossRef] [PubMed]
48. Saraiva, J.; Nobre, R.J.; De Almeida, L.P. Gene therapy for the CNS using AAVs: The impact of systemic delivery by AAV9. *J. Control. Release* **2016**, *241*, 94–109. [CrossRef]
49. Hudry, E.; Andres-Mateos, E.; Lerner, E.P.; Volak, A.; Cohen, O.; Hyman, B.T.; Maguire, C.A.; Vandenberghe, L.H. Efficient Gene Transfer to the Central Nervous System by Single-Stranded Anc80L65. *Mol. Ther. Methods Clin. Dev.* **2018**, *10*, 197–209. [CrossRef]
50. Hösel, M.; Huber, A.; Bohlen, S.; Lucifora, J.; Ronzitti, G.; Puzzo, F.; Boisgerault, F.; Hacker, U.T.; Kwanten, W.; Klöting, N.; et al. Autophagy determines efficiency of liver-directed gene therapy with adeno-associated viral vectors. *Hepatology* **2017**, *66*, 252–265. [CrossRef]
51. Wang, H.; Richter, M.; Psatha, N.; Li, C.; Kim, J.; Liu, J.; Ehrhardt, A.; Nilsson, S.K.; Cao, B.; Palmer, N.; et al. A Combined In Vivo HSC Transduction/Selection Approach Results in Efficient and Stable Gene Expression in Peripheral Blood Cells in Mice. *Mol. Ther. Methods Clin. Dev.* **2017**, *8*, 52–64. [CrossRef] [PubMed]
52. Langou, K.; Moumen, A.; Pellegrino, C.; Aebischer, J.; Medina, I.; Aebischer, P.; Raoul, C. AAV-mediated expression of wild-type and ALS-linked mutant VAPB selectively triggers death of motoneurons through a Ca^{2+}-dependent ER-associated pathway. *J. Neurochem.* **2010**, *114*, 795–809. [CrossRef] [PubMed]
53. Wang, H.C.; Zhang, T.; Kuerban, B.; Jin, Y.L.; Le, W.; Hara, H.; Fan, D.S.; Wang, Y.J.; Tabira, T.; Chui, D.H. Autophagy is involved in oral rAAV/Abeta vaccine-induced Abeta clearance in APP/PS1 transgenic mice. *Neurosci. Bull.* **2015**, *31*, 491–504. [CrossRef] [PubMed]
54. Niu, H.; Shen, L.; Li, T.; Ren, C.; Ding, S.; Wang, L.; Zhang, Z.; Liu, X.; Zhang, Q.; Geng, D.; et al. Alpha-synuclein overexpression in the olfactory bulb initiates prodromal symptoms and pathology of Parkinson's disease. *Transl. Neurodegener.* **2018**, *7*, 25. [CrossRef] [PubMed]
55. Zheng, Q.; Zheng, X.; Zhang, L.; Luo, H.; Qian, L.; Fu, X.; Liu, Y.; Gao, Y.; Niu, M.; Meng, J.; et al. The Neuron-Specific Protein TMEM59L Mediates Oxidative Stress-Induced Cell Death. *Mol. Neurobiol.* **2016**, *54*, 4189–4200. [CrossRef]
56. Mancuso, R.; Martínez-Muriana, A.; Leiva, T.; Gregorio, D.; Ariza, L.; Morell, M.; Esteban-Pérez, J.; García-Redondo, A.; Calvo, A.; Atencia-Cibreiro, G.; et al. Neuregulin-1 promotes functional improvement by enhancing collateral sprouting in SOD1G93A ALS mice and after partial muscle denervation. *Neurobiol. Dis.* **2016**, *95*, 168–178.
57. Kato, S.; Sugawara, M.; Kobayashi, K.; Kimura, K.; Inoue, K.-I.; Takada, M.; Kobayashi, K. Enhancement of the transduction efficiency of a lentiviral vector for neuron-specific retrograde gene delivery through the point mutation of fusion glycoprotein type E. *J. Neurosci. Methods* **2019**, *311*, 147–155. [CrossRef]
58. Artusi, S.; Miyagawa, Y.; Goins, W.F.; Cohen, J.; Glorioso, J.C. Herpes Simplex Virus Vectors for Gene Transfer to the Central Nervous System. *Disease* **2018**, *6*, 74. [CrossRef]
59. Ansorena, E.; Casales, E.; Aranda, A.; Tamayo, E.; Garbayo, E.; Smerdou, C.; Blanco-Prieto, M.J.; Aymerich, M.S. A simple and efficient method for the production of human glycosylated glial cell line-derived neurotrophic factor using a Semliki Forest virus expression system. *Int. J. Pharm.* **2013**, *440*, 19–26. [CrossRef]
60. Berns, K.I.; Muzyczka, N. AAV: An Overview of Unanswered Questions. *Hum. Gene Ther.* **2017**, *28*, 308–313. [CrossRef]
61. Ehrengruber, M.U. Alphaviral Vectors for Gene Transfer into Neurons. *Mol. Neurobiol.* **2002**, *26*, 183–202. [CrossRef]
62. Ehrengruber, M.U.; Lundstrom, K.; Schweitzer, C.; Heuss, C.; Schlesinger, S.; Gähwiler, B.H. Recombinant Semliki Forest virus and Sindbis virus efficiently infect neurons in hippocampal slice cultures. *Proc. Natl. Acad. Sci. USA* **1999**, *96*, 7041–7046. [CrossRef] [PubMed]

63. VandenDriessche, T.; Thorrez, L.; Acosta-Sanchez, A.; Petrus, I.; Wang, L.; Ma, L.; De Waele, L.; Iwasaki, Y.; Gillijns, V.; Wilson, J.M.; et al. Efficacy and safety of adeno-associated viral vectors based on serotype 8 and 9 vs. lentiviral vectors for hemophilia B gene therapy. *J. Thromb. Haemost.* **2006**, *5*, 16–24. [CrossRef] [PubMed]
64. Hoy, S.M. Onasemnogene Abeparvovec: First Global Approval. *Drugs* **2019**, *79*, 1255–1262. [CrossRef] [PubMed]
65. Miller, T.M.; Pestronk, A.; David, W.; Rothstein, J.; Simpson, E.; Appel, S.H.; Andres, P.L.; Mahoney, K.; Allred, P.; Alexander, K.; et al. An antisense oligonucleotide against SOD1 delivered intrathecally for patients with SOD1 familial amyotrophic lateral sclerosis: A phase 1, randomised, first-in-man study. *Lancet Neurol.* **2013**, *12*, 435–442. [CrossRef]
66. Auricchio, A.; Smith, A.; Ali, R.R. The Future Looks Brighter After 25 Years of Retinal Gene Therapy. *Hum. Gene Ther.* **2017**, *28*, 982–987. [CrossRef]
67. Grimm, D.; Zhou, S.; Nakai, H.; Thomas, C.E.; Storm, T.A.; Fuess, S.; Matsushita, T.; Allen, J.; Surosky, R.; Lochrie, M.; et al. Preclinical in vivo evaluation of pseudotyped adeno-associated virus vectors for liver gene therapy. *Blood* **2003**, *102*, 2412–2419. [CrossRef]
68. Paulk, N.; Pekrun, K.; Zhu, E.; Nygaard, S.; Li, B.; Xu, J.; Chu, K.; Leborgne, C.; Dane, A.P.; Haft, A.; et al. Bioengineered AAV Capsids with Combined High Human Liver Transduction In Vivo and Unique Humoral Seroreactivity. *Mol. Ther.* **2017**, *26*, 289–303. [CrossRef]
69. White, S.; Nicklin, S.A.; Buüning, H.; Brosnan, M.J.; Leike, K.; Papadakis, E.D.; Hallek, M.; Baker, A.H. Targeted Gene Delivery to Vascular Tissue In Vivo by Tropism-Modified Adeno-Associated Virus Vectors. *Circulation* **2004**, *109*, 513–519. [CrossRef]
70. Work, L.M.; Büning, H.; Pustulka-Hunt, E.; Nicklin, S.A.; Denby, L.; Britton, N.; Leike, K.; Odenthal, M.; Drebber, U.; Hallek, M.; et al. Vascular bed-targeted in vivo gene delivery using tropism-modified adeno-associated viruses. *Mol. Ther.* **2006**, *13*, 683–693. [CrossRef]
71. Halbert, C.L.; Allen, J.M.; Miller, A.D. Adeno-Associated Virus Type 6 (AAV6) Vectors Mediate Efficient Transduction of Airway Epithelial Cells in Mouse Lungs Compared to That of AAV2 Vectors. *J. Virol.* **2001**, *75*, 6615–6624. [CrossRef] [PubMed]
72. Harper, S.Q.; Hauser, M.A.; DelloRusso, C.; Duan, D.; Crawford, R.W.; Phelps, S.F.; Harper, H.A.; Robinson, A.S.; Engelhardt, J.F.; Brooks, S.V.; et al. Modular flexibility of dystrophin: Implications for gene therapy of Duchenne muscular dystrophy. *Nat. Med.* **2002**, *8*, 253–261. [CrossRef] [PubMed]
73. Tervo, D.G.R.; Hwang, B.-Y.; Viswanathan, S.; Gaj, T.; Lavzin, M.; Ritola, K.D.; Lindo, S.; Michael, S.; Kuleshova, E.; Ojala, D.; et al. A Designer AAV Variant Permits Efficient Retrograde Access to Projection Neurons. *Neuron* **2016**, *92*, 372–382. [CrossRef] [PubMed]
74. Asokan, A.; Schaffer, D.V.; Samulski, R.J. The AAV Vector Toolkit: Poised at the Clinical Crossroads. *Mol. Ther.* **2012**, *20*, 699–708. [CrossRef] [PubMed]
75. Hartmann, J.; Thalheimer, F.B.; Höpfner, F.; Kerzel, T.; Khodosevich, K.; García-González, D.; Monyer, H.; Diester, I.; Büning, H.; Carette, J.E.; et al. GluA4-Targeted AAV Vectors Deliver Genes Selectively to Interneurons while Relying on the AAV Receptor for Entry. *Mol. Ther. Methods Clin. Dev.* **2019**, *14*, 252–260. [CrossRef]
76. Qing, K.; Mah, C.; Hansen, J.; Zhou, S.; Dwarki, V.; Srivastava, A. Human fibroblast growth factor receptor 1 is a co-receptor for infection by adeno-associated virus 2. *Nat. Med.* **1999**, *5*, 71–77. [CrossRef]
77. Di Pasquale, G.; Davidson, B.L.; Stein, C.S.; Martins, I.; Scudiero, D.; Monks, A.; Chiorini, J.A. Identification of PDGFR as a receptor for AAV-5 transduction. *Nat. Med.* **2003**, *9*, 1306–1312. [CrossRef]
78. Kashiwakura, Y.; Tamayose, K.; Iwabuchi, K.; Hirai, Y.; Shimada, T.; Matsumoto, K.; Nakamura, T.; Watanabe, M.; Oshimi, K.; Daida, H. Hepatocyte Growth Factor Receptor Is a Coreceptor for Adeno-Associated Virus Type 2 Infection. *J. Virol.* **2005**, *79*, 609–614. [CrossRef]
79. Asokan, A.; Hamra, J.B.; Govindasamy, L.; Agbandje-McKenna, M.; Samulski, R.J. Adeno-associated virus type 2 contains an integrin alpha5beta1 binding domain essential for viral cell entry. *J. Virol.* **2006**, *80*, 8961–8969. [CrossRef]
80. Summerford, C.; Bartlett, J.S.; Samulski, R.J. AlphaVbeta5 integrin: A co-receptor for adeno-associated virus type 2 infection. *Nat. Med.* **1999**, *5*, 78–82. [CrossRef]
81. Walters, R.W.; Pilewski, J.M.; Chiorini, J.A.; Zabner, J. Secreted and Transmembrane Mucins Inhibit Gene Transfer with AAV4 More Efficiently than AAV5. *J. Boil. Chem.* **2002**, *277*, 23709–23713. [CrossRef] [PubMed]

82. Akache, B.; Grimm, D.; Pandey, K.; Yant, S.R.; Xu, H.; Kay, M.A. The 37/67-kilodalton laminin receptor is a receptor for adeno-associated virus serotypes 8, 2, 3, and 9. *J. Virol.* **2006**, *80*, 9831–9836. [CrossRef] [PubMed]
83. Vana, K.; Zuber, C.; Pflanz, H.; Kolodziejczak, D.; Zemora, G.; Bergmann, A.-K.; Weiss, S. LRP/LR as an alternative promising target in therapy of prion diseases, Alzheimer's disease and cancer. *Infect. Disord. Drug Targets* **2009**, *9*, 69–80. [CrossRef]
84. Cearley, C.N.; Vandenberghe, L.H.; Parente, M.K.; Carnish, E.R.; Wilson, J.M.; Wolfe, J.H. Expanded Repertoire of AAV Vector Serotypes Mediate Unique Patterns of Transduction in Mouse Brain. *Mol. Ther.* **2008**, *16*, 1710–1718. [CrossRef] [PubMed]
85. Hildinger, M.; Auricchio, A.; Gao, G.; Wang, L.; Chirmule, N.; Wilson, J.M. Hybrid Vectors Based on Adeno-Associated Virus Serotypes 2 and 5 for Muscle-Directed Gene Transfer. *J. Virol.* **2001**, *75*, 6199–6203. [CrossRef] [PubMed]
86. Gao, G.-P.; Alvira, M.R.; Wang, L.; Calcedo, R.; Johnston, J.; Wilson, J.M. Novel adeno-associated viruses from rhesus monkeys as vectors for human gene therapy. *Proc. Natl. Acad. Sci. USA* **2002**, *99*, 11854–11859. [CrossRef]
87. Li, W.; Asokan, A.; Wu, Z.; Van Dyke, T.; DiPrimio, N.; Johnson, J.; Govindaswamy, L.; Agbandje-McKenna, M.; Leichtle, S.; Redmond, D.; et al. Engineering and Selection of Shuffled AAV Genomes: A New Strategy for Producing Targeted Biological Nanoparticles. *Mol. Ther.* **2008**, *16*, 1252–1260. [CrossRef]
88. Maheshri, N.; Koerber, J.T.; Kaspar, B.K.; Schaffer, D.V. Directed evolution of adeno-associated virus yields enhanced gene delivery vectors. *Nat. Biotechnol.* **2006**, *24*, 198–204. [CrossRef]
89. Körbelin, J.; Sieber, T.; Michelfelder, S.; Lunding, L.; Spies, E.; Hunger, A.; Alawi, M.; Rapti, K.; Indenbirken, D.; Müller, O.J.; et al. Pulmonary Targeting of Adeno-associated Viral Vectors by Next-generation Sequencing-guided Screening of Random Capsid Displayed Peptide Libraries. *Mol. Ther.* **2016**, *24*, 1050–1061. [CrossRef]
90. Chen, Y.H.; Chang, M.; Davidson, B.L. Molecular signatures of disease brain endothelia provide new sites for CNS-directed enzyme therapy. *Nat. Med.* **2009**, *15*, 1215–1218. [CrossRef]
91. Tordo, J.; O'Leary, C.; Antunes, A.; Palomar, N.; Aldrin-Kirk, P.; Basche, M.; Bennett, A.; D'Souza, Z.; Gleitz, H.; Godwin, A.; et al. A novel adeno-associated virus capsid with enhanced neurotropism corrects a lysosomal transmembrane enzyme deficiency. *Brain* **2018**, *141*, 2014–2031. [CrossRef] [PubMed]
92. Zinn, E.; Pacouret, S.; Khaychuk, V.; Turunen, H.T.; Carvalho, L.S.; Andres-Mateos, E.; Shah, S.; Shelke, R.; Maurer, A.C.; Plovie, E.; et al. In Silico Reconstruction of the Viral Evolutionary Lineage Yields a Potent Gene Therapy Vector. *Cell Rep.* **2015**, *12*, 1056–1068. [CrossRef] [PubMed]
93. Münch, R.C.; Janicki, H.; Völker, I.; Rasbach, A.; Hallek, M.; Büning, H.; Buchholz, C.J. Displaying High-affinity Ligands on Adeno-associated Viral Vectors Enables Tumor Cell-specific and Safe Gene Transfer. *Mol. Ther.* **2012**, *21*, 109–118. [CrossRef] [PubMed]
94. Davidsson, M.; Wang, G.; Aldrin-Kirk, P.; Cardoso, T.; Nolbrant, S.; Hartnor, M.; Mudannayake, J.; Parmar, M.; Björklund, T. A systematic capsid evolution approach performed in vivo for the design of AAV vectors with tailored properties and tropism. *Proc. Natl. Acad. Sci. USA* **2019**, *116*, 27053–27062. [CrossRef] [PubMed]
95. Osaki, T.; Shin, Y.; Sivathanu, V.; Campisi, M.; Kamm, R.D. In Vitro Microfluidic Models for Neurodegenerative Disorders. *Adv. Healthc. Mater.* **2017**, *7*, 1700489. [CrossRef] [PubMed]
96. Park, J.; Lee, B.K.; Jeong, G.S.; Hyun, J.K.; Lee, C.J.; Lee, S.-H. Three-dimensional brain-on-a-chip with an interstitial level of flow and its application as an in vitro model of Alzheimer's disease. *Lab Chip* **2015**, *15*, 141–150. [CrossRef]
97. Zurina, I.M.; Shpichka, A.; Saburina, I.N.; Kosheleva, N.V.A.; Gorkun, A.A.; Grebenik, E.; Kuznetsova, D.S.; Zhang, D.A.; Rochev, Y.; Butnaru, D.V.; et al. 2D/3D buccal epithelial cell self-assembling as a tool for cell phenotype maintenance and fabrication of multilayered epithelial linings in vitro. *Biomed. Mater.* **2018**, *13*, 054104. [CrossRef]
98. Soriani, M.; Yi, W.; He, A.; Ml, S. Faculty of 1000 evaluation for Microfluidic blood-brain barrier model provides in vivo-like barrier properties for drug permeability screening. *F1000 Post Publ. Peer Rev. Biomed. Lit.* **2017**, *114*, 184–194.
99. Dauth, S.; Maoz, B.M.; Sheehy, S.P.; Hemphill, M.A.; Murty, T.; Macedonia, M.K.; Greer, A.M.; Budnik, B.; Parker, K.K. Neurons derived from different brain regions are inherently different in vitro: A novel multiregional brain-on-a-chip. *J. Neurophysiol.* **2017**, *117*, 1320–1341. [CrossRef]

100. Merkel, S.F.; Andrews, A.; Lutton, E.M.; Mu, D.; Hudry, E.; Hyman, B.T.; Maguire, C.A.; Ramirez, S.H. Trafficking of adeno-associated virus vectors across a model of the blood-brain barrier; a comparative study of transcytosis and transduction using primary human brain endothelial cells. *J. Neurochem.* **2016**, *140*, 216–230. [CrossRef]
101. Donadon, I.; Bussani, E.; Riccardi, F.; Licastro, D.; Romano, G.; Pianigiani, G.; Pinotti, M.; Konstantinova, P.; Evers, M.; Lin, S.; et al. Rescue of spinal muscular atrophy mouse models with AAV9-Exon-specific U1 snRNA. *Nucleic Acids Res.* **2019**, *47*, 7618–7632. [CrossRef] [PubMed]
102. Hughes, M.; Smith, D.A.; Morris, L.; Fletcher, C.; Colaço, A.; Huebecker, M.; Tordo, J.; Palomar, N.; Massaro, G.; Henckaerts, E.; et al. AAV9 intracerebroventricular gene therapy improves lifespan, locomotor function and pathology in a mouse model of Niemann–Pick type C1 disease. *Hum. Mol. Genet.* **2018**, *27*, 3079–3098. [CrossRef] [PubMed]
103. Osman, E.Y.; Bolding, M.R.; Villalón, E.; Kaifer, K.A.; Lorson, Z.C.; Tisdale, S.; Hao, Y.; Conant, G.C.; Pires, J.C.; Pellizzoni, L.; et al. Functional characterization of SMN evolution in mouse models of SMA. *Sci. Rep.* **2019**, *9*, 9472. [CrossRef]
104. Amado, D.; Rieders, J.M.; Diatta, F.; Hernandez-Con, P.; Singer, A.; Mak, J.T.; Zhang, J.; Lancaster, E.; Davidson, B.L.; Chen-Plotkin, A.S. AAV-Mediated Progranulin Delivery to a Mouse Model of Progranulin Deficiency Causes T Cell-Mediated Toxicity. *Mol. Ther.* **2019**, *27*, 465–478. [CrossRef]
105. Commisso, B.; Ding, L.; Varadi, K.; Gorges, M.; Bayer, D.; Boeckers, T.M.; Ludolph, A.C.; Kassubek, J.; Müller, O.J.; Roselli, F. Stage-dependent remodeling of projections to motor cortex in ALS mouse model revealed by a new variant retrograde-AAV9. *eLife* **2018**, *7*, 7. [CrossRef] [PubMed]
106. Pastor, P.D.H.; Du, X.; Fazal, S.; Davies, A.N.; Gomez, C.M. Targeting the CACNA1A IRES as a Treatment for Spinocerebellar Ataxia Type 6. *Cerebellum* **2018**, *17*, 72–77. [CrossRef] [PubMed]
107. Yang, C.; Kang, F.; Wang, S.; Han, M.; Zhang, Z.; Li, J. SIRT1 Activation Attenuates Bone Cancer Pain by Inhibiting mGluR1/5. *Cell Mol. Neurobiol.* **2019**, *39*, 1165–1175. [CrossRef]
108. Lee, S.H.; Kim, S.; Lee, N.; Lee, J.; Yu, S.S.; Kim, J.-H.; Kim, S. Intrathecal delivery of recombinant AAV1 encoding hepatocyte growth factor improves motor functions and protects neuromuscular system in the nerve crush and SOD1-G93A transgenic mouse models. *Acta Neuropathol. Commun.* **2019**, *7*, 96. [CrossRef]
109. Wen, D.; Cui, C.; Duan, W.; Wang, W.; Wang, Y.; Liu, Y.; Li, Z.; Li, C. The role of insulin-like growth factor 1 in ALS cell and mouse models: A mitochondrial protector. *Brain Res. Bull.* **2019**, *144*, 1–13. [CrossRef]
110. Gong, Y.; Berenson, A.; Laheji, F.; Gao, G.; Wang, D.; Ng, C.; Volak, A.; Kok, R.; Kreouzis, V.; Dijkstra, I.M.; et al. Intrathecal Adeno-Associated Viral Vector-Mediated Gene Delivery for Adrenomyeloneuropathy. *Hum. Gene Ther.* **2019**, *30*, 544–555. [CrossRef]
111. Janda, E.; LaScala, A.; Carresi, C.; Parafati, M.; Aprigliano, S.; Russo, V.; Savoia, C.; Ziviani, E.; Musolino, V.; Morani, F.; et al. Parkinsonian toxin-induced oxidative stress inhibits basal autophagy in astrocytes via NQO2/quinone oxidoreductase 2: Implications for neuroprotection. *Autophagy* **2015**, *11*, 1063–1080. [CrossRef] [PubMed]
112. Rocha, E.M.; Smith, G.; Park, E.; Cao, H.; Brown, E.; Hayes, M.A.; Beagan, J.; McLean, J.R.; Izen, S.C.; Perez-Torres, E.; et al. Glucocerebrosidase gene therapy prevents α-synucleinopathy of midbrain dopamine neurons. *Neurobiol. Dis.* **2015**, *82*, 495–503. [CrossRef] [PubMed]
113. Hunt, J.B.; Nash, K.R.; Placides, D.; Moran, P.; Selenica, M.-L.B.; Abuqalbeen, F.; Ratnasamy, K.; Slouha, N.; Rodriguez-Ospina, S.; Savlia, M.; et al. Sustained Arginase 1 Expression Modulates Pathological Tau Deposits in a Mouse Model of Tauopathy. *J. Neurosci.* **2015**, *35*, 14842–14860. [CrossRef] [PubMed]
114. Zhong, L.; Xu, Y.; Zhuo, R.; Wang, T.; Wang, K.; Huang, R.; Wang, D.; Gao, Y.; Zhu, Y.; Sheng, X.; et al. Soluble TREM2 ameliorates pathological phenotypes by modulating microglial functions in an Alzheimer's disease model. *Nat. Commun.* **2019**, *10*, 1365. [CrossRef]
115. Stoker, T.B.; Torsney, K.M.; Barker, R.A. Emerging Treatment Approaches for Parkinson's Disease. *Front. Mol. Neurosci.* **2018**, *12*, 693. [CrossRef]
116. Gutekunst, C.-A.; Tung, J.K.; McDougal, M.E.; Gross, R.E. C3 transferase gene therapy for continuous conditional RhoA inhibition. *Neurosci.* **2016**, *339*, 308–318. [CrossRef]
117. Jin, D.; Muramatsu, S.-I.; Shimizu, N.; Yokoyama, S.; Hirai, H.; Yamada, K.; Liu, H.-X.; Higashida, C.; Hashii, M.; Higashida, A.; et al. Dopamine release via the vacuolar ATPase V0 sector c-subunit, confirmed in N18 neuroblastoma cells, results in behavioral recovery in hemiparkinsonian mice. *Neurochem. Int.* **2012**, *61*, 907–912. [CrossRef]

118. Higashida, H.; Yokoyama, S.; Tsuji, C.; Muramatsu, S.-I. Neurotransmitter release: Vacuolar ATPase V0 sector c-subunits in possible gene or cell therapies for Parkinson's, Alzheimer's, and psychiatric diseases. *J. Physiol. Sci.* **2016**, *67*, 11–17. [CrossRef]
119. Lu, M.-H.; Ji, W.-L.; Xu, D.-E.; Yao, P.-P.; Zhao, X.-Y.; Wang, Z.-T.; Fang, L.-P.; Huang, R.; Lan, L.-J.; Chen, J.-B.; et al. Inhibition of sphingomyelin synthase 1 ameliorates alzheimer-like pathology in APP/PS1 transgenic mice through promoting lysosomal degradation of BACE1. *Exp. Neurol.* **2019**, *311*, 67–79. [CrossRef]
120. Stoica, L.; Sena-Esteves, M. Adeno Associated Viral Vector Delivered RNAi for Gene Therapy of SOD1 Amyotrophic Lateral Sclerosis. *Front. Mol. Neurosci.* **2016**, *9*, 1971. [CrossRef]
121. Frakes, A.E.; Braun, L.; Ferraiuolo, L.; Guttridge, D.C.; Kaspar, B.K. Additive amelioration of ALS by co-targeting independent pathogenic mechanisms. *Ann. Clin. Transl. Neurol.* **2017**, *4*, 76–86. [CrossRef] [PubMed]
122. Lu, J.-M.; Liu, D.-D.; Li, Z.; Ling, C.; Mei, Y.-A. Neuritin Enhances Synaptic Transmission in Medial Prefrontal Cortex in Mice by Increasing CaV3.3 Surface Expression. *Cereb. Cortex* **2017**, *27*, 3842–3855. [CrossRef] [PubMed]
123. Evers, M.; Miniarikova, J.; Juhas, S.; Vallès, A.; Bohuslavova, B.; Juhasova, J.; Skalnikova, H.K.; Vodicka, P.; Valekova, I.; Brouwers, C.; et al. AAV5-miHTT Gene Therapy Demonstrates Broad Distribution and Strong Human Mutant Huntingtin Lowering in a Huntington's Disease Minipig Model. *Mol. Ther.* **2018**, *26*, 2163–2177. [CrossRef] [PubMed]
124. Cideciyan, A.V.; Sudharsan, R.; Dufour, V.L.; Massengill, M.T.; Iwabe, S.; Swider, M.; Lisi, B.; Sumaroka, A.; Marinho, L.F.; Appelbaum, T.; et al. Mutation-independent rhodopsin gene therapy by knockdown and replacement with a single AAV vector. *Proc. Natl. Acad. Sci. USA* **2018**, *115*, E8547–E8556. [CrossRef] [PubMed]
125. Lee, S.; Ahn, H.J. Anti-EpCAM-conjugated adeno-associated virus serotype 2 for systemic delivery of EGFR shRNA: Its retargeting and antitumor effects on OVCAR3 ovarian cancer in vivo. *Acta Biomater.* **2019**, *91*, 258–269. [CrossRef]
126. Zharikov, A.; Bai, Q.; De Miranda, B.R.; Van Laar, A.; Greenamyre, J.T.; Burton, E.A. Long-term RNAi knockdown of α-synuclein in the adult rat substantia nigra without neurodegeneration. *Neurobiol. Dis.* **2019**, *125*, 146–153. [CrossRef]
127. Li, H.; Wan, C.; Li, F. Recombinant adeno-associated virus-, polyethylenimine/plasmid- and lipofectamine/carboxyfluorescein-labeled small interfering RNA-based transfection in retinal pigment epithelial cells with ultrasound and/or SonoVue. *Mol. Med. Rep.* **2015**, *11*, 3609–3614. [CrossRef]
128. Komor, A.C.; Badran, A.H.; Liu, D.R. CRISPR-Based Technologies for the Manipulation of Eukaryotic Genomes. *Cell* **2017**, *169*, 559. [CrossRef]
129. Reyon, D.; Tsai, S.Q.; Khayter, C.; Foden, J.A.; Sander, J.D.; Joung, J.K. FLASH assembly of TALENs for high-throughput genome editing. *Nat. Biotechnol.* **2012**, *30*, 460–465. [CrossRef]
130. Komor, A.C.; Kim, Y.B.; Packer, M.S.; Zuris, J.A.; Liu, D.R. Programmable editing of a target base in genomic DNA without double-stranded DNA cleavage. *Nature* **2016**, *533*, 420–424. [CrossRef]
131. Raikwar, S.; Thangavel, R.; Dubova, I.; Selvakumar, G.P.; Ahmed, M.E.; Kempuraj, D.; Zaheer, S.A.; Iyer, S.S.; Zaheer, A. Targeted Gene Editing of Glia Maturation Factor in Microglia: A Novel Alzheimer's Disease Therapeutic Target. *Mol. Neurobiol.* **2018**, *56*, 378–393. [CrossRef] [PubMed]
132. Gaj, T.; Ojala, D.S.; Ekman, F.K.; Byrne, L.C.; Limsirichai, P.; Schaffer, D.V. In vivo genome editing improves motor function and extends survival in a mouse model of ALS. *Sci. Adv.* **2017**, *3*, eaar3952. [CrossRef] [PubMed]
133. Ekman, F.K.; Ojala, D.S.; Adil, M.M.; Lopez, P.A.; Schaffer, D.V.; Gaj, T. CRISPR-Cas9-Mediated Genome Editing Increases Lifespan and Improves Motor Deficits in a Huntington's Disease Mouse Model. *Mol. Ther. Nucleic Acids* **2019**, *17*, 829–839. [CrossRef] [PubMed]
134. Huang, P.; Sun, J.; Wang, F.; Luo, X.; Feng, J.; Gu, Q.; Liu, T.; Sun, X. MicroRNA Expression Patterns Involved in Amyloid Beta–Induced Retinal Degeneration. *Investig. Opthalmol. Vis. Sci.* **2017**, *58*, 1726–1735. [CrossRef] [PubMed]
135. Ma, X.; Liu, L.; Meng, J. MicroRNA-125b promotes neurons cell apoptosis and Tau phosphorylation in Alzheimer's disease. *Neurosci. Lett.* **2017**, *661*, 57–62. [CrossRef]

136. Miyazaki, Y.; Adachi, H.; Katsuno, M.; Minamiyama, M.; Jiang, Y.-M.; Huang, Z.; Doi, H.; Matsumoto, S.; Kondo, N.; Iida, M.; et al. Viral delivery of miR-196a ameliorates the SBMA phenotype via the silencing of CELF2. *Nat. Med.* **2012**, *18*, 1136–1141. [CrossRef]
137. Boissonneault, V.; Plante, I.; Rivest, S.; Provost, P. MicroRNA-298 and microRNA-328 regulate expression of mouse beta-amyloid precursor protein-converting enzyme 1. *J. Boil. Chem.* **2008**, *284*, 1971–1981. [CrossRef]
138. Tuszynski, M.H.; Yang, J.H.; Barba, D.; Hoi-Sang, U.; Bakay, R.A.; Pay, M.M.; Masliah, E.; Conner, J.M.; Kobalka, P.; Roy, S.; et al. Nerve Growth Factor Gene Therapy: Activation of Neuronal Responses in Alzheimer Disease. *JAMA Neurol.* **2015**, *72*, 1139–1147. [CrossRef]
139. Quintino, L.; Avallone, M.; Brännstrom, E.; Kavanagh, P.; Lockowandt, M.; Jareño, P.G.; Breger, L.S.; Lundberg, C. GDNF-mediated rescue of the nigrostriatal system depends on the degree of degeneration. *Gene Ther.* **2018**, *26*, 57–64. [CrossRef]
140. Islamov, R.; Rizvanov, A.; Mukhamedyarov, M.; Salafutdinov, I.; Garanina, E.; Fedotova, V.; Solovyeva, V.V.; Mukhamedshina, Y.; Safiullov, Z.; Izmailov, A.; et al. Symptomatic Improvement, Increased Life-Span and Sustained Cell Homing in Amyotrophic Lateral Sclerosis After Transplantation of Human Umbilical Cord Blood Cells Genetically Modified with Adeno-Viral Vectors Expressing a Neuro-Protective Factor and a Neural Cell Adhesion Molecule. *Curr. Gene Ther.* **2015**, *15*, 266–276.
141. Schubert, T.; Wissinger, B. Restoration of synaptic function in sight for degenerative retinal disease. *J. Clin. Investig.* **2015**, *125*, 2572–2575. [CrossRef] [PubMed]
142. Chen, Y.-C.; Ma, N.-X.; Pei, Z.-F.; Wu, Z.; Do-Monte, F.H.; Keefe, S.; Yellin, E.; Chen, M.S.; Yin, J.-C.; Lee, G.; et al. A NeuroD1 AAV-Based Gene Therapy for Functional Brain Repair after Ischemic Injury through In Vivo Astrocyte-to-Neuron Conversion. *Mol. Ther.* **2020**, *28*, 217–234. [CrossRef] [PubMed]
143. Ginn, S.L.; Amaya, A.K.; Alexander, I.E.; Edelstein, M.; Abedi, M.R. Gene therapy clinical trials worldwide to 2017: An update. *J. Gene Med.* **2018**, *20*, e3015. [CrossRef]
144. Tan, J.-K.Y.; Sellers, D.L.; Pham, B.; Pun, S.H.; Horner, P.J. Non-Viral Nucleic Acid Delivery Strategies to the Central Nervous System. *Front. Mol. Neurosci.* **2016**, *9*, 41. [CrossRef] [PubMed]
145. Huang, X.-P.; Ding, H.; Lu, J.-D.; Tang, Y.-H.; Deng, B.-X.; Deng, C.-Q. Autophagy in cerebral ischemia and the effects of traditional Chinese medicine. *J. Integr. Med.* **2015**, *13*, 289–296. [CrossRef]

MDPI
St. Alban-Anlage 66
4052 Basel
Switzerland
Tel. +41 61 683 77 34
Fax +41 61 302 89 18
www.mdpi.com

Viruses Editorial Office
E-mail: viruses@mdpi.com
www.mdpi.com/journal/viruses

www.ingramcontent.com/pod-product-compliance
Lightning Source LLC
LaVergne TN
LVHW070125170726
843515LV00007B/1753

9783036524900